Mechanical Equipment Mechanism Design
100 Cases

Hao Zhang
Fellow & Fellow Assessor of The Institution of Engineering and Technology,
London, UK

Adrian David Cheok
Member of The Order of Australia,
Australia

Yujun Zhang
Studying at College of Artificial Intelligence and Automation,
Hohai University, China

Tianyong Zhang
CAO of Hangzhou Huajian Intelligent Technology Research Institute,
Hangzhou, China

CRC Press
Taylor & Francis Group
Boca Raton London New York

CRC Press is an imprint of the
Taylor & Francis Group, an **informa** business
A SCIENCE PUBLISHERS BOOK

First edition published 2026
by CRC Press
2385 NW Executive Center Drive, Suite 320, Boca Raton FL 33431

and by CRC Press
4 Park Square, Milton Park, Abingdon, Oxon, OX14 4RN

CRC Press is an imprint of Taylor & Francis Group, LLC

Library of Congress Cataloging-in-Publication Data (applied for)

ISBN: 978-1-032-82705-6 (hbk)
ISBN: 978-1-032-85573-8 (pbk)
ISBN: 978-1-003-51876-1 (ebk)

DOI: 10.1201/9781003518761

Typeset in Times New Roman
by Prime Publishing Services

Preface

This book is dedicated to technological innovation in mechanical items, and years of technical accumulation contributed to its success. The goal of writing this book is to first summarize previous mechanical design work; second, the ever-changing mechanical structures serve as a catalyst for future innovation in mechanical structures; and finally, based on the description of some classic structural designs, readers can form a design concept in future mechanical structural design and be adept at accepting new innovative designs.

100 cases of mechanical equipment mechanism design have been carefully selected and classified. The book is divided as such, Chapter 1 addresses Classic Structural Design of Non-Standard Mechanical Equipment; Chapter 2 focuses on Packaging Machinery; Chapter 3 examines Automated Assembly Machinery; Chapter 4 discusses Parts Processing Equipment; Chapter 5 covers Testing Instruments and Equipment; Chapter 6 pertains to Soldering Machinery; and Chapter 7 explores Lifting and Transfer Devices. Readers can refer to relevant chapters according to the relevant industries, adopt relevant structures, and form their own designs. The authors hope that their work will benefit readers.

The most notable aspect of this book is that each instance described has been implemented. It is not simply talk on paper, and the result in operation may not be optimal, but it can generally match the design criteria. This book's case structure design is not ideal. If readers can improve on this foundation, it will benefit their own design skills. We hope that readers will criticize and amend it in a timely manner.

Each case presents structural details to the maximum extent possible, but it may still be difficult to see clearly. Therefore, all structures in this book are provided with 3D source files, which readers can download and carefully study.

This book is suitable for all mechanical engineers, mechanics, automation engineers, and technical workers to read, and is suitable as a teaching reference for all college students in mechanical related majors. Due to the wide range of industries covered, readers are highly likely to come up with brilliant ideas and achieve unexpected results while reading.

Due to the authors' limited experience, the book may contain errors and omissions. We hope you will not hesitate to provide feedback and improve the accuracy and reliability of future updates and publishing, resulting in a classic reference book for mechanical equipment structure design.

Engineering Council

ZHANG Hao
CEng FRSA FIET
Chartered Engineer
Registrant 563384 of the Engineering Council

Contents

Chapter 1

Classic Structural Design of Non-Standard Mechanical Equipment

Case 1: Descartes Robot

I. Case Description

Descartes Robot, also known as Cartesian Coordinate Robot, is a commonly used industrial robot mainly employed in computer numerical control (CNC) machine tools. It can also be employed as a milling machine or a drawing machine. The main structure of the Descartes Robot is illustrated in Fig. 1.1, comprising components such as electric motor (1, 5, 8), the first robotic arm (2), horizontal guiding rail (3), frame (4), the second robotic arm (6), vertical guiding rail (7). The primary advantage of the Descartes Robot lies in the linearity of all its control axes, as opposed to rotational, greatly simplifying the arm calculations of the robot.

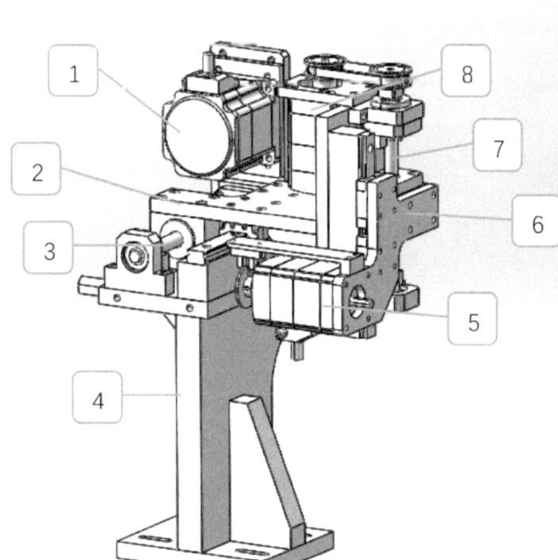

Fig. 1.1 Descartes Robot
1, 5, 8-Electric Motor; 2-First Robotic Arm; 3-Horizontal Guiding Rail; 4-Frame; 6-Second Robotic Arm; 7-Vertical Guiding Rail

II. Working Principle

The basic form of a Cartesian robot consists of three 'arms.' Each robotic arm can only move along a two-dimensional axis. The first robotic arm is positioned in the horizontal plane and can only move backward or forward, while the second robotic arm is vertical and can move upward or downward. Electric motors are used to control the robot arms to accomplish predefined tasks [1]. To program the robot for specific tasks, the programmer must be able to write code that guides the robot along its control axes to achieve various required positions. Solving problems with this arm solution requires calculations, and because linear calculations for finding the appropriate locations relative to the robot's control axes are easier, programmers may conduct these calculations in a closed form. [2]

III. Main Structure Introduction

The first robotic arm, as shown in Fig. 1.2, operates with the electric motor (5) driving the horizontal guiding rail (2) to rotate through the pulley (1). The surface of the horizontal guiding rail (2) is threaded to engage with the internal surface of the horizontal slider (3). The rotation direction and distance of the first robotic arm (4) are adjusted by the electric motor (5), controlling the translation direction and distance of the horizontal slider (3).

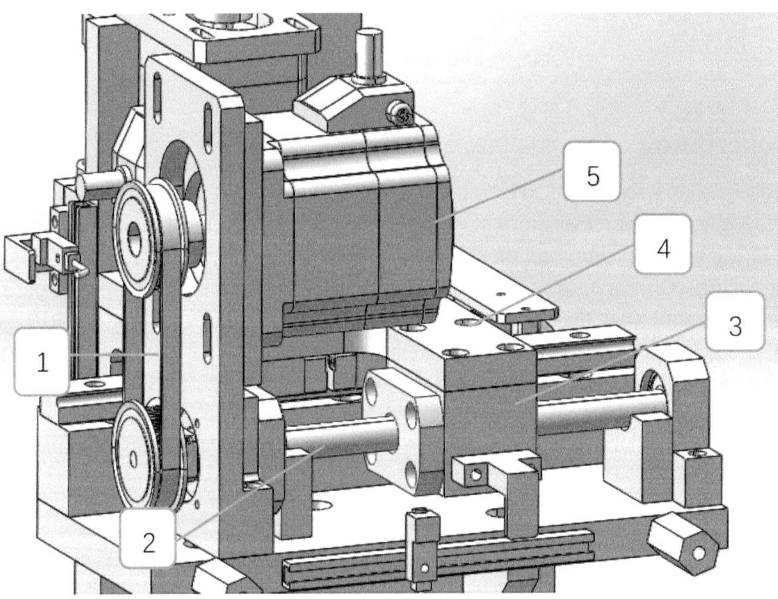

Fig. 1.2 First Robotic Arm
1 - Pulley; 2 - Horizontal Guiding Rail; 3 - Horizontal Slider; 4 - First Robotic Arm; 5 - Electric Motor

The second robotic arm, as shown in Fig. 1.3, operates with the electric motor (5) driving the vertical guiding rail (2) to rotate through the pulley (1). The surface of the vertical guiding rail (2) is threaded to engage with the internal surface of the vertical slider (3). The rotation direction and distance of the second robotic arm (4) are adjusted by the electric motor (5), controlling the translation direction and distance of the vertical slider (3).

The electric motor of the Cartesian robot, as shown in Fig. 1.4, can have various operational devices installed beneath it, depending on the robot's purpose. When the robot is in operation, the electric motor (1), along with the installed operational devices, rotates around the rotation axis (2) in the vertical plane, achieving the preset objectives.

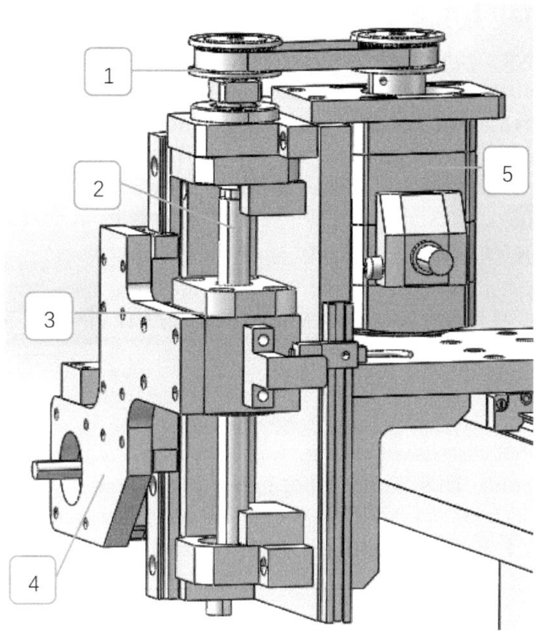

Fig. 1.3 Second Robotic Arm
1 - Pulley; 2 - Vertical Guiding Rail; 3 - Vertical Slider; 4 - Second Robotic Arm; 5 - Electric Motor

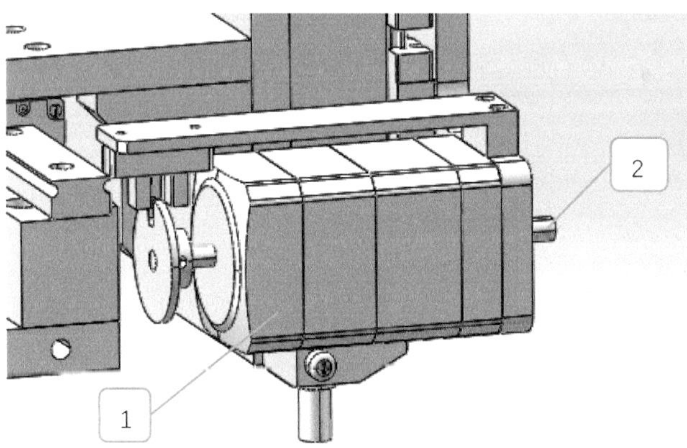

Fig. 1.4 Electric Motor
1 - Electric Motor; 2 - Rotation Axis

IV. Specific Parameters (see Table 1.1)

Table 1.1 Cartesian Robot Parameters

Working Range:	50 × 70 mm
Ambient Temperature:	0~60°C
Permissible Humidity:	10%~95%
Equipment Accuracy:	0.06 mm
Load Capacity:	5 kg
Equipment Dimensions:	355 mm(H) × 190 mm(L) × 230 mm(W)

V. Mechanical Design Highlights

Due to the Cartesian robot's capability to easily reach different points in three-dimensional space, its most common application is a numerical control (CNC) machine tool. CNC machine tools utilize computer programs to extract necessary commands, instructing the robot to work in the desired manner, and then load these commands into the robot. They enable the precise movements of the robot, making the Cartesian robot suitable for various drawing and plotting functions [3]. When utilized in this fashion, the Cartesian robot can set or lift tools from surfaces as it moves along the X and Y planes, allowing for the fabrication of specialized patterns.

Case 2: 5-Axis Industrial Robot

I. Case Description

The 5-axis industrial robot, as shown in Fig. 1.5, is primarily composed of four robotic arms, a base, and four servo motors. This 5-axis robot may replace human labor in a variety of difficult working circumstances, as well as long-duration, high-intensity repetitive activities in high-risk and hazardous environments. It enables intelligent, multifaceted, and flexible automated production.

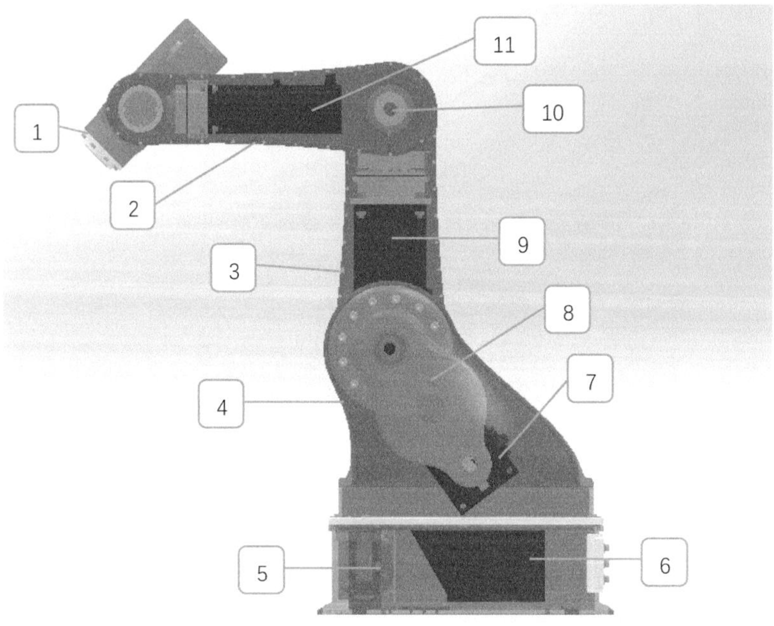

Fig. 1.5 5-Axis Industrial Robot
1 - First Robotic Arm; 2 - Second Robotic Arm; 3 - Third Robotic Arm; 4 - Fourth Robotic Arm; 5 - Base; 6, 7, 9, 11 - Electric Motors; 8 - First Transmission Mechanism; 10 - Second Transmission Mechanism

II. Working Principle

The basic working principle of the 5-axis robot is teaching and playback. With this method, the user guides the robot through each step of the actual task. The robot automatically memorizes the position, posture, motion parameters, and process parameters of each action during the teaching process. It then generates a program to execute all the operations continuously [4]. After the teaching process is complete, simply providing a start command to the robot will make it accurately replicate the taught actions, step by step, to complete the entire operation [5].

III. Main Structure Introduction

The robot base structure, as shown in Fig. 1.6, is typically fixed to the ground or a device/instrument. The base (1) contains an internal servo motor (3) that drives the fourth robotic arm to rotate horizontally.

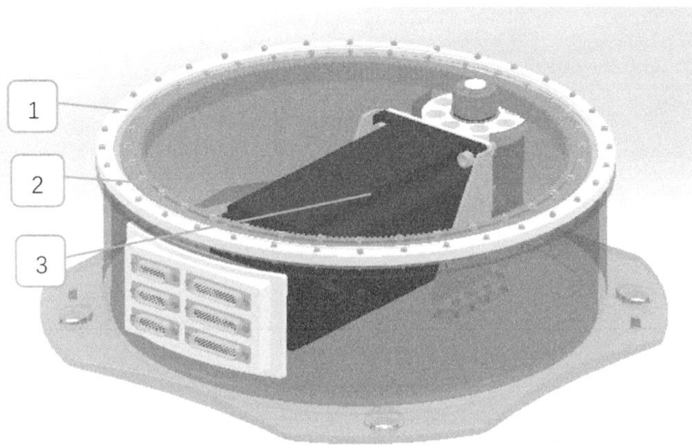

Fig. 1.6 Robot Base
1 - Base; 2 - Internal Gear; 3 - Electric Motor

The first transmission mechanism, as shown in Fig. 1.7, controls the third robotic arm (1) through the fourth robotic arm (2) via the first transmission mechanism (6).

Fig. 1.7 Robot First Transmission Mechanism
1 - Third Robotic Arm; 2 - Fourth Robotic Arm; 3 - Base; 4, 5, 7 - Electric Motors; 6 - First Transmission Mechanism

The second transmission mechanism, as shown in Fig. 1.8, controls the second robotic arm (1) through the third robotic arm (3) via the second transmission mechanism (5).

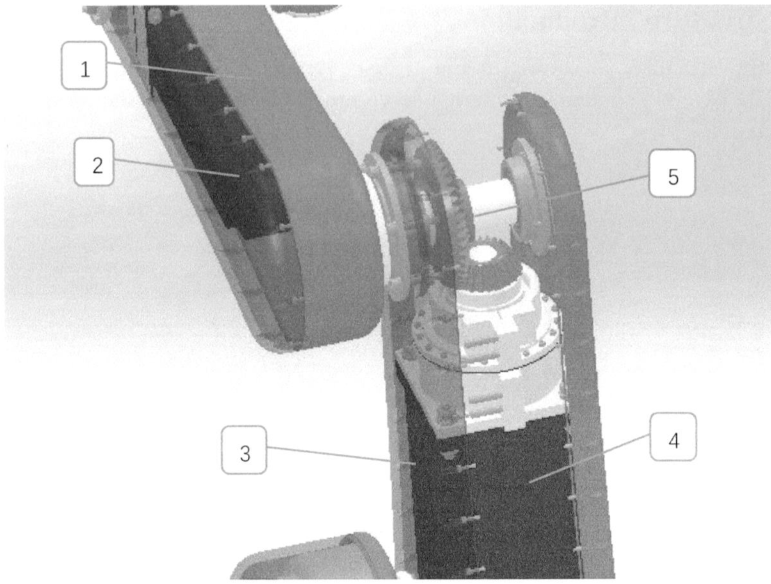

Fig. 1.8 Robot Second Transmission Mechanism
1 - Second Robotic Arm; 3 - Third Robotic Arm; 2, 4 - Electric Motors; 5 - Second Transmission Mechanism

IV. Specific Parameters (see Table 1.2)

Table 1.2 Industrial Robot Parameters

Robotic Arm Travel:	X:700 mm Y:700 mm Z:100 mm
Transmission Method:	XY Axis: Servo Motor + Synchronous Belt; Z Axis: Servo Motor + Lead Screw
Moving Speed:	X:500 mm/s Y:400 mm/s Z:300 mm/s
Z-Axis Load Capacity:	35 kg
Repeatability:	0.05 mm
Drive Mode:	Servo Motor
Control Mode:	PLC (Programmable Logic Controller)/Motion Control Card
Mode:	Timed Mode; Continuous Mode; Quantitative Mode; Cyclic Mode;

V. Mechanical Design Highlights

Industrial robots typically consist of three main parts: the mechanical arm assembly, the control system, and the teaching system. The mechanical arm is the component used to grasp the workpiece and comes in various structural forms, such as gripping, supporting, and suction, depending on the shape, size, weight, material, and operational requirements of the object being grasped [5]. The motion mechanism enables the mechanical arm to perform various rotations, movements, or compound motions to achieve specified actions, altering the position and posture of the grasped object. The independent motion modes, such as lifting, extending, rotating, etc., of the motion mechanism are referred to as the degrees of freedom of the mechanical arm. The more degrees of freedom, the greater the flexibility and versatility of the mechanical arm, but the more complex its structure becomes.

The control system controls the motors of each degree of freedom of the mechanical arm to perform specific actions. It also receives feedback information from sensors, forming a stable closed-loop control. The core of the control system is typically composed of microcontrollers or digital signal processing (DSP) chips, programmed to achieve the desired functions [6].

The teaching system instructs the robot on how to perform actions either through a "teaching box" or manually by a person. This information is then stored in the controller's memory, and the robot repeats the taught actions continuously. The teaching system essentially serves as a specialized intelligent terminal.

Case 3: Palletizing Robot

I. Case Description

Palletizing robots can be configured based on different material packaging, stacking sequences, layers, and other requirements to achieve palletizing tasks for various types of materials. The overall structure, as shown in Fig. 1.9, mainly consists of the end effector (1), the first robotic arm (2), the second robotic arm (3), servo motors (4, 5), the base (6), hydraulic cylinder (7), the first transmission mechanism (8), and the second transmission mechanism (9). Palletizing robots have a wide range of applications, improving work efficiency, enhancing packaging quality, reducing labor costs, and optimizing working environments.

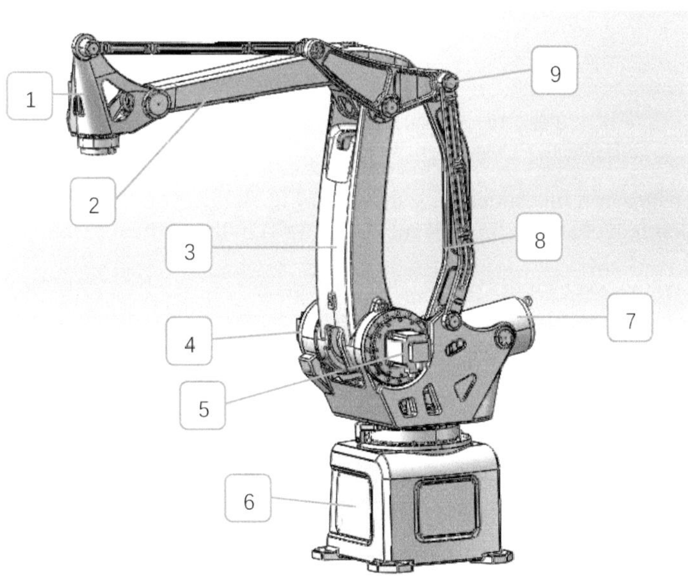

Fig. 1.9 Palletizing Robot
1-End Effector; 2-First Robotic Arm; 3-Second Robotic Arm; 4, 5-Servo Motors; 6-Base; 7-Hydraulic Cylinder; 8-First Transmission Mechanism; 9-Second Transmission Mechanism

II. Working Principle

The basic working principle of a palletizing robot is teaching and playback. To instruct the robot on how to perform a task, programmers use a controller to guide the robotic arm through the entire set of actions. The robot accurately stores the sequence of actions in its memory, and subsequently, whenever products are conveyed on the assembly line, it repeatedly executes this set of actions.

The mechanical arm of the robot is responsible for moving the end effector [7]. The end effector typically includes a built-in pressure sensor to prevent objects in the robot's hand from falling or being crushed [8].

III. Main Structure Introduction

The driving mechanism of the second robotic arm, as shown in Fig. 1.10, is primarily driven by a hydraulic cylinder (1) to control the rotation movement of the second robotic arm (4) in the vertical plane.

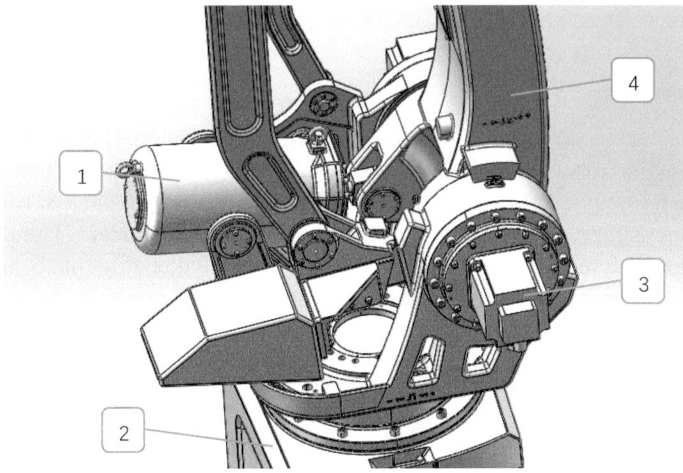

Fig. 1.10 Second Robotic Arm Driving Mechanism
1 - Hydraulic Cylinder; 2 - Base; 3 - Servo Motor; 4 - Second Robotic Arm

The first transmission mechanism, as shown in Fig. 1.11, is driven by the servo motor (2) to control the rotation movement of the first robotic arm (3) in the vertical plane.

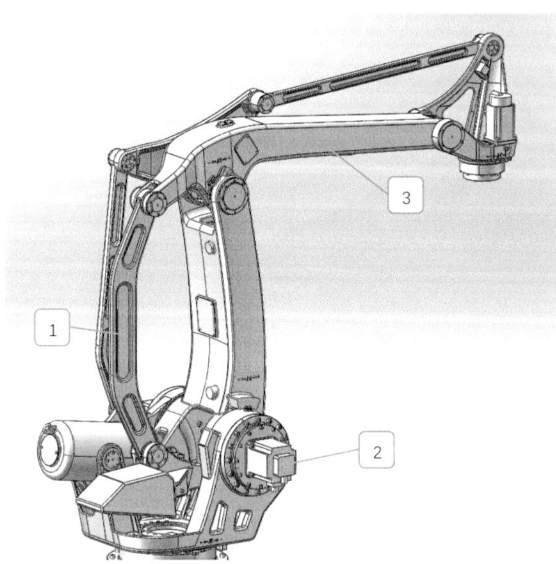

Fig. 1.11 First Transmission Mechanism
1 - First Transmission Mechanism; 2 - Servo Motor; 3 - First Robotic Arm

The second transmission mechanism, as shown in Fig. 1.12, is driven by the servo motor (3) to control the rotation movement of the end effector (1) in the vertical plane.

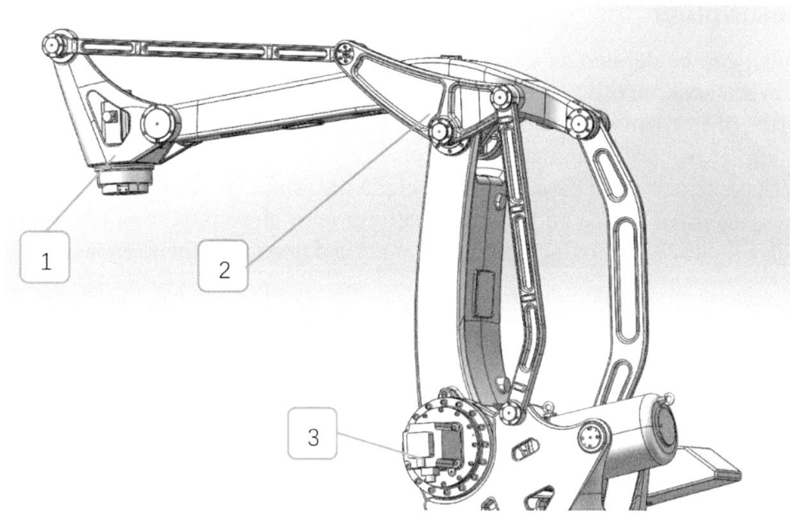

Fig. 1.12 Second Transmission Mechanism
1 - End Effector; 2 - Second Transmission Mechanism; 3 - Servo Motor

IV. Specific Parameters (see Table 1.3)

Table 1.3 Palletizing Robot Parameters

Working Efficiency:	Loading 600 to 1200 packaged items per hour.
Load Capacity:	35 kg
Repeatability:	0.06 mm
Drive Mode:	Servo Motor
Control Mode:	PLC (Programmable Logic Controller)/Motion Control Card
Mode:	Timed Mode; Continuous Mode; Quantitative Mode; Cyclic Mode;

V. Precautions

(1) Handling material conditions. To adapt to the robot's work, it is required that the handled items must be in boxes or bags [9]. This allows the robot to transport the items onto the conveyor. Additionally, items loaded manually should not undergo any changes in their state after being placed [10].

(2) Shape of handled items. One of the working conditions for palletizing robots is that the shape of the handled items should be regular for efficient packing [11]. Irregularly shaped items such as cylinders, cans made of glass, iron, aluminum, as well as rod-shaped, cylindrical, and ring-shaped items, require special gripping devices due to their irregular shapes [12]. Items suitable for the robot's operation include cardboard boxes, wooden boxes, paper bags, burlap bags, cloth bags, etc.

Case 4: Parallel Robot

I. Case Description

A parallel robot can be defined as a closed-loop mechanism where the moving platform and the fixed platform are connected by at least two independent motion chains. The mechanism has two or more degrees of freedom and is driven in a parallel manner [13]. The overall structural diagram is shown in Fig. 1.13, consisting mainly of a control system, a drive system, and a transmission system. The characteristics of a parallel robot include zero cumulative error, high precision, and the ability to place the drive device on the fixed platform or in close proximity to the fixed platform. This results in a lightweight moving part, high speed, and good dynamic response [14].

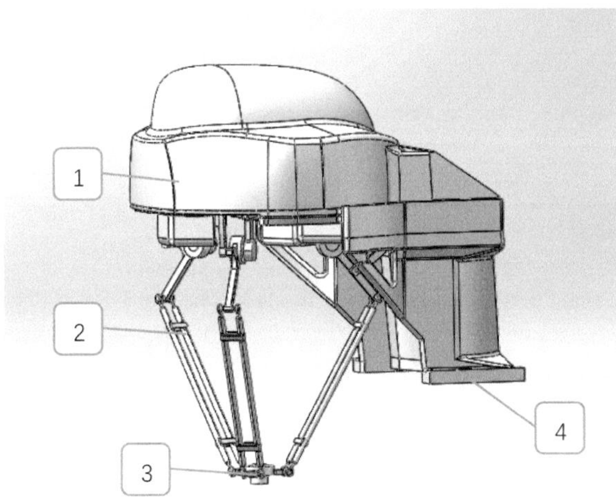

Fig. 1.13 Parallel Robot
1 - Frame 2 - Parallel Manipulator 3 - End Effector 4 - Base

II. Working Principle

The entire control system of the parallel robot is centered around an industrial control computer, with a PLC (Programmable Logic Controller) serving as the main control unit for servo control and digital signal control. It includes the servo control of the robot body, control of the drive system, control of the transmission system, and control of various auxiliary facilities. During the operation of the robot, the on-off state of two electromagnetic valves is controlled by the digital signal from the controller [14]. When the suction cup is directly above the object to be picked up, the vacuum generator produces a vacuum as the electromagnetic valve opens, allowing the suction cup to grip the object. When reaching the placement position, the vacuum-breaking electromagnetic valve opens, and the pressure under the suction cup becomes higher than atmospheric pressure, causing the object to be released. The transmission system mainly consists of an electric motor and two belts. When the robot is in motion, the controller provides an enable signal to drive the electric motor, which moves the belts. Encoders on the belts provide real-time feedback on the belt's speed to the controller.

III. Main Structure Introduction

The parallel robot, as shown in Fig. 1.14, is driven and controlled by a servo motor installed inside the frame (1). The servo motor controls the elevation of the parallel robot (2). The end effector (3) can take various structural forms based on the shape, size, weight, material, and operational requirements of the object to be gripped.

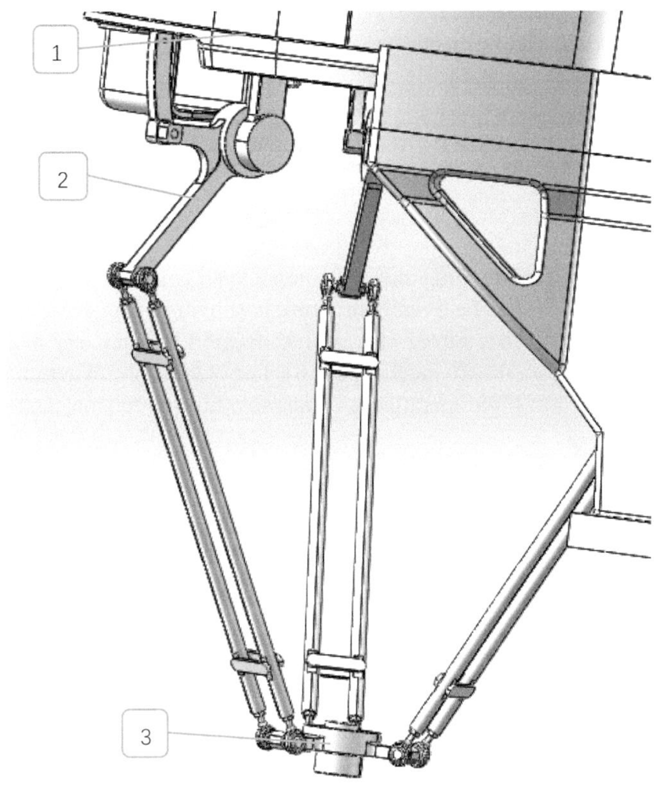

Fig. 1.14 Parallel Robot
1 - Frame 2 - Parallel Manipulator 3 - End Effector

IV. Specific Parameters (see Table 1.4)

Table 1.4 Parallel Robot Parameters

Structural Form:	Parallel Configuration
Drive Mode:	Servo Motor Driven
Load Capacity:	10 kg
Repeat Positioning Accuracy:	0.02 mm
Maximum Movement Speed:	100 mm/s
Overall Weight:	≤70 kg
Operating Mode:	Teach and Repeat/Programming
Equipment Appearance Dimensions:	500 mm(H) × 450 mm(L) × 420 mm(W)

V. Mechanical Design Highlights

The entire control system of the parallel robot is centered around an industrial control computer, with a PLC (Programmable Logic Controller) serving as the main control unit for servo control and digital signal control. It includes servo control of the robot body, control of the vision system, control of the pneumatic system, control of the transmission system, and control of various auxiliary facilities. The robot's vision system captures images using an image capture device and then passes them to the processing unit. The system distinguishes dimensions, shapes, colors, and so forth using digital processing based on pixel distribution, brightness, color, and other data. The discrimination results are then used to regulate the on-site equipment operations, increasing the robot's intelligence.

Case 5: Clamping and Automatic Positioning Unit

I. Case Explanation

The clamping and automatic positioning unit is a compact and cost-effective unit with clamping and automatic positioning functions. The overall structure is shown in Fig. 1.15, and it mainly consists of components such as an electric motor (1), product frame (2), planetary gearbox (3), clamping mechanism (4), spindle (5), and brake mechanism (6). The clamping and automatic positioning unit features a compact structure, high operational efficiency, high precision, and low manufacturing costs [15].

Fig. 1.15 Clamping and Automatic Positioning Unit
1-Electric Motor 2-Product Frame 3-Planetary Gearbox 4-Clamping Mechanism 5-Spindle 6-Brake Mechanism

II. Working Principle

When the clamping and automatic positioning unit is in operation, the clamping mechanism secures the workpiece to be processed. When the required processing parameters are entered into the equipment and the parameter setup is completed, the electric motor operates, causing the planetary gearbox to revolve. The gearbox connects to the spindle, causing it to rotate. The device is equipped with a brake system to ensure the precision of workpiece repositioning. When the workpiece reaches the preset angle, the brake mechanism stops the further rotation of the spindle [16], completing the automatic repositioning operation for the workpiece.

III. Main Structure Introduction

The planetary gear mechanism, as shown in Fig. 1.16, is driven by the electric motor (1), causing the planetary gears (2) to rotate and, in turn, driving the sun gear (3) and the spindle (4). The planetary gear mechanism utilizes the interaction of two spur gears (2) and (3) to transmit force vertically to the spindle (4) under the power drive, achieving the purpose of reduction.

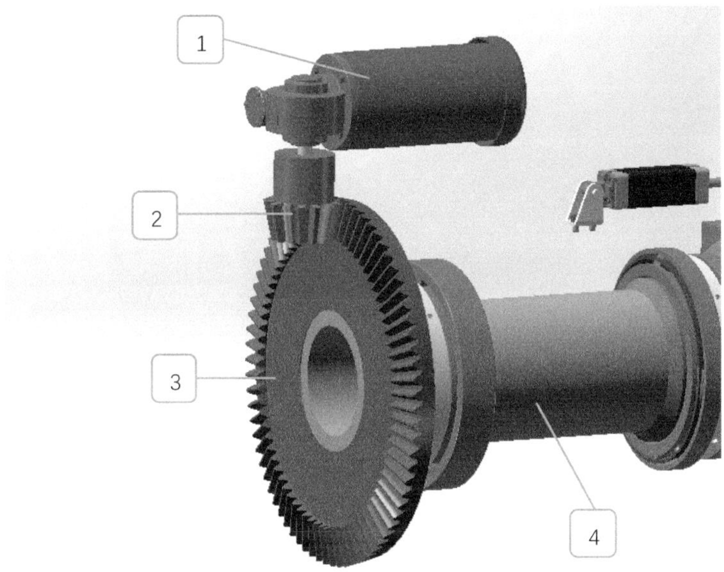

Fig. 1.16 Planetary Gear Mechanism
1 - Electric Motor 2 - Planetary Gears 3 - Sun Gear 4 - Spindle

The Brake M echanism as shown in Fig. 1.17, when the workpiece to be processed rotates to the preset angle, cylinder (1) starts, pressing down brake pad (3), thereby preventing the spindle (2) from continuing to rotate.

Fig. 1.17 Brake Mechanism
1 - Cylinder 2 - Spindle 3 - Brake Pad

The Clamping M echanism, as shown in Fig. 1.18, works by placing the workpiece into the spring chuck (2). Start the machine tool, and the piston (3) of the locking hydraulic cylinder (1) moves to the left to clamp the workpiece.

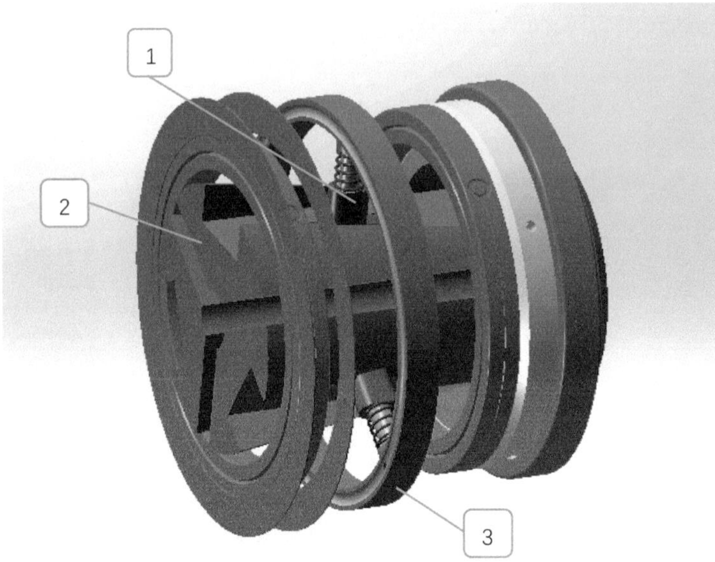

Fig. 1.18 Clamping Mechanism
1 - Locking Hydraulic Cylinder 2 - Spring Chuck 3 - Piston

IV. Specific Parameters (see Table 1.5)

Table 1.5 Clamping and Automatic Positioning Unit Parameters

Positioning Accuracy:	0.005 mm
Repeatability Accuracy:	0.002 mm
Clamping Force:	≥ 14000 N
Repositioning:	4 × 90°
Ambient Temperature:	0~55°C
Permissible Humidity:	1%~95%
Device Dimensions:	700 mm(H) × 860 mm(L) × 580 mm(W)

V. Mechanical Design Highlights

When processing more than two slots or faces on rotary parts, it is necessary to reposition the workpiece. If there are requirements for symmetry and dynamic balance, accurate indexing is needed. In the machining of such parts, indexing heads or special indexing fixtures are usually used, or multiple clamping processes are employed. However, these methods have the following drawbacks: multiple clamping or manual indexing is required, and it is challenging to ensure dimensional and positional tolerances, as well as dynamic balance. This often leads to a high scrap rate. Furthermore, equipment investment costs are expensive, and production efficiency is low.

The clamping and automatic indexing unit is designed for the machining of rotary parts, allowing the completion of the machining of more than two slots or faces with just one clamping.

It eliminates the positioning errors associated with multiple clamping, ensuring the symmetry and dynamic balance requirements for high-speed rotating products such as woodworking tools. With a high degree of automation, it significantly improves production efficiency, achieving more than six times the productivity of ordinary milling machines.

Case 6: Cutting Machine

I. Case Explanation

The flat press cutting machine is suitable for the mass cutting of wood, and its overall structure is depicted in Fig. 1.19. It consists of the following components: cutting board table (1), cutting board (2), guide rail (3), frame (4), handwheel (5), pad (6), height adjustment mechanism (7), and cutting pressure mechanism (8). The cutting pressure mechanism includes an eccentric wheel mechanism, electric motor, and other components. The cutting machine features a simple structure, easy maintenance, user-friendly operation, and precise die-cutting.

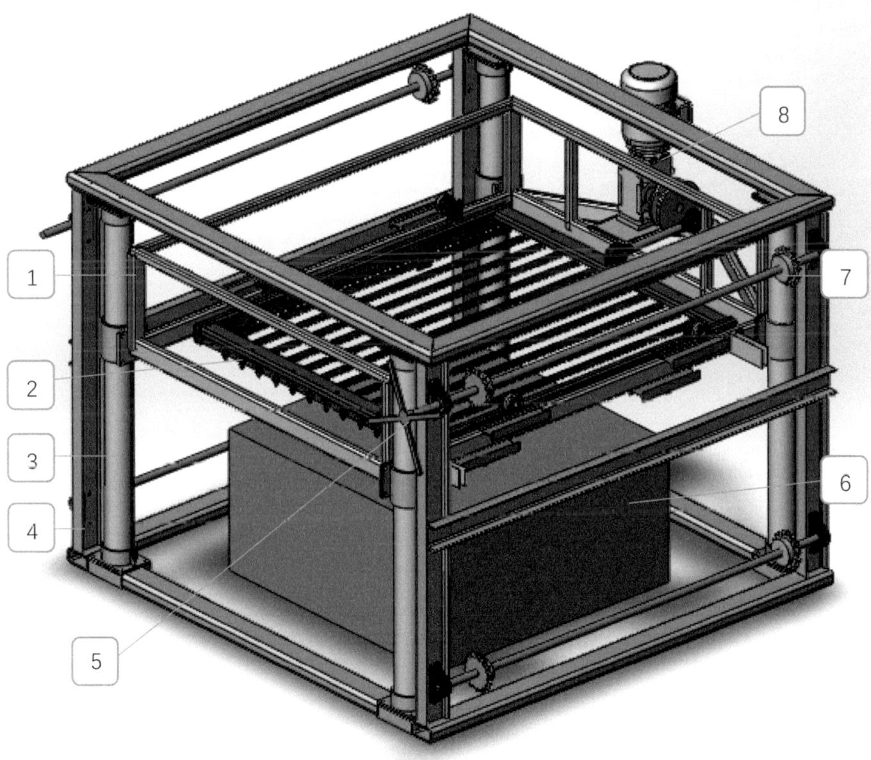

Fig. 1.19 Cutting Machine
1 - Cutting Board Table; 2 - Cutting Board; 3 - Guide Rail; 4 - Frame; 5 - Handwheel; 6 - Pad; 7 - Height Adjustment Mechanism; 8 - Cutting Pressure Mechanism

II. Working Principle

The cutting board table and the cutting pressure mechanism are the main devices responsible for the cutting action. Before starting the cutting machine, secure the cutting board to the flat cutting board table. Place the wood to be processed above the pad, and use the handwheel to adjust the height adjustment mechanism to ensure the cutting board is in close contact with the wood to be processed. Start the cutting machine, and the cutting board table slowly presses down. The electric motor drives the eccentric wheel mechanism to rotate. The eccentric wheel mechanism is connected to the rocker arm, which, through the motion of the rocker arm, drives the cutting board to reciprocate. This continuous engagement and disengagement between the cutting board and the wood being processed achieves the cutting process [17].

III. Main Structure Introduction

The cutting pressure mechanism, as shown in Fig. 1.20, mainly consists of the eccentric wheel mechanism (1) and an electric motor (2). The electric motor (2) provides the power needed for the rotation of the eccentric wheel mechanism (1). The eccentric wheel mechanism (1) is composed of a wheel disc with an eccentric hole and a rocker arm. When the shaft rotates, the outer edge of the wheel disc drives the rocker arm, causing it to undergo reciprocating motion.

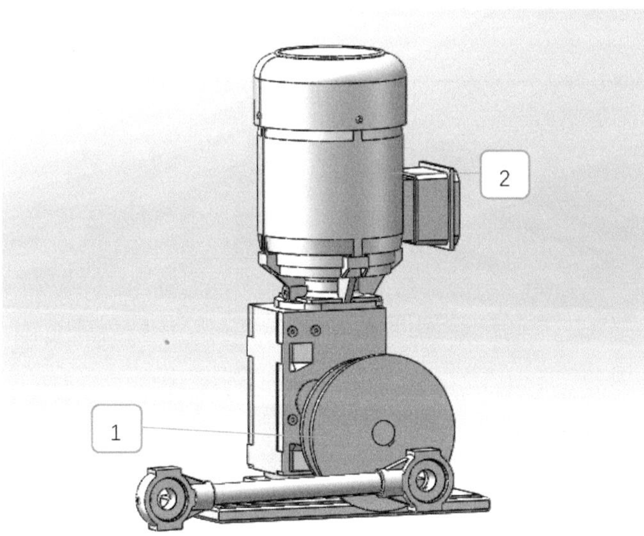

Fig. 1.20 Pressure Cutting Mechanism
1- Eccentric Wheel Mechanism 2- Electric Motor

IV. Specific Parameters (see Table 1.6)

Table 1.6 Cutting Machine Parameters

Motor Power:	3 kW
Operating Temperature:	0~65°C
Maximum Processing Material Size:	1220 × 1220 mm
Production Efficiency:	4~6 t/h
Blade Material:	Alloy Steel
Equipment Size:	1660 mm(H) × 2600 mm(L) × 2180 mm(W)

V. Mechanical Design Highlights

The cutting machine improves the accuracy of the cutting tool for sheet cutting by adding the pressing mechanism and the cutting board platform. In the pressing mechanism, the use of springs allows the pressing block to buffer excessive force on the sheet, preventing damage caused by excessive downward pressure. The installation of the cutting board platform on the cutting tool mount ensures greater stability during the downward process, enhancing the stability of the cutting tool and contributing to improved cutting precision.

VI. Precautions

(1) Before use, carefully inspect the performance of the equipment to ensure the integrity of each component.
(2) Conduct a detailed check on the power switch and the tightness of the saw blade. The operating platform must be stable, and there should be sufficient lighting and brightness for night operations.
(3) The power line must be safe and reliable. It is strictly prohibited to randomly pull or place the power line carelessly to avoid breakage. Before use, carefully inspect the performance of the equipment to ensure the integrity of each component.
(4) Strong cutting operations are not allowed. Wait for the electric motor to reach full speed before cutting.
(5) If any abnormal noise occurs, stop immediately for inspection. Power must be cut off before repairing or replacing parts, and wait for the machine to completely stop working.
(6) If the equipment shakes or experiences other malfunctions, stop it immediately for repairs. Wearing gloves during operation is strictly prohibited. If dust is generated during operation, wear a mask or face shield.
(7) After processing is complete, turn off the power and clean the equipment and the surrounding area.

Case 7: Pipe Cutting Machine

I. Case Description

The pipe cutting machine is typically used for cutting stainless steel pipes. The overall structure is shown in Fig. 1.21, and a zoomed-in view is provided in Fig. 1.22. It consists mainly of the feeding mechanism (1), the first clamping mechanism (2), the feeding mechanism (3), the second clamping mechanism (4), and the cutting mechanism (5). The pipe cutting machine features a high level of automation, low cutting errors, and easy operation, making it suitable for cutting steel pipes of different diameters.

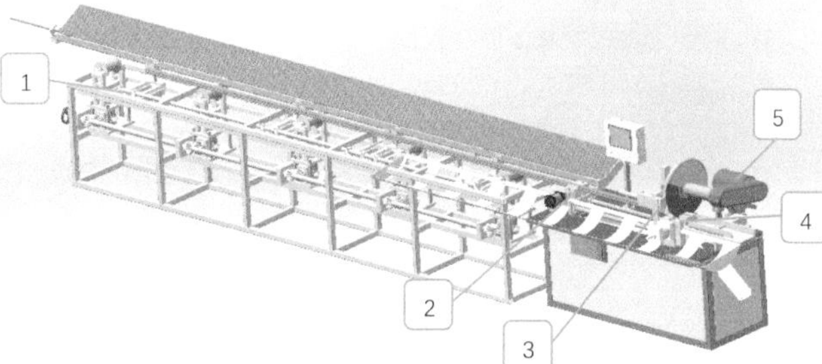

Fig. 1.21 Automatic Pipe Cutting Machine
1 - Feeding Mechanism 2 - First Clamping Mechanism 3 - Feeding Mechanism 4 - Second Clamping Mechanism 5 - Cutting Mechanism

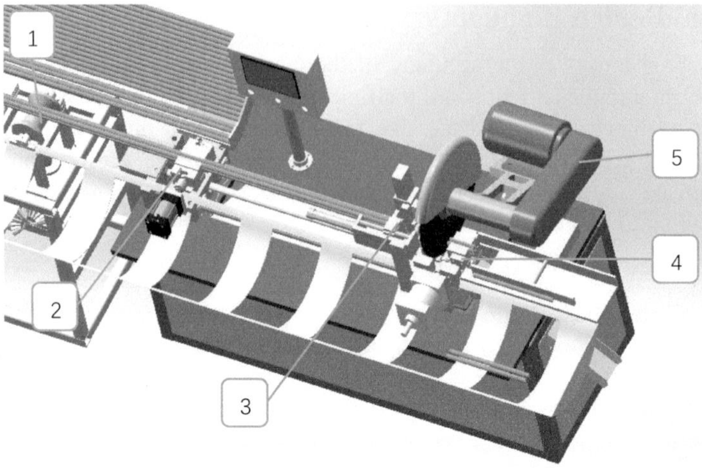

Fig. 1.22 Enlarged View of the Pipe Cutting Machine
1 - Feeding Mechanism 2 - First Clamping Mechanism 3 - Feeding Mechanism 4 - Second Clamping Mechanism
5 - Cutting Mechanism

II. Working Principle

When the pipe cutting machine is in operation, adjust the height of the feeding mechanism to match the height of the clamping mechanism. Stainless steel pipes, arranged in groups of four, are fed into the pipe cutting machine by the feeding mechanism. When a stainless steel pipe reaches below the cutting machine and exceeds the specified length of the saw blade, the clamping mechanism secures the stainless steel pipe, and the cutting machine is activated. The saw blade descends to cut the stainless steel pipe. After cutting, the clamping mechanism releases the stainless steel pipe, and the feeding mechanism continues to grab the pipe, delivering it under the cutting machine. The finished pipe is then pushed out of the clamping mechanism. The pipe cutting machine repeats these operations to batch-cut pipes of equal lengths [18].

III. Main Structure Introduction

The first clamping mechanism, as shown in Fig. 1.23, is driven by cylinder (2) to clamp the pipe, ensuring that the pipe will not swing violently during cutting, thus preventing accidents.

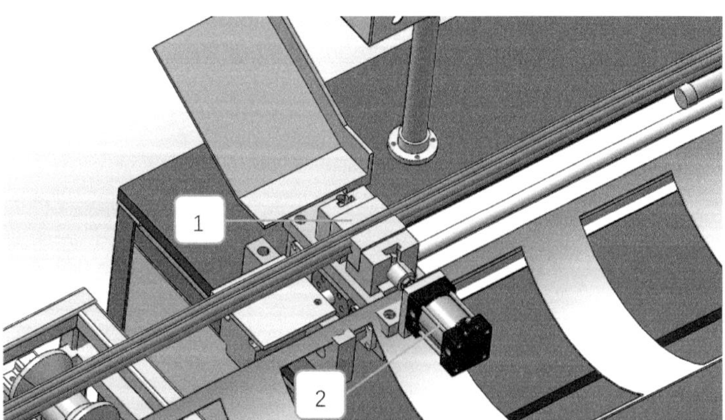

Fig. 1.23 Clamping Mechanism
1 - Fixed Block 2 - Cylinder

The second clamping mechanism, as shown in Fig. 1.24, is driven by a cylinder (2) to clamp the pipe, ensuring that the pipe will not swing violently during cutting, thus preventing accidents. Additionally, it ensures that the pipe will not shift during cutting, maintaining the precision of the cutting of stainless steel pipes.

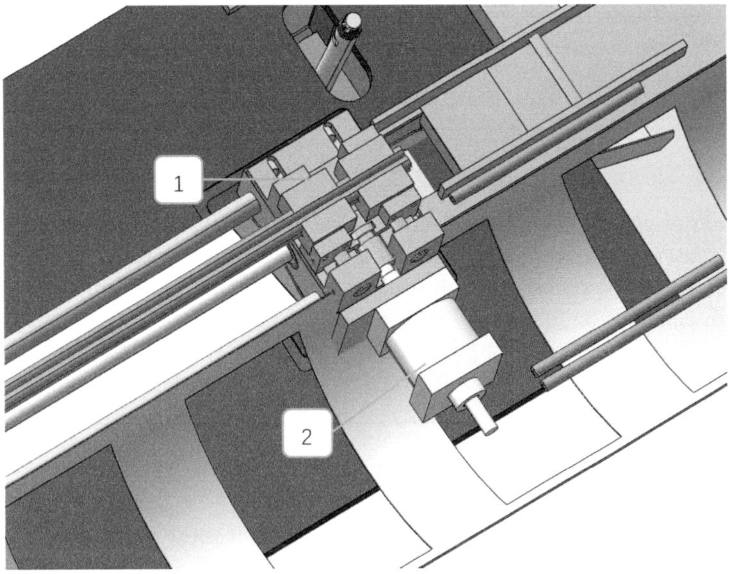

Fig. 1.24 Clamping Mechanism
1 - Fixed Block 2 - Cylinder

The cutting machine, as shown in Fig. 1.25, is driven by an electric motor (3), which rotates the saw blade (1) through the pulley (2) to cut stainless steel pipes.

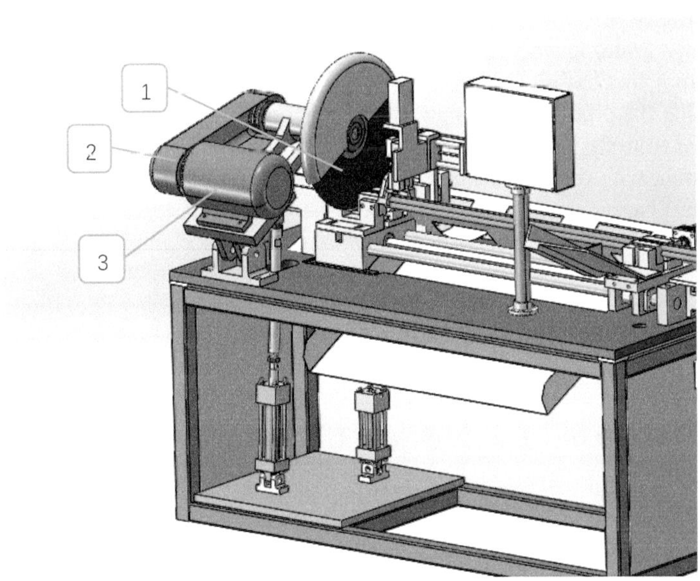

Fig. 1.25 Cutting Machine
1 - Saw Blade 2 - Pulley 3 - Electric Motor

IV. Specific Parameters (see Table 1.7)

Table 1.7 Saw Pipe Machine Parameters

Motor Speed:	4000 r/min
Saw Blade Center Distance:	62 mm
Saw Blade Diameter:	140 mm
Maximum Cutting Wall Thickness:	Steel: 6 mm Stainless Steel: 5 mm
Equipment dimensions:	1250 mm(H) × 6930 mm(L) × 950 mm(W)

V. Mechanical Design Highlights

In the production of certain equipment, there is a need for a substantial amount of stainless steel pipe parts to be cut, requiring not only ensuring the cutting dimensions but also preventing surface scratches during the sawing process. Additionally, it should facilitate easy adjustment of dimensions. This places high demands on the overall structure of the material rack and the positioning method of the locating fixture. The automatic pipe cutting machine, by setting dimensions and automating the cutting process, reduces costs for cutting orders with small batches and complex shapes of steel pipes, demonstrating high economic efficiency.

VI. Precautions

(1) The dust removal mechanism of the pipe cutting machine must be intact before cutting.

(2) Spindle speed adjustment must be done after the machine has come to a complete stop. During speed adjustment, gears should fully engage. If any abnormalities are detected in the machine, stop immediately for inspection.

(3) When loading long pipes onto the material rack or releasing bundled wires , measures should be taken to prevent the pipes from rolling, impacting, or injuring individuals.

(4) During machine rotation, no part of the human body should come into contact with the transmission components. When operating, cuffs should be securely fastened, and working with gloves is strictly prohibited. The human head should be kept away from the cutting direction.

(5) When using a grinding wheel pipe cutting machine, check the grinding wheel for any damage, cracks, or moisture before use, and ensure the reliability of the power supply.

(6) When cutting the pipe with the pipe cutting machine, precautions should be taken to prevent the pipe end from flying out and causing injury.

(7) Before cutting with the pipe cutting machine, adjust the tools and clamp the workpiece. The length of the clamping area should not be less than 50 millimeters. The stop plate must be fixed, and work can only commence after sequential testing for clamping, loosening, forward, and backward movements.

(8) People should not stand in the direction of the workpiece entering or exiting.

(9) When changing tools, measuring workpieces, lubricating, or cleaning the pipe end, the machine must be stopped.

Case 8: Double Saw Blade Cutting Machine

I. Case Explanation

In the cutting process of standard profiles, manual labor is primarily involved in feeding, manually operating the machine for cutting, and then unloading. This reliance on manual labor demands a high level of skill from the workers, often leading to lower processing efficiency. Additionally, the production yield is influenced by human factors, making it challenging to ensure high-quality output. The high rate of defects results in significant material waste, ultimately increasing production costs. The double saw blade cutting machine is suitable for the rapid cutting of standard profiles. The overall structure is shown in Fig. 1.26, consisting mainly of a frame (1), Flat belt (2), cutting motor (3), saw blades (4), output shaft (5), and feeding mechanism (6, 7). The double saw blade cutting machine offers advantages such as high cutting efficiency and high product yield.

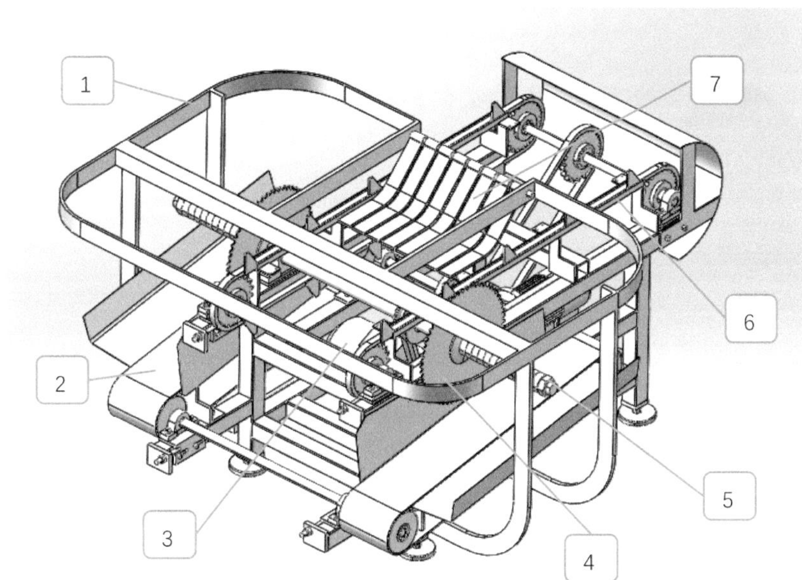

Fig. 1.26 Double Saw Blade Cutting Machine
1-Frame 2-Flat Belt 3-Cutting Motor 4-Saw Blade 5-Output Shaft 6,7-Feeding Mechanism

II. Working Principle

Before operating the double saw blade cutting machine, adjust the position of the saw blades on the output shaft according to the cutting requirements. During operation, place the profile to be processed on the feeding mechanism, start the cutting machine, and the cutting motor drives the saw blades to rotate at high speed. The feeding mechanism continuously advances, pushing the profile to be processed to the cutting position, completing the cutting of the profile. The waste material after cutting is then conveyed out by the conveyor belt. After each cut, the feeding mechanism rotates back continuously, bringing the next profile to be processed to the cutting position for sequential processing [19].

III. Main Structure Introduction

The cutting mechanism, as shown in Fig. 1.27, consists of the cutting motor (1) driving the output shaft (4) to rotate through the pulley (3), thereby causing the saw blades (2) to rotate at high speed.

The feeding mechanism is shown in Fig. 1.28. It is driven by the motor (3), which rotates the pulley (2), thus driving the feeding belt (1). The feeding belt (1) is equipped with stoppers, allowing continuous feeding of the plates to the cutting mechanism. At the same time, the flat belt (4) is also driven synchronously by the motor (3) to transport the waste material out of the cutting machine.

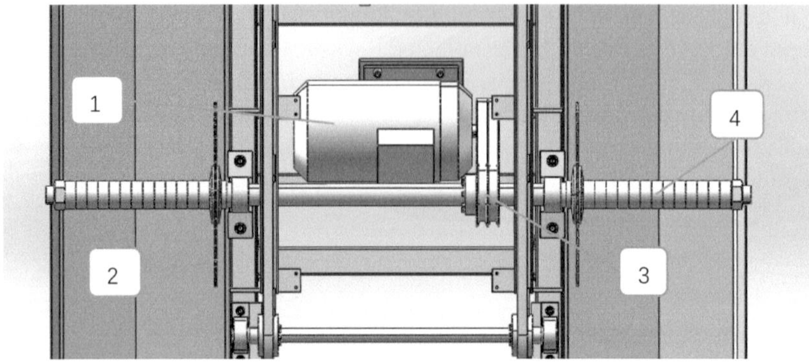

Fig. 1.27 Cutting Mechanism
1 - Cutting Motor 2 - Saw Blades 3 - Pulley 4 - Output Shaft

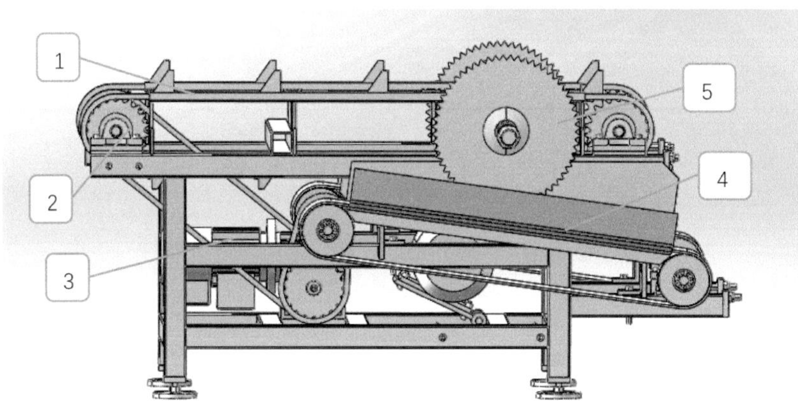

Fig. 1.28 Feeding Mechanism
1 - Feeding belt 2 - Pulley 3 - Motor 4 - Flat belt 5 - Saw blade

IV. Specific Parameters (see Table 1.8)

Table 1.8 Parameters of Double Saw Blade Cutting Machine

Yield Rate:	≥98%
Ambient Temperature:	–10~65°C
Permissible Humidity:	10%~90%
Maximum Processing Length of Raw Material:	1400 mm
Equipment Dimensions:	970 mm(H) × 1700 mm(L) × 1875 mm(W)

V. Mechanical Design Highlights

The double saw blade cutting machine, designed with a sequential feeding mechanism (6) and a waste material conveyor belt, operates with synchronized movement driven by a single electric motor, enhancing the automation and efficiency of the equipment. Two saw blades are positioned on the output shaft, allowing for adjustable distances between them to meet the cutting requirements for different lengths. The double saw blade cutting machine enables continuous feeding without stopping the machine, and one person can operate multiple machines, reducing labor requirements.

Moreover, the product quality is not influenced by manual factors, and the machine's simple and stable structure ensures high processing accuracy.

VI. Precautions

(1) Before use, carefully inspect the equipment's performance to ensure the integrity of each component.
(2) Thoroughly check the power switch, blade tension, and stability of the operating platform; ensure there is sufficient lighting during nighttime operations.
(3) Ensure the power supply lines are safe and reliable; avoid improper handling or pulling of power cords, and be cautious in placing them to prevent accidental disconnection. Prior to use, thoroughly check the equipment's performance to ensure all components are intact.
(4) Avoid performing forceful cutting operations; wait for the electric motor to reach full speed before initiating the cutting process.
(5) If any abnormal sounds occur, immediately stop and inspect the machine; before repairing or replacing parts, disconnect the power and wait for the machine to completely stop working.
(6) In the event of equipment vibration or other malfunctions, stop the machine immediately for repair; wearing gloves during operation is strictly prohibited. If dust is generated during operation, wear a mask or face shield.
(7) After completing the processing, turn off the power and ensure the equipment and surrounding area are cleaned.

Case 9: Paper Cutter

I. Case Explanation

The paper cutter is a paper processing equipment, mainly used for processing single sheets of paper in paper mills and shaping printed materials in printing companies. The overall structure is shown in Fig. 1.29, mainly composed of the main machine (1), paper pushing mechanism (2), cutting mechanism (3), workbench (4), paper pressing mechanism (5), and control panel (6). The paper pushing mechanism (2) is used to push and position the paper, while the paper pressing mechanism (5) presses the paper firmly to ensure that the original positioning accuracy is not compromised during cutting. The cutting mechanism (3) is used for cutting the paper, and the workbench (4)

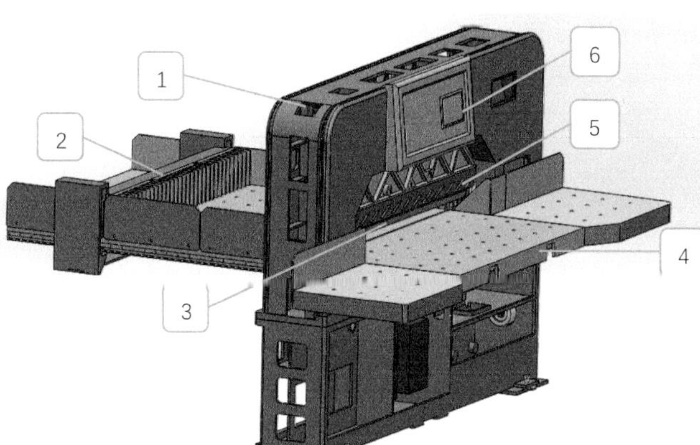

Fig. 1.29 Paper Cutter
1-Main Machine 2-Paper Pushing Mechanism 3-Cutting Mechanism 4-Workbench 5-Paper Pressing Mechanism 6-Control Panel

provides support. The paper cutter features short preparation time, high cutting precision, low labor intensity, and safe operation.

II. Working Principle

The preparation work before cutting paper involves aligning the left and right side panels, the paper-pushing mechanism, and the workbench of the paper cutter to form a 90° angle, ensuring that the specifications of the cut paper remain consistent. After starting the paper cutter, inputting parameters into the control panel, and initiating the machine, the paper-pushing mechanism automatically advances the paper to a specified length. At this point, the paper pressing mechanism descends to secure the paper, and the cutting mechanism diagonally descends to complete the cutting process [20].

III. Main Structure Introduction

The paper-pushing mechanism, as shown in Fig. 1.30, is driven by a cylinder (2) and continuously pushes the paper in the direction of the workbench (1).

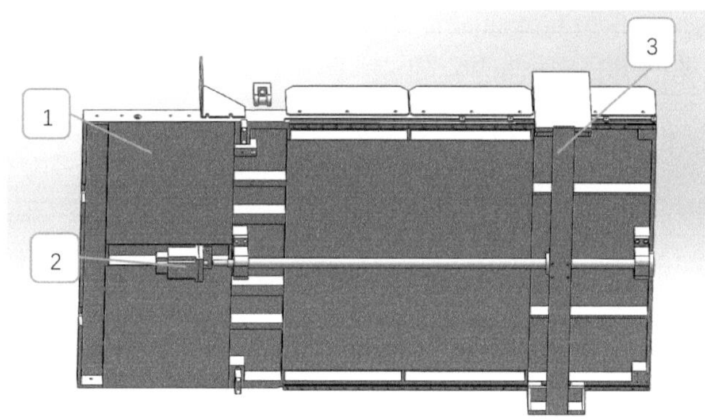

Fig. 1.30 Paper-Pushing Mechanism
1 - Workbench 2 - Cylinder 3 - Paper-Pushing Mechanism

The cutting mechanism and the paper-pressing mechanism are shown in Fig. 1.31. The paper-pressing mechanism (5) is driven by an electric motor (1), and through pulley (1), it controls the downward pressure to fix the paper. The cutting mechanism (4) is driven by an electric motor (3), and through pulley (2), it controls the oblique downward cutting of the paper [21].

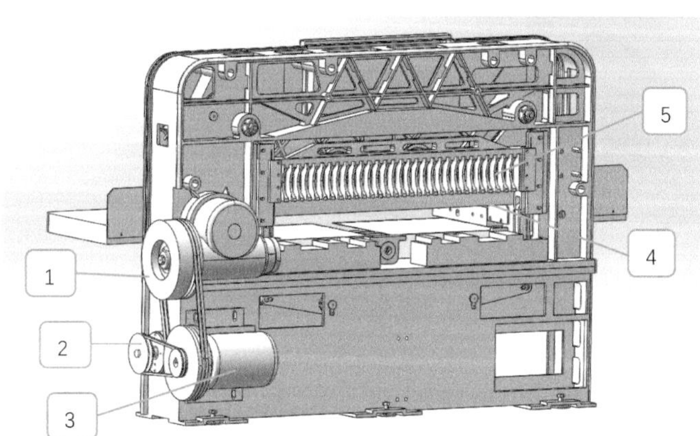

Fig. 1.31 Cutting Mechanism and Paper-pressing Mechanism
1, 2 - Pulleys 3 - Electric Motor 4 - Cutting Mechanism 5 - Paper-pressing Mechanism

IV. Specific Parameters (see Table 1.9)

Table 1.9 Paper Cutting Machine Parameters

Maximum Cutting Width:	1300 mm
Maximum Cutting Length:	1300 mm
Maximum Cutting Height:	120 mm
Length in Front of the Worktable:	700 mm
Worktable Height:	930 mm
Cutting Speed:	42 times/min
Machine Weight:	2100 kg
Equipment Dimensions:	1600 mm(H) × 2700 mm(L) × 2900 mm(W)

V. Mechanical Design Highlights

The paper cutter is equipped with a paper pressing mechanism that can secure the paper cutter and the paper, facilitating the fixation of the paper and cutting through the paper cutter. The control panel is connected electrically to the linear displacement sensor and the paper cutter, enabling automatic control of the cutting speed. This prevents the occurrence of curled edges when feeding paper quickly and enhances the uniformity of the cutting edge [22].

VI. Precautions

During actual usage, the paper cutter has to deal with a diverse range of cutting materials, including not only paper and cardboard but also leather, plastic, cork flooring, and more. Due to the variety of materials to be cut, it is necessary to adjust the pressure of the paper pressing mechanism and the angle of the cutting blade accordingly for different materials. This ensures the production of high-quality cut products.

Case 10: UV Curing Oven

I. Case Explanation

The UV curing oven is capable of igniting ultraviolet lamps, emitting ultraviolet light with a main peak of around 360nm, under the excitation of instantaneous high voltage. This light is directed onto the ink layer, initiating the curing of the ink through the cross-linking of acrylic resin. It is mainly used for the curing of ink layers after UV coating on products such as paper, PVC, and plastics [23]. The overall structure is shown in Fig. 1.32, consisting mainly of the exhaust fan unit (1), light shield (2), conveyor belt (3), and ultraviolet light source (4). The UV curing oven can significantly save resources and improve work efficiency.

II. Working Principle

To initiate polymerization, cross-linking, and grafting reactions in the resin, a light initiator (or photosensitizer) is added to a special formulation of resin. After absorbing high-intensity ultraviolet light in the UV curing equipment, the light initiator generates active free radicals. This process transforms the resin from a liquid to a solid state within a few seconds [24]. When using the equipment, power up by opening the control door, turning on the power and air switches, and adjusting the speed of the conveyor belt. Turn on the UV lamps, and when the temperature rises to 40~50°C, activate the exhaust fan unit to cool the curing oven [25].

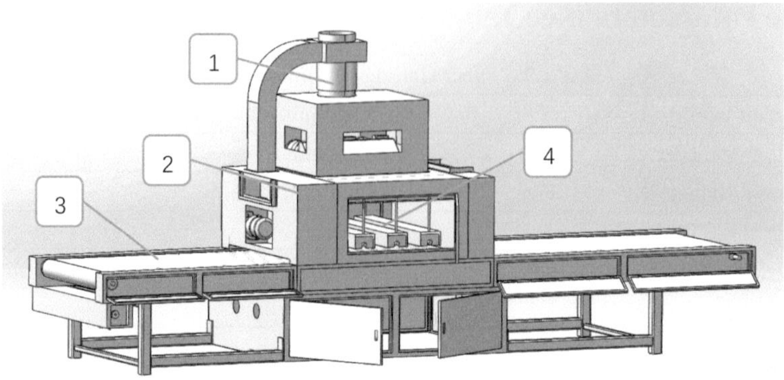

Fig. 1.32 UV Curing Oven
1 - Exhaust Fan Unit 2 - Light Shield 3 - Conveyor Belt 4 - Ultraviolet Light Source

III. Main Structure Introduction

Exhaust fan unit, as shown in Fig. 1.33, consists of the top-side fan unit (1), left-side fan unit (2), and right-side fan unit (3). When the internal temperature of the curing oven becomes too high, these fans automatically activate to cool it down.

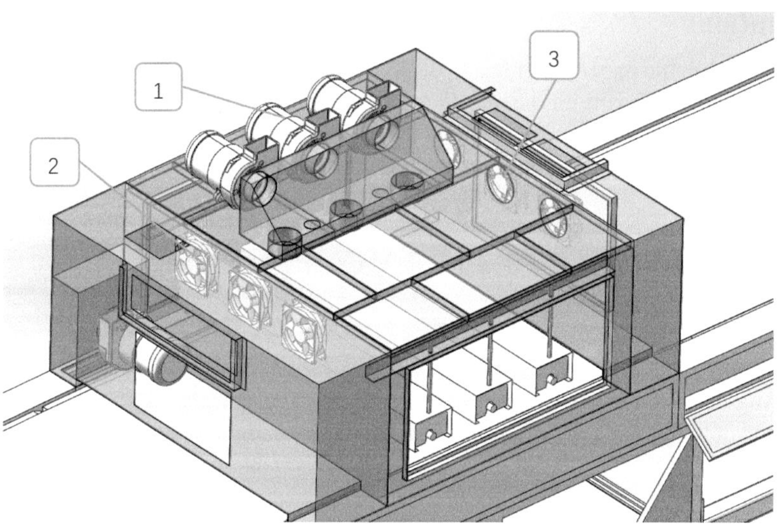

Fig. 1.33 Exhaust Fan Unit
1 - Top-side Fan Unit 2 - Left-side Fan Unit 3 - Right-side Fan Unit

IV. Specific Parameters (see Table 1.10)

Table 1.10 UV Curing Oven Parameters

UV Light Sources:	2 sets of 3 kW/1 set of 2 kW
Belt Width:	400 mm
Belt Height above Ground:	750 mm
Belt Material:	Teflon-coated fiberglass

(*Contd.*)

Belt Speed:	1~5 m/min
Temperature:	Controlled within 80°C
Material:	Outer shell made of cold-rolled steel plate, bent and surface high-temperature powder coated
Lamp Power:	2000 W
Peak Wavelength:	365 nm
Irradiance:	200 mw/cm^2
Lamp Life:	1000 hours (affected by the number of switch cycles)
Total Power:	2150 W
Total Power:	Leakage protection circuit breaker, motor overload protection

V. Mechanical Design Highlights

By installing displacement sensors on the side of the UV curing oven, when the displacement sensor receives information about the product entering the UV curing oven, the UV oven automatically opens the door using an automatic door opener. The curing time is calculated by a timer, and once cured, the conveyor belt automatically outputs the finished product, effectively improving production efficiency.

VI. Precautions

(1) Wear labor protective equipment.
(2) Check the electrical insulation and grounding of the equipment, and ensure the integrity of the metal casing protection.
(3) Inspect the equipment for any damage and check if there is any leakage of harmful ultraviolet radiation.
(4) Open the oven door, check for foreign objects inside, and ensure proper ventilation.
(5) Following the equipment operation sequence, gradually turn on the power to the UV lamps and then slowly adjust the energy of the UV lamps to the maximum.
(6) After preheating the UV lamps for 3-5 minutes, turn on the exhaust fan to complete the "lighting" process.
(7) Adjust the conveyor belt speed to ensure proper energy for UV coating.
(8) Place the samples or workpieces to be cured in the central position of the conveyor belt to ensure sufficient and uniform exposure.
(9) When stopping the oven, turn off the UV lamp power first, then adjust the lamp energy to the lowest setting.
(10) Press the stop button, and the exhaust system will automatically stop after the UV lamps have cooled. After the exhaust stops, disconnect the equipment's main power.
(11) After the UV lamps in the UV curing oven are turned off, do not restart immediately. Wait for the UV lamps to cool down before restarting.

Case 11: Automatic Sugar Beet Harvester

I. Case Explanation

The automatic sugar beet harvester belongs to agricultural harvesting equipment and is used for the large-scale automatic harvesting of sugar beets. The overall structure is shown in Fig. 1.34, mainly consisting of the clamping and lifting device (1), soil filtering and debris removal screw roller (2), crawler (3), lifting wheel assembly (4), and soil filtering conveyor chain assembly (5). The automatic sugar beet harvester has advantages such as high harvesting efficiency and minimal damage to sugar beets.

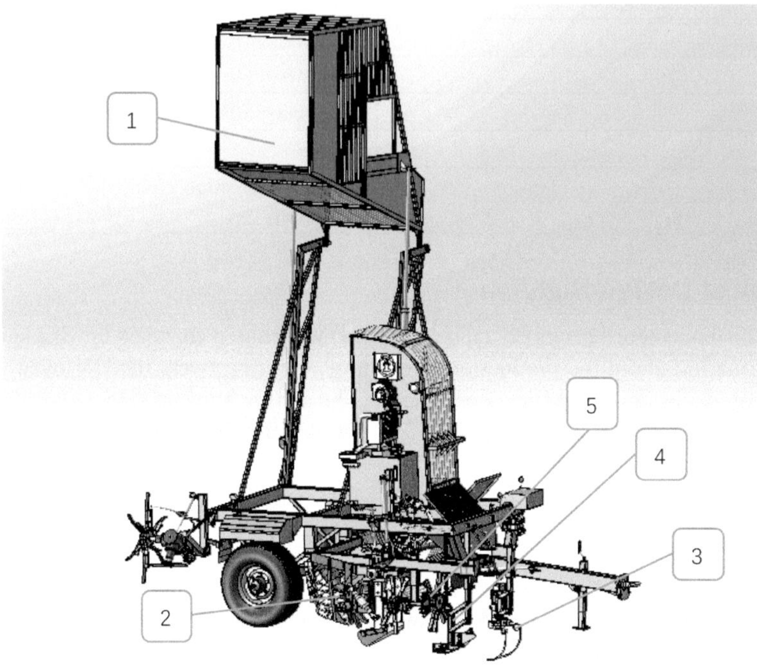

Fig. 1.34 Automatic Sugar Beet Harvester
1 - Clamping and Lifting Device 2 - Soil Filtering and Debris Removal Screw Roller 3 - Crawler 4 - Lifting Wheel Assembly 5 - Soil Filtering Conveyor Chain Assembly

II. Working Principle

The automatic sugar beet harvester operates by advancing the entire machine. Under the control of the crawler, the lifting wheel assembly pulls out sugar beets along with soil moisture and debris. After the first soil filtering and debris removal by the soil filtering conveyor chain assembly, the sugar beets are conveyed to the rotating soil filtering and debris removal screw roller for a second round of soil filtering and debris removal. Finally, the clamping and lifting device lifts the sugar beets to complete the unloading operation [26].

III. Main Structure Introduction

The crawler and lifting wheel assembly are shown in Fig. 1.35. The crawler unit (1) is used to detect the position of the sugar beets, while the lifting wheel assembly (2) pulls out the sugar beets and transports them to the soil filtering and debris removal screw roller for soil filtering and debris removal processing.

In the Filtering and Debris Removing Spiral Roller, as shown in Fig. 1.36, the sugar beets with soil and debris are fed into the spiral roller (2) for filtering and debris removal. The spiral roller (2) is inclined at a certain angle relative to the horizontal plane, and the rotating centrifugal force is used to throw out the soil and debris from the roots of the sugar beets [27].

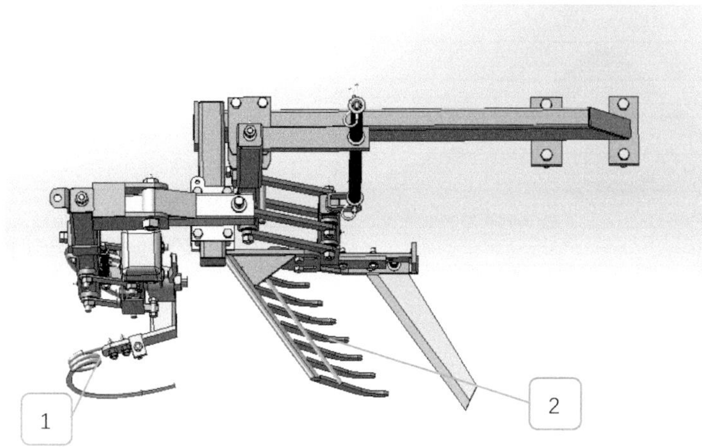

Fig. 1.35 Crawling Unit and Lifting Wheel Assembly
1 - Crawling Unit 2 - Lifting Wheel Assembly

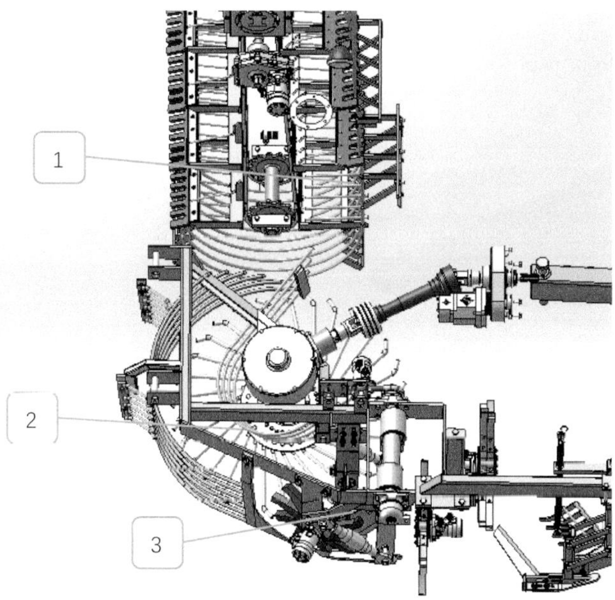

Fig. 1.36 Filter and Debris Removal Spiral Roller
1 - Vegetable Unloading Conveyor Assembly 2 - Filter and Debris Removal Spiral Roller 3 - Filter Soil Conveyor Assembly

IV. Specific Parameters (see Table 1.11)

Table 1.11 Automatic Sugar Beet Harvester Parameters

Applicable Tractor:	45-80 Horsepower
Number of Rows:	2 rows
Row Width:	50-66 cm

(Contd.)

Total Weight:	2000 kg
Tire Size:	7.50-16
Excavation Method:	Lift and pull wheel
Unloading Method:	Direct feeding
Harvesting Tank Payload:	0.7 t
Unloading Height:	2.5 m
Unloading Height:	2900 mm(H) × 5300 mm(L) × 1600 mm(W)

V. Mechanical Design Highlights

The principle of the automatic sugar beet harvester involves conveying the excavated sugar beets to the soil and fruit separation rollers via the front conveyor chain. During this process, a large amount of excavated soil falls through the gaps in the conveyor chain. The counter-rotating soil and fruit separation rollers primarily remove the soil adhering to the sugar beet tubers. Finally, the sugar beets are collected in the fruit collection box by the clamping and lifting mechanism. The cleaning mechanism of the automatic sugar beet harvester can clean the sugar beets of mud and debris from multiple directions and angles. Additionally, these cleaning components can serve as transportation parts, facilitating the centralized loading of sugar beets [28]. The inclined angle of the soil and fruit separation screw roller (2) can be adjusted as needed to prevent sugar beets from being squeezed and damaged due to impact forces.

Chapter 2

Packaging Machinery

Case 12: Automatic Liquid Filling Machine with Filling Head

I. Case Description

The filling machine is suitable for the mass production of products that require liquid filling, such as water injections in glass bottles and plastic bottles, as well as the filling of biochemical medicine bottles. It consists of the liquid filling main machine, conveyor belt, and electrical control box. The overall structure is shown in Fig. 2.1. The liquid filling main machine includes multiple pumps (1), liquid suction cylinders (2), and filling heads (9). The conveyor belt consists of a flat-top conveyor

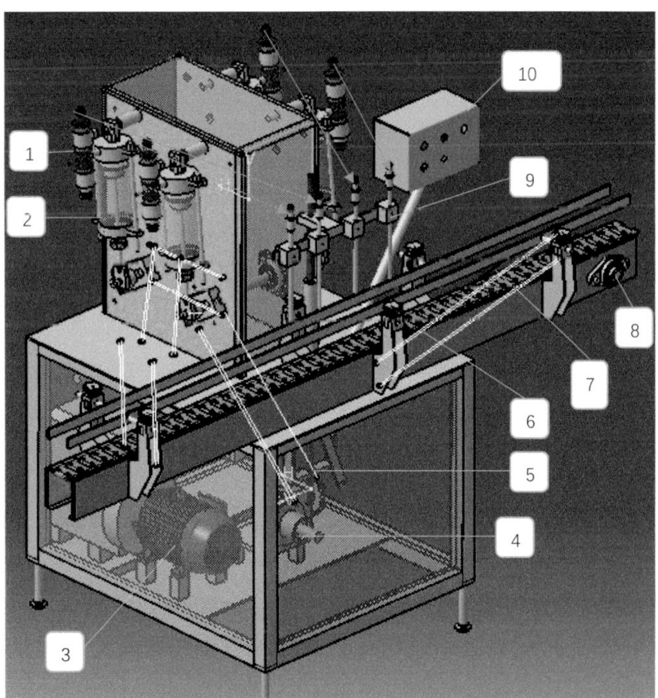

Fig. 2.1 Automatic Liquid Filling Machine with Heads
1-Multi-head Pump 2-Liquid Drawing Cylinder 3-Electric Motor 4-Speed Reduction Mechanism 5-Positioning Electromagnet 6-Flat-top Conveyor Chain 7-Guardrail 8-Electric Motor 9-Filling Head 10-Control Panel

chain (6), guardrails (7), an electric motor (8), and corresponding control circuits. The electrical components include an electric motor (3) and its drive power supply, variable speed reduction mechanism (4), and positioning electromagnetic iron (5) [29]. This filling machine is characterized by extremely high measurement accuracy and good repeatability. It has a wide range of filling capacities, and adjustments can be easily made to accommodate various container sizes.

II. Working Principle

The liquid filling main machine and conveyor belt are the primary mechanisms for completing the filling action. The main machine is equipped with 4 filling heads, and the solution is extracted by the multi-head pumps and liquid drawing cylinders on both sides, with each drawing cylinder corresponding to a filling head [30]. Each filling head has a miniature cylinder that drives a small piston to open and close the nozzle, acting as a valve. Before the filling machine operates, empty bottles are transported into the filling machine by the flat-top conveyor chain until they are positioned under the filling heads. At this point, the positioning electromagnet moves down, inserting the nozzles of the filling heads into the bottles. The valve opens at a certain distance from the bottom of the bottle to start filling the liquid. The filling heads move up while filling until the liquid filling is complete. Then, the nozzles are closed, and the heads move up to the bottle opening position, completing the liquid filling process [31].

III. Main Structure Introduction

The filling head, as shown in Fig. 2.2, is controlled by the positioning electromagnet (1) during the operation of the filling machine to adjust the height of the nozzle (3). The filling head is equipped with a miniature cylinder that drives a small piston to act as a valve, allowing for the precise opening and closing according to the set filling volume and achieving quantitative filling of the solution.

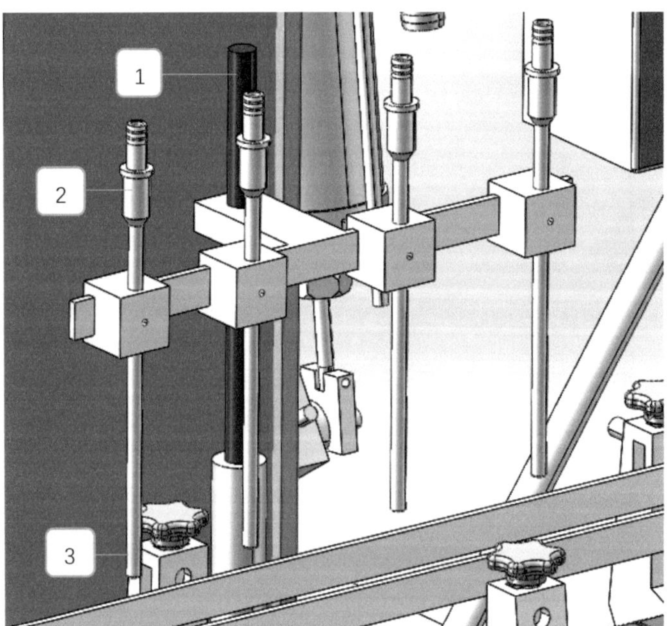

Fig. 2.2 Filling Head
1 - Positioning Electromagnet 2 - Flow Control Valve 3 - Nozzle

The power system of the filling machine is shown in Fig. 2.3. In this filling machine, the output shaft of the electric motor is connected to the drive shaft of the liquid extraction cylinder. This setup allows the electric motor to serve as both the power source for filling liquids and to precisely control the operation of the liquid extraction cylinder by controlling the running steps of the motor. This arrangement enables precise metering.

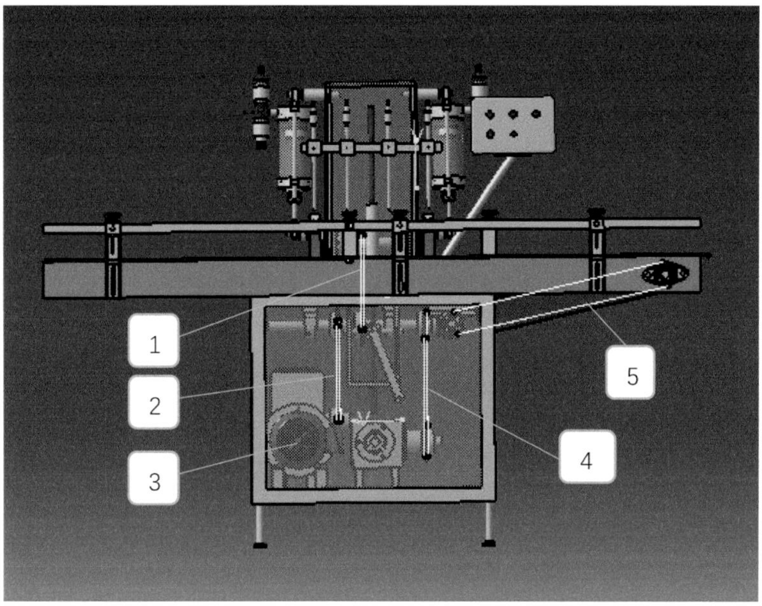

Fig. 2.3 Filling Machine Power System
1, 2, 4, 5 - Conveyor Belt 3 - Electric Motor

IV. Specific Parameters (see Table 2.1)

Table 2.1 Filling Machine Parameters

Working Efficiency:	1200~1500 bottles/hour (for standard capacity)
Filling Volume Range:	100~500 ml
Filling Accuracy:	≤±1 ~3% (depending on the volume and liquid)
Operating Temperature:	0~55°C
Air Pressure:	0.4-0.6 MPa

V. Mechanical Design Highlights

The 4-head automatic liquid filling machine is designed reasonably, and suitable for the automatic filling of various bottle shapes, with a filling range of 30 ml to 1 L. It is suitable for the filling of most liquid products, making the entire packaging process environmentally friendly, precise, and safe. It reduces labor costs for businesses, liberates people from the production line, and achieves packaging automation.

Case 13: Winding Machine

I. Case Description

Winding machine is a specialized machinery that forms a certain winding shape for large-area sheet-like materials according to a specific scheme. The overall structure is shown in Fig. 2.4. It mainly consists of auxiliary pressure roller mechanism (1), winding mechanism (2), tape feeding mechanism (3), cutting mechanism (4), frame (5), guiding mechanism (6), and control console (7). The winding machine has the advantages of simple operation, high winding accuracy, low equipment failure rate, and easy maintenance [32].

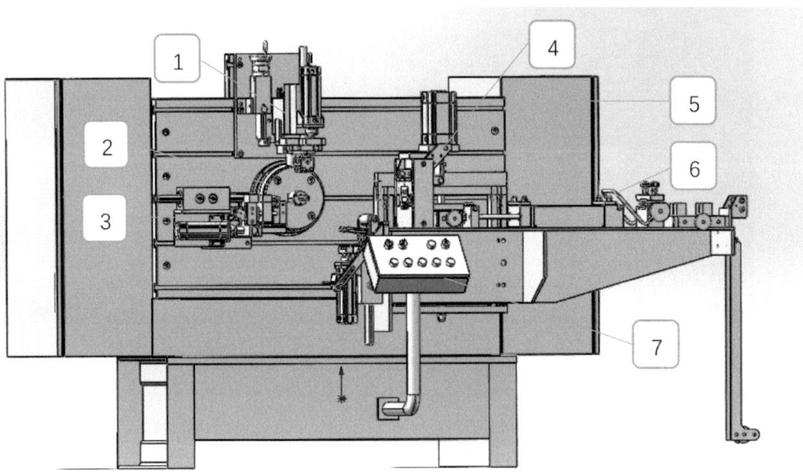

Fig. 2.4 Winding Machine
1-Auxiliary Pressure Roller Mechanism 2-Winding Mechanism 3-Tape Feeding Mechanism 4-Cutting Mechanism 5-Frame 6-Guiding Mechanism 7-Control Console

II. Working Principle

The winding mechanism and the guiding plate are the main mechanisms responsible for the winding action. When using the winding machine, the core of the wound material is placed on the winding needle, and the winding mechanism is connected to rotate the winding needle. The wound material is fed into the winding mechanism through the guiding mechanism, while the auxiliary pressure roller applies pressure to ensure the smooth operation of the winding machine. When the wound material reaches the set length, the cutting mechanism automatically cuts the material, and the winding mechanism continues to rotate, winding the remaining material onto the roll. At this point, the tape feeding mechanism applies the termination tape to the wound material, completing the entire winding process [33].

III. Main Structure Introduction

The guiding mechanism is shown in Fig. 2.5. When the winding machine is in operation, the sheet material is introduced from the right side of the guiding track (2). It passes through the transition wheel (3), transition wheel (4), elastic pressure plate (5), and transition wheel (1) in sequence. This process fully expands the sheet material. By adjusting the elastic pressure plate (5), the tension of the sheet material can be adjusted to prevent material damage due to excessive tension.

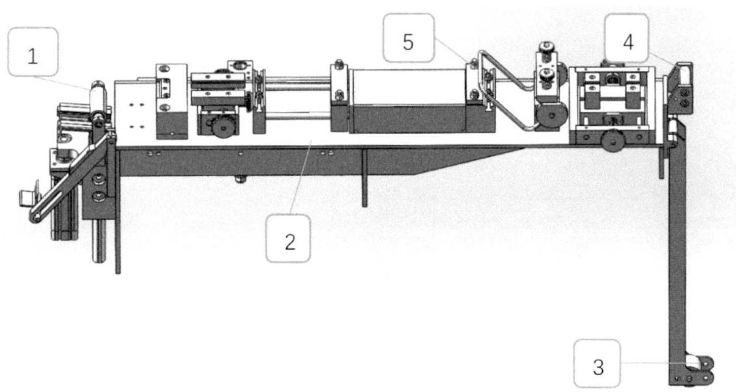

Fig. 2.5 Guiding Mechanism
1, 3, 4 - Transition wheels 2 - Guiding track 5 - Elastic pressure plate

The cutting mechanism, as shown in Fig. 2.6, mainly consists of cylinder (1), mounting bracket (2), and blade (3). When the wound material reaches the set length, cylinder (1) activates, and blade (3) descends to cut the material.

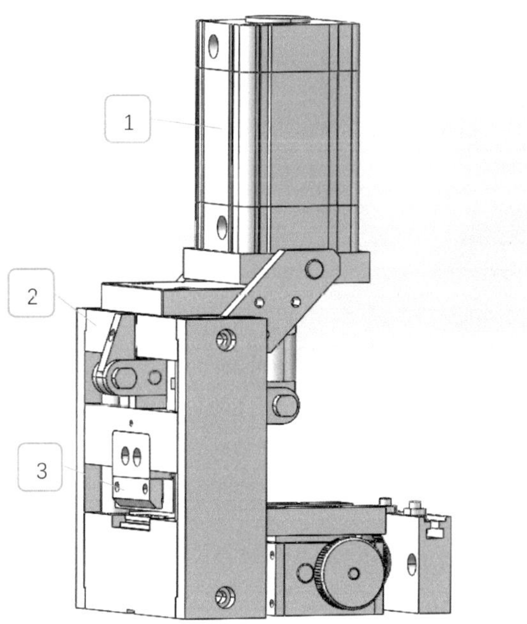

Fig. 2.6 Cutting Mechanism
1 - Cylinder 2 - Mounting Bracket 3 - Blade

The winding mechanism is shown in Fig. 2.7, mainly consisting of auxiliary pressure roller mechanism (1), cylinder (4), tape feeding mechanism (5), cylinder (7), and winding shaft (8). The winding shaft (8) continuously rotates under the power provided by cylinders (4) and (7). The auxiliary pressure roller mechanism (1) is controlled by cylinder (2), ensuring that roller (6) always adheres to the wound material, preventing it from becoming thicker and offsetting. After the winding is completed, the tape feeding mechanism (5) applies the finishing tape to the wound material.

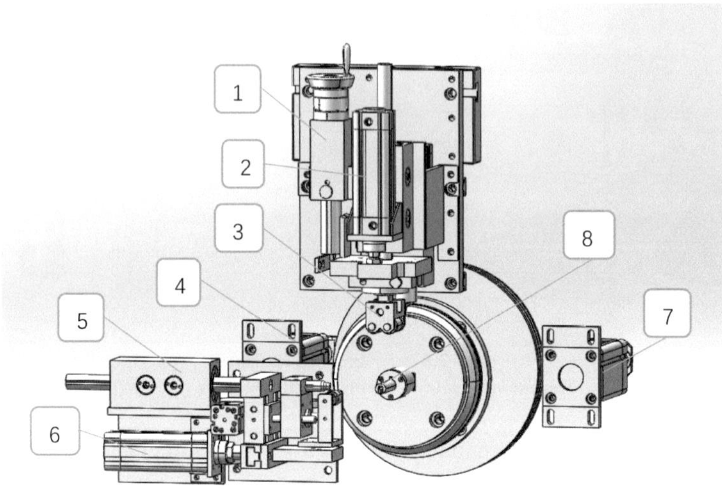

Fig. 2.7 Winding Mechanism
1 - Auxiliary Pressure Roller Mechanism 2, 3, 4, 7 - Cylinders 5 - Tape Feeding Mechanism 6 - Pressure Roller 8 - Winding Needle Image 2-7 Winding Mechanism

IV. Specific Parameters (see Table 2.2)

Table 2.2 Winding Machine Parameters

Speed:	2-4 t/h
Winding Accuracy (Overlap):	±0.5~0.7 mm
Utilization Rate:	≥93%
Yield Rate:	≥98%
Total Power of Equipment:	1.0 kW
Air Source Pressure:	0.4~0.7 MPa
Weight:	800 kg
Environmental Requirements:	Temperature: 1~45°C
Humidity:	<75%

V. Mechanical Design Highlights

The general winding machine's transition wheel cannot adjust the pressure required, making it prone to material damage due to tension. In this winding machine, an elastic pressure plate is added to the guiding mechanism. When the material passes over the surface of the outer tube [34], if the pressure is too high, the sheet-like material will compress the outer tube, further compressing the elastic pressure plate to adaptively adjust the tension of the material, making the winding process smoother.

VI. Precautions

(1) When hoisting the winding shaft, it should be done slowly and steadily, and no rough actions that could harm the equipment are permitted.

(2) The winding machine must be stopped when threading, to prevent the risk of hands being caught during threading.
(3) It is not permitted to enter the running area of the winding machine while it is in operation.

Case 14: One-line Sealing Machine

I. Case Description

The one-word sealing machine can complete the one-word sealing of the carton in one go, as shown in the overall structure in Fig. 2.8. It consists of the frame (1), sealing mechanism (2), conveying mechanism (3), width adjustment mechanism (4), tape holder (5), power switch (6), height adjustment mechanism (7), and tape holder (8). The sealing mechanism (2) is primarily composed of pressure rollers, pressure adjustment mechanisms, and tape-cutting mechanisms. The height adjustment mechanism (7) consists of a vertical rocking handle and a lifting frame. The width adjustment mechanism (4) is mainly composed of guide wheels and conveyor clamp rods [35]. The one-word sealing method is currently widely used in the market. It can operate as a standalone machine or be connected to an automated production line. Its characteristics include a broad range of applicable sizes and simple operation.

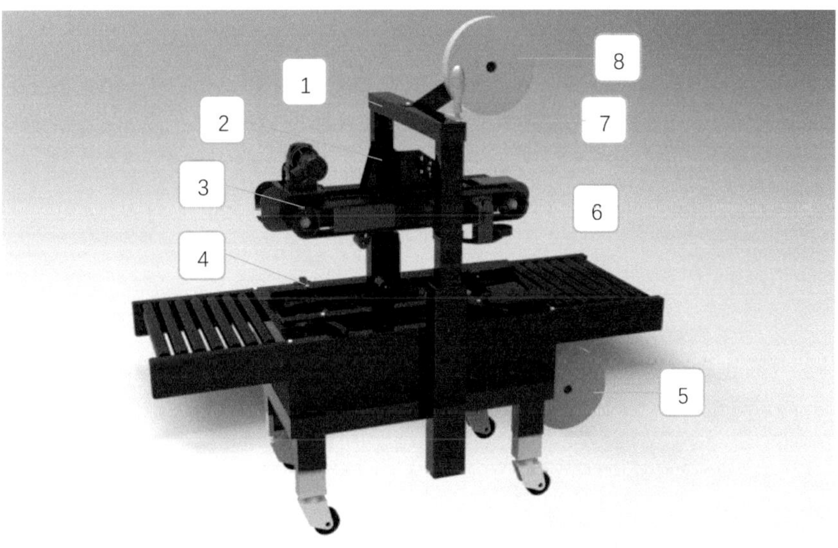

Fig. 2.8 One-word Sealing Machine
1 - Frame 2 - Sealing Mechanism 3 - Conveying Mechanism 4 - Width Adjustment Mechanism 5, 8 - Tape Holder
6 - Power Switch 7 - Height Adjustment Mechanism

II. Working Principle

During the initial debugging, the box to be sealed is placed into the conveying mechanism, with the length being approximately one-third of the box length. Adjust the height of the upper sealing mechanism by rocking the vertical rocking handle based on the size of the box, lowering it until it touches the box. Adjust the width adjustment mechanism's guide wheel and conveying clamp to ensure that they tightly adhere to the sides of the box and secure with screws. After debugging is complete, continuous operation can be performed [36]. During operation, press the power switch, push the box into the sealing machine, hand it over to the conveying mechanism, and let the guide wheel and conveying clamp grip the sides of the box. The conveyor belt drives the box through the sealing mechanism, automatically completing the upper and lower sealing of the cardboard box as well as the cutting of the tape [37].

III. Main Structure Introduction

The sealing mechanism, as shown in Fig. 2.9, is the core of the entire sealing machine, responsible for sealing and cutting the tape on the cardboard boxes. After turning on the power switch, the box to be sealed enters the sealing machine through the conveyor belt. As it passes through the sealing mechanism, the front pressure roller (4) applies the tape to the surface of the box. The box continues to move, and the front pressure roller flattens the tape on the box surface. After passing through the front pressure roller (4), the mechanism rebounds, and the tape cutting mechanism (3) on the machine cuts the tape. The rear pressure roller (2) then smoothe s the tape on the back of the box, completing the sealing operation.

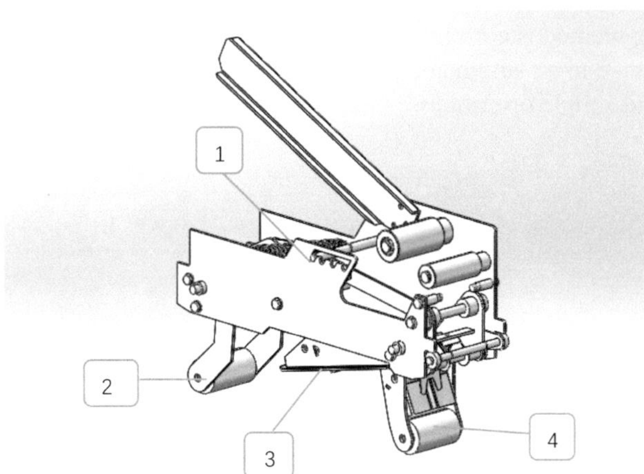

Fig. 2.9 Sealing Mechanism
1 - Pressure Adjustment Mechanism 2 - Rear Pressure Roller 3 - Tape Cutting Mechanism 4 - Front Pressure Roller

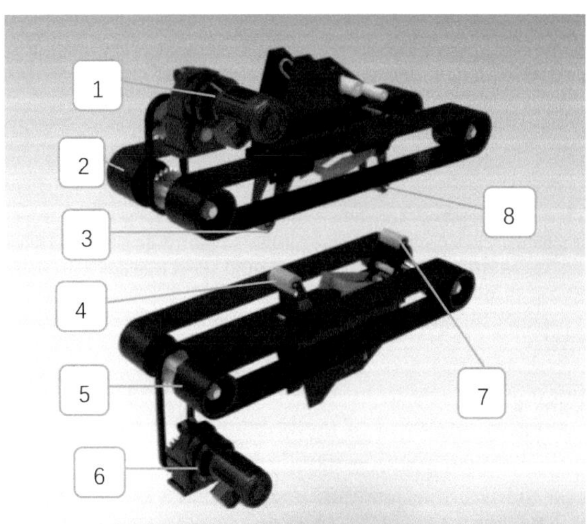

Fig. 2.10 Linked Sealing Structure
1, 6 - Electric Motor 2, 5 - Flat Belt 7, 8 - Front Pressure Roller 3, 4 - Rear Pressure Roller

The linked sealing structure is shown in Fig. 2.10. Each sealing machine has upper and lower sets of sealing mechanisms. When the box is conveyed into the sealing machine through the conveying mechanism, Front Pressure Roller (7) and Front Pressure Roller (8) apply tape to the box. The box continues to move, and the pressure rollers are pressed below the tabletop. After passing through the linkage mechanism, Rear Pressure Roller (3) and Rear Pressure Roller (4) apply tape to the back of the box, completing the sealing of both sides of the box.

IV. Specific Parameters (see Table 2.3)

Table 2.3 One-Sided Carton Sealing Machine Parameters

Conveying Speed:	0-20 m/min
Power Consumption:	1.8 KVA
Applicable Tape:	W48 mm/60 mm/75 mm Kraft Paper Tape, BOPP Tape
Temperature:	1~40°C
Humidity:	<75%
Machine Dimensions:	1350 mm(H) × 1020 mm(L) × 850 mm(W) (excluding front and rear roller stands)
Machine Weight:	120 kg

V. Mechanical Design Highlights

(1) Stable performance, reliable quality, high sealing efficiency, strong applicability, and long service life.
(2) Suitable for sealing lighter cartons after product packaging, providing a smooth, standardized, and aesthetically pleasing sealing effect.
(3) Adopts instant tape sealing, economically fast, easy to adjust, and can complete the sealing action in one go.
(4) Precision and durable mechanical components, tight structural design, vibration-free operation, ensuring stability and reliability during operation.
(5) Equipped with a blade protection device to prevent accidental injuries during operation, ensuring safe and efficient packaging.

Case 15: Strapping Machine

I. Case Description

Strapping machine is specialized equipment used for packaging electronic components into dedicated carrier tapes. The overall structure is shown in Fig. 2.11, with a detailed enlarged view in Fig. 2.12. It mainly consists of a display screen (1), tape conveying mechanism (2), frame (3), tension adjustment mechanism (4), cover tape disc (5), detection mechanism (6), and sealing mechanism (7) [38]. The strapping machine features a simple structure, easy operation and maintenance, and fast packaging speed.

II. Working Principle

The strapping machine works by extracting the tape from the cover tape disc, adjusting the tension through the tension adjustment mechanism, and feeding it into the tape conveying mechanism. At this point, the tape with electronic components is introduced into the tape conveying mechanism

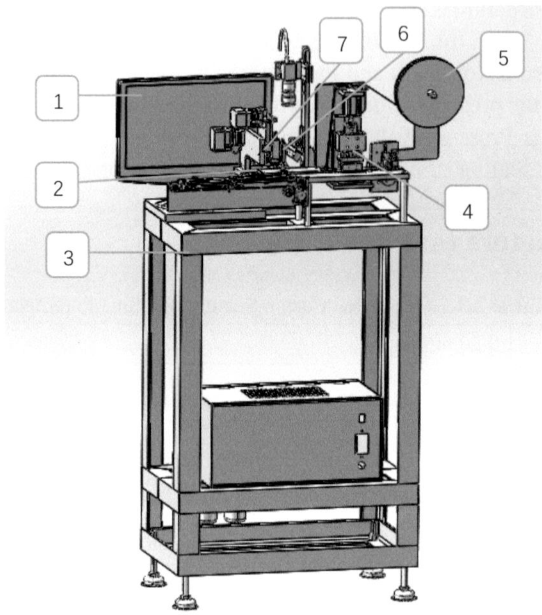

Fig. 2.11 Strapping Machine
1-Display Screen 2-Tape Conveying Mechanism 3-Frame 4-Tension Adjustment Mechanism 5-Cover Tape Disc 6-Detection Mechanism 7-Sealing Mechanism

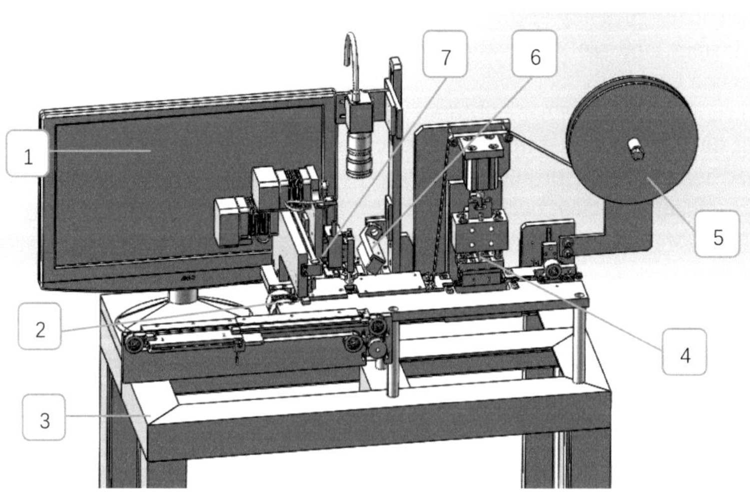

Fig. 2.12 Enlarged View of Strapping Machine
1-Display Screen 2-Tape Conveying Mechanism 3-Frame 4-Tension Adjustment Mechanism 5-Cover Tape Disc 6-Detection Mechanism 7-Sealing Mechanism

[39]. The tape conveying mechanism, driven by the electric motor, transports the tape and cover tape forward to the sealing mechanism. The sealing mechanism heats and presses the tape, completing the packaging process. Simultaneously, the detection mechanism monitors the real-time sealing condition of the tape to ensure a complete seal.

III. Main Structure Introduction

The tape conveying mechanism is shown in Fig. 2.13, where the tape is placed in the tape groove (1) and is continuously transported to the sealing mechanism by the electric motor (2).

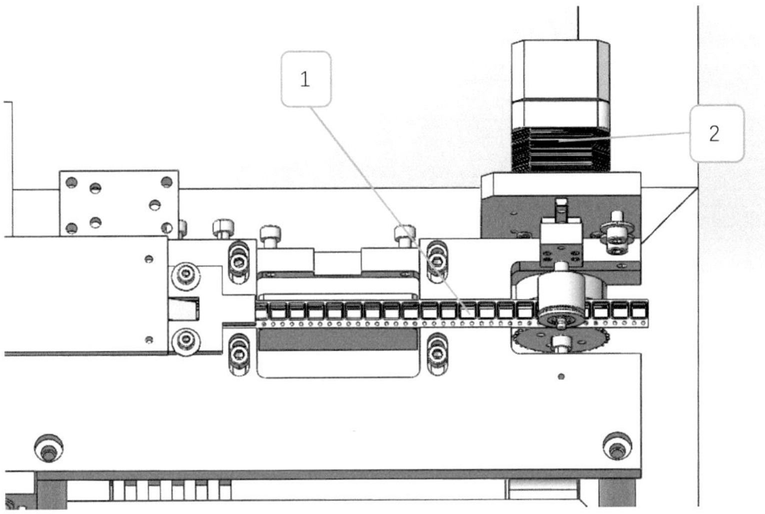

Fig. 2.13 Tape Conveying Mechanism
1 - Tape Groove 2 - Electric Motor

Detection mechanism and sealing mechanism are shown in Fig. 2.14. The detection mechanism utilizes a camera (3) to capture images and transmit them to the computer system for assessing the sealing condition of the tape. The sealing mechanism controls the lifting of the sealing head through electric motor (1) and the lateral displacement of the sealing head through electric motor (2). Both motors work in coordination to repeatedly seal the tape [39].

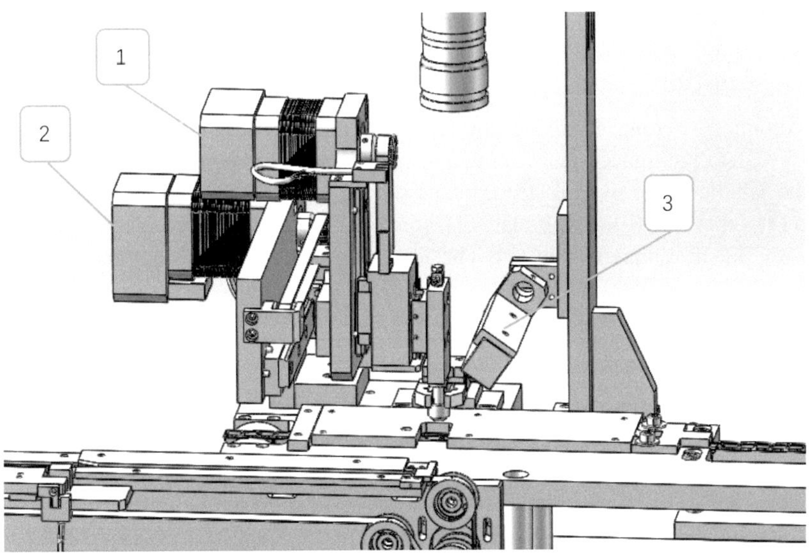

Fig. 2.14 Shows the detection mechanism and sealing mechanism.
1, 2 - Electric motors 3 - Camera

IV. Specific Parameters (see Table 2.4)

Table 2.4 Belt Bundling Machine Parameters

Packaging Method:	Automatic sealing, automatic material collection, manual boxing
Packaging Speed:	1000-5000 pcs/h, freely adjustable according to requirements
Operation Rate:	≥93%
Good Product Rate:	≥98%
Packaging Width:	8~15 mm
Air Pressure:	4~5 kg/cm^2
Temperature Adjustment Range:	Room temperature~300°C
Sealing Method:	Stepless speed regulation for belt movement, reciprocating sealing, and continuous sealing, suitable for cold sealing self-adhesive belts and hot sealing belts
Equipment Dimensions:	1355 mm(H) × 785 mm(L) × 460 mm(W)

V. Mechanical Design Highlights

When the belt bundling machine is in operation, the tape is pulled out from the belt collecting disc. After adjusting the tension with the tension adjustment mechanism, it is fed into the tape conveying mechanism. The electronic components to be packaged are placed in the tape. Under the drive of the motor inside the conveying mechanism, the tape moves forward. Subsequently, it passes through the sealing component, where the tape is heated and press-sealed. After that, it goes around the tape roller and winds onto the collecting disc [40]. By installing an anti-rotation component on the carrier tape disc, the collecting disc can only rotate in a specific direction, preventing automatic rotation. This simple and convenient structure effectively addresses the issue of automatic rotation of the collecting disc, making it highly practical.

Case 16: Fully Automatic Vertical Round Bottle Labeling Machine

I. Case Description

The Fully Automatic Vertical Round Bottle Labeling Machine is suitable for labeling round bottles in industries such as pharmaceuticals, food, and cosmetics. It can be used for full-wrap labeling and semi-circular labeling. The overall structure, as shown in Fig. 2.15, consists mainly of adjustment mechanism (1), label feed (2), fixing plate (3), conveyor belt (4), material receiving rack (5), display screen (6), and labeling mechanism (7). The machine features a wide application range, flexible usage, high stability, excellent labeling quality, and durability [41].

II. Working Principle

Before the fully automatic vertical round bottle labeling machine starts working, adjust the adjustment mechanism to the appropriate height and angle according to the labeling requirements for round bottles. After the adjustment, start the conveyor belt to transport the round bottle to the labeling mechanism. The labeling belt on the fixing plate is driven by an electric motor to control and rotate the round bottle, thereby attaching the label to the designated position on the round bottle, completing the labeling process [42].

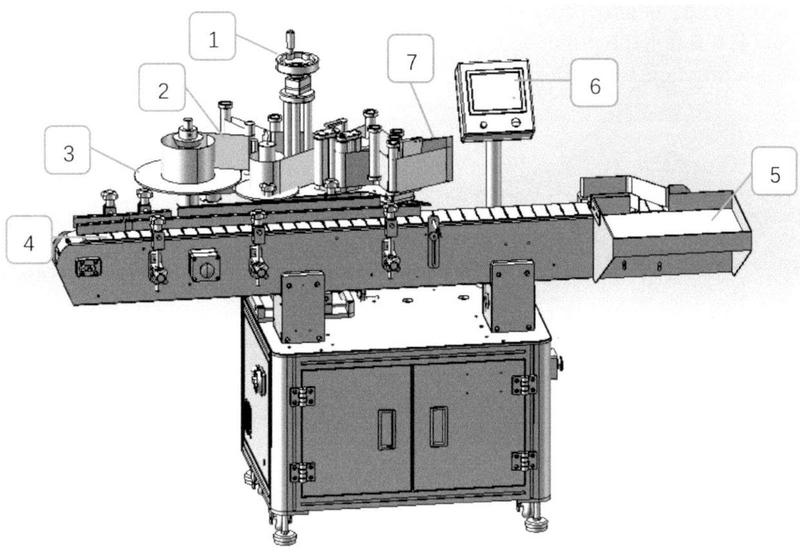

Fig. 2.15 Fully Automatic Vertical Round Bottle Labeling Machine
1-Adjustment Mechanism 2-Label Feed 3-Fixing Plate 4-Conveyor Belt 5-Material Receiving Rack 6-Display Screen 7-Labeling Mechanism

III. Main Structure Introduction

Adjustment mechanism, as shown in Fig. 2.16, adjusts the height of the fixing plate by shaking handle (1), the horizontal lateral displacement of the fixing plate by shaking handle (2), the horizontal vertical displacement of the fixing plate by shaking handle (3), and the tilt angle of the fixing plate by shaking handle (4).

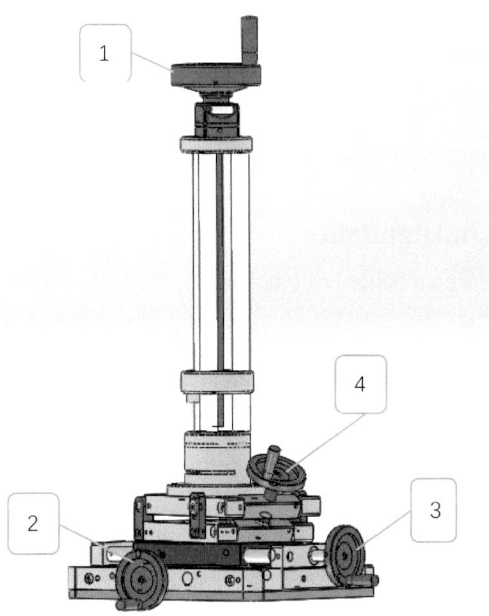

Fig. 2.16 Adjustment Mechanism
1, 2, 3, 4-Handles

The fixing plate, as shown in Fig. 2.17, is driven by the electric motor (1) and rotates the rotating shaft (4) through the pulley (3). This rotation causes the labeling tape to move, attaching the label on the labeling tape to the designated position on the round bottle [43].

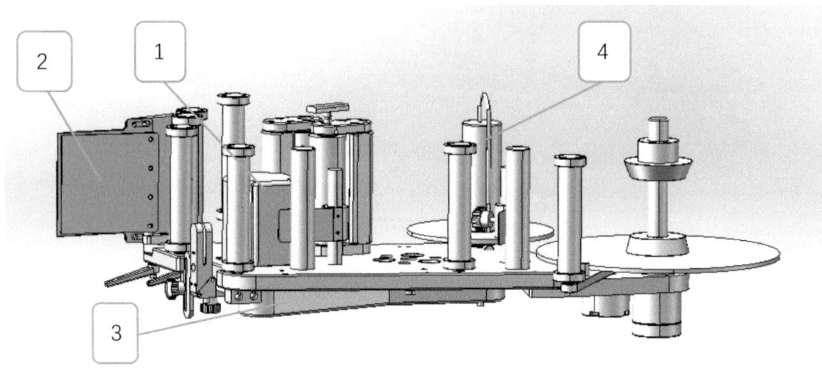

Fig. 2.17 Fixed Plate
1 - Electric motor 2 - Labeling mechanism 3 - Pulley 4 - Rotating shaft

IV. Specific Parameters (see Table 2.5)

Table 2.5 Parameters of Fully Automatic Vertical Round Bottle Labeling Machine

Applicable Bottle Diameter (for round bottles):	ϕ25 mm ~ ϕ100 mm
Applicable Bottle Height:	25~230 mm
Labeling Accuracy:	±1 mm
Labeling Speed:	40~120 pcs/min (Dependent on bottle and label size)
Weight:	Approximately 185 kg
Power:	980 W
Machine Dimensions:	1300 mm(H) × 1950 mm(L) × 1100 mm(W)

V. Mechanical Design Highlights

(1) Wide Applicability: The machine can meet the full-circle or semi-circle labeling requirements for round bottles. The switch between labeling of front and rear bottles is simple, and adjustments are convenient.

(2) High Label Overlapping Precision: The labeling belt uses a correction mechanism to ensure that the belt does not deviate. There are eight degrees of freedom for adjustment in the x/y/z directions and inclination, allowing precise adjustments with no dead angles and ensuring high label overlapping precision [44].

(3) Superior Labeling Quality: The use of elastic pressure-covered labeling belts results in smooth and wrinkle-free labels, enhancing packaging quality.

(4) Flexible Application: The machine labels bottles in a standing position, and it is equipped with automatic bottle separation functionality. It can be used for standalone production or integrated into a production line.

(5) Intelligent Control: The machine features automatic photoelectric tracking, ensuring that labels are applied to every item. It also has functions for automatic correction in the absence of items and automatic label detection to prevent missed labeling and label wastage [45].
(6) Easy Adjustment: Labeling speed, conveyor speed, and bottle separation speed can be continuously adjusted according to needs, allowing for versatile adjustments.
(7) Sturdy and Durable: The machine adopts a three-rod adjustment mechanism, making full use of the stability of the triangle. The overall structure is robust and durable.

VI. Applicability

(1) Applicable Labels: Self-adhesive labels, adhesive films, electronic supervision codes, barcodes, etc.
(2) Applicable Products: Products that require labeling or film attachment on the circumferential surface.
(3) Applicable Industries: Widely used in the food, pharmaceutical, cosmetics, daily chemical, electronics, hardware, plastic, and other industries.
(4) Application Examples: Labeling of round bottles, labeling of plastic bottles, labeling of food cans, etc.

Case 17: Non-standard Inspection and Adhesive Packaging Integrated Machine

I. Case Description

Non-standard inspection and adhesive packaging integrated machines are used for tasks such as product flatness inspection and adhesive packaging. The overall structure is illustrated in Fig. 2.18, with an enlarged view provided in Fig. 2.19. It mainly consists of a conveyor belt (1), frame (2), adhesive labeling module (3), packaging module (4), flatness detection module (5), and handling mechanism (6). The detection module primarily comprises a photoelectric coupler detector.

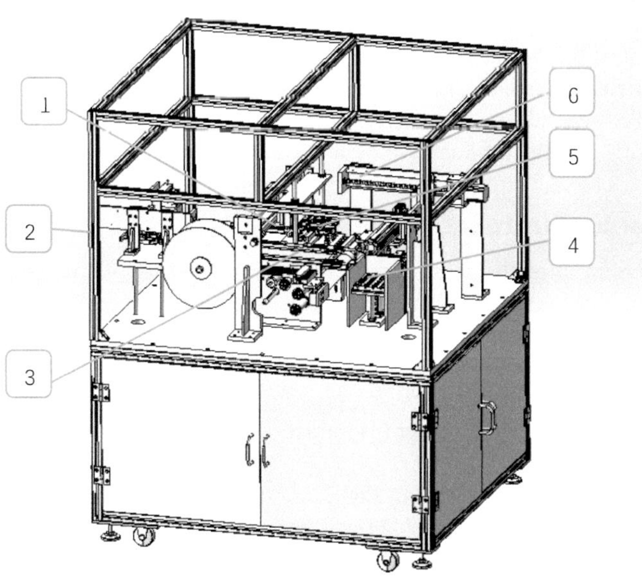

Fig. 2.18 Non-standard inspection and adhesive packaging integrated machine
1-Conveyor belt 2-Frame 3-Adhesive labeling module 4-Packaging module 5-Flatness detection module 6-Handling mechanism

This integrated machine automates the steps of detection, adhesive application, and packaging, avoiding manual inspection or adhesive application that could cause secondary deformation to the product, thus affecting product quality [46]. By integrating these processes into one machine, it reduces the need for multiple devices, thereby increasing production efficiency, saving space occupied by equipment, and significantly reducing labor costs.

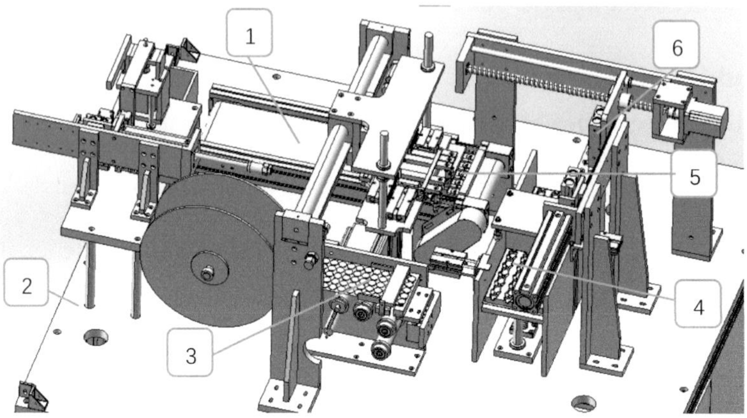

Fig. 2.19 Partial enlarged view of the non-standard inspection and adhesive packaging integrated machine
1-Conveyor belt 2-Frame 3-Adhesive labeling module 4-Packaging module 5-Flatness detection module 6-Handling mechanism

II. Working Principle

The non-standard inspection and adhesive packaging integrated machine operates by conveying the products to be inspected to the flatness detection module using the conveyor belt. The products are then examined by the photoelectric coupler detector, which determines the inspection results. If the product is found to be non-compliant, the first handling mechanism retrieves the product. If the product is deemed good, the first handling mechanism transfers the product to the adhesive labeling module [46]. Once the first handling mechanism moves away, the cylinder extends, and the pressure block presses against the product for 2-3 seconds, ensuring tight adhesion between the insulating paper and the shielding cover. After the adhesive labeling is complete, the cylinder retracts the pressure block. The second handling mechanism picks up the product with the attached adhesive paper and transports it to the packaging module, where the product is packaged and output.

III. Main Structure Introduction

The adhesive labeling module, as shown in Fig. 2.20, consists of the adhesive tape wheel (1), which supplies the insulating paper for the module. During the operation of the adhesive labeling and packaging integrated machine, the first handling mechanism picks up the good products and transports them to the flattened surface of the adhesive paper. At this point, cylinder (4) is activated, causing pressure block (3) to extend upward, pressing against the product for 2-3 seconds. Once the adhesive labeling is completed, cylinder (4) retracts, pulling back pressure block (3), thus finishing the adhesive labeling process.

The packaging module, as shown in Fig. 2.21, receives the product after adhesive application from the second handling mechanism (1) and accurately places it in the packaging box (2) for packaging. After packaging is complete, the product is output.

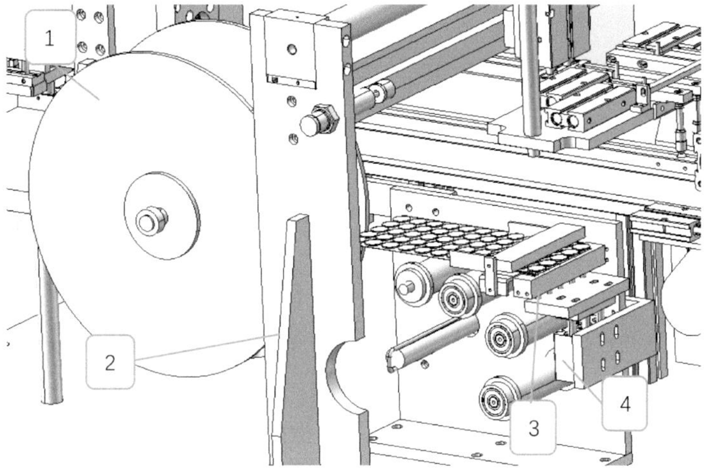

Fig. 2.20 Adhesive Labeling Module
1 - Adhesive Tape Wheel 2 - First Handling Mechanism 3 - Pressure Block 4 - Cylinder

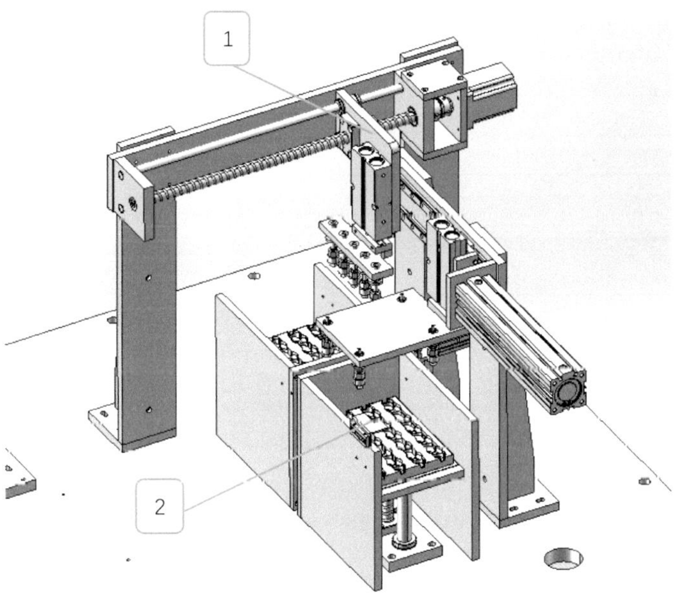

Fig. 2.21 Packaging Module
1 - Second Handling Mechanism 2 - Packaging Box

The handling mechanism is shown in Fig. 2.22, mainly composed of the first handling mechanism (1) and the second handling mechanism (8). The first handling mechanism (1) is controlled by cylinder (2) to raise and lower the suction cup (3). Each suction cup of suction cup (3) is controlled by a cylinder and is used to sort out the unqualified products and transport the good products to the glue coating module. The second handling mechanism (8) is controlled by motor (7), cylinder (5), and cylinder (6), enabling the transportation of products in three-dimensional space. It is used to pick up the products after glue application and transport them to the packaging module.

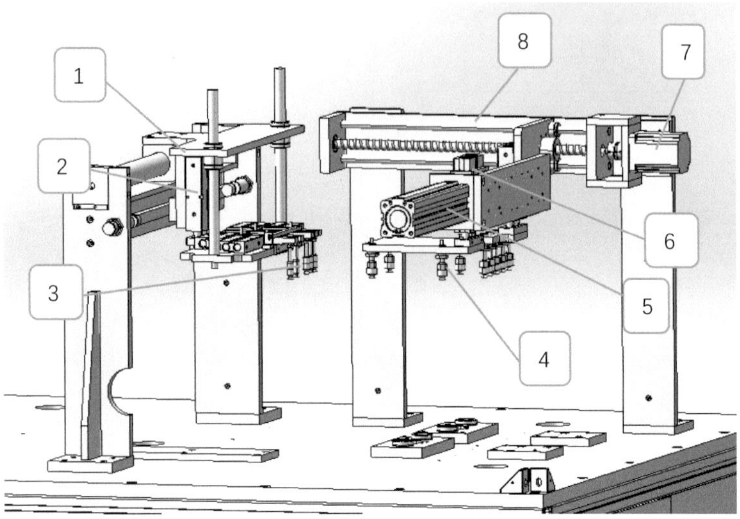

Fig. 2.22 Conveying Mechanism
1 - First Conveying Mechanism 2, 5, 6 - Cylinders 3, 4 - Suction Nozzles 7 - Electric Motor 8 - Second Conveying Mechanism

IV. Specific Parameters (see Table 2.6)

Table 2.6 Non-standard Inspection Gluing and Packaging Integrated Machine Parameters

Operating Efficiency:	2000-2500 pcs/h
Error Rate:	≤0.2%
Operating Temperature:	0~55°C
Allowable Humidity:	20%~95%
Equipment Dimensions:	1600 mm(H) × 1300 mm(L) × 1140 mm(W)

Case 18: Non-standard Label Printer

I. Case Description

Non-standard label printers are used to print labels on the surface of products. The overall structure is shown in Fig. 2.23, with a close-up view provided in Fig. 2.24. It mainly consists of the frame (1), workbench (2), control box (3), high-precision slide table (4), exhaust port (5), and printing mechanism (6). The high-precision slide table (4) includes swing blocks, fixed blocks, and moving blocks. Non-standard label printers are characterized by stable operation, high reliability, and high work efficiency, ensuring effective label printing quality on products.

II. Working Principle

The non-standard labeling machine operates by fixing the product on the workbench. The high-precision slide platform transports the workbench below the labeling mechanism and aligns the product with the printing head of the labeling mechanism, thereby imprinting labels on the product.

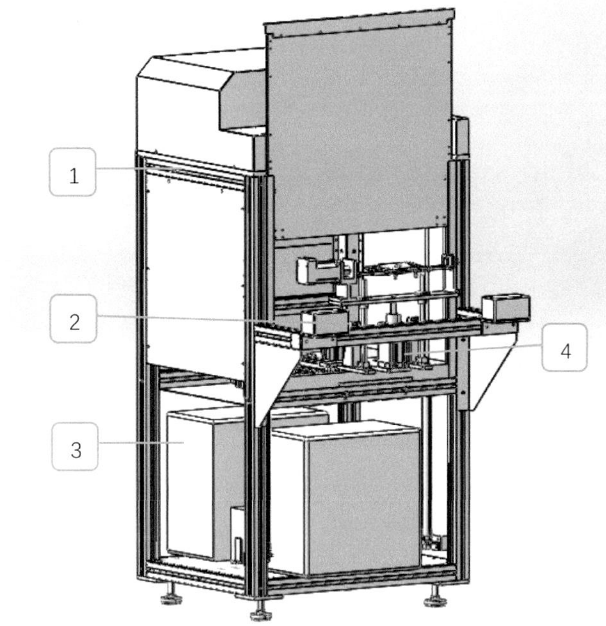

Fig. 2.23 Non-standard Labeling Machine
1 - Frame 2 - Workbench 3 - Control Box 4 - High-precision Slide Platform

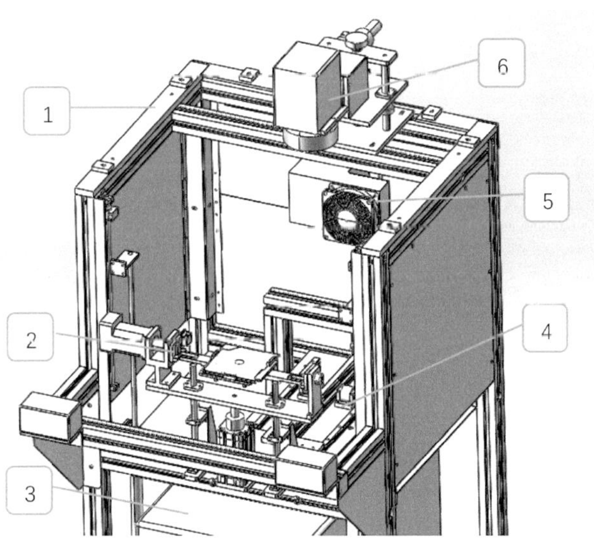

Fig. 2.24 Partial Enlarged View of Non-standard Labeling Machine
1 - Frame 2 - Workbench 3 - Control Box 4 - High-precision Slide Platform 5 - Exhaust Port 6 - Labeling Mechanism

III. Main Structure Introduction

The printing and labeling mechanism, as shown in Fig. 2.25, can adjust the height of the printing head (1) using the vertical lifting rod (2) to accommodate different specifications of labeled products.

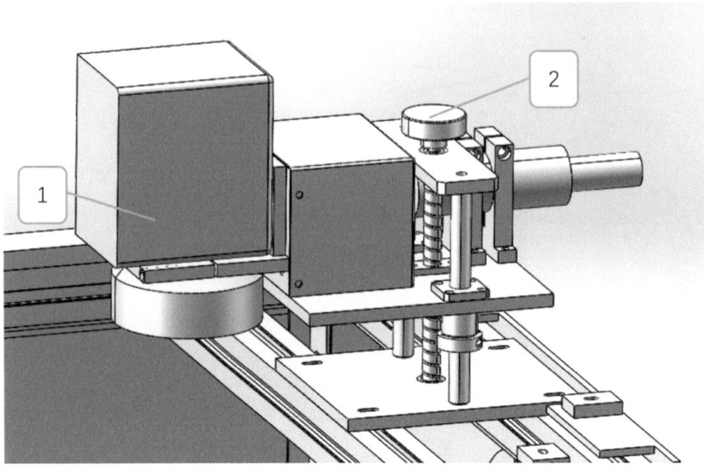

Fig. 2.25 Printing and Labeling Mechanism
1 - Printing Head 2 - Vertical Lifting Rod

The high-precision sliding table, as shown in Fig. 2.26, is controlled by cylinder (1) to achieve horizontal displacement of the workbench (2) and by cylinder (3) to achieve vertical displacement of the workbench (2). Meanwhile, the high-precision sliding table (4) can slide on the track (4), and a fixed block (5) is provided at the end of the track to prevent disconnection failure of the workbench (2).

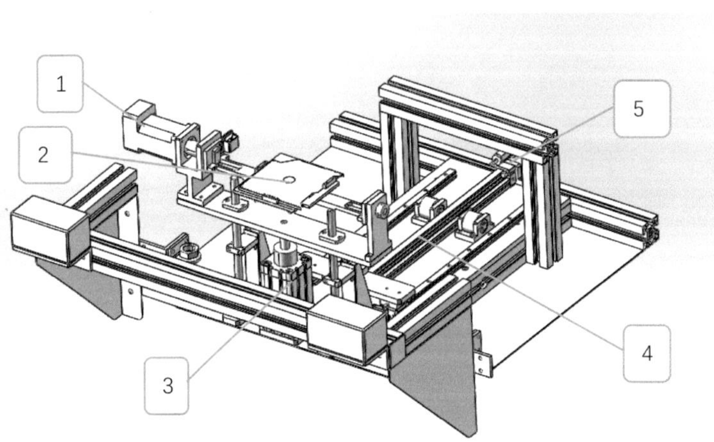

Fig. 2.26 High-precision Sliding Table
1, 3 - Cylinders 2 - Workbench 4 - Guide Rail 5 - Fixed Block

IV. Specific Parameters (see Table 2.7)

Table 2.7 Non-standard Labeling Machine Parameters

Yield Rate:	≥98%
Repeatability:	0.05 mm
Workbench Stroke:	X:200 mm Y: 480 mm Z:300 mm

Contd.

Temperature:	0~55°C
Humidity:	<75%
Equipment Dimensions:	1530 mm(H) × 740 mm(L) × 1170 mm(W)

V. Mechanical Design Highlights

The high-precision slide table (4) of the non-standard labeling machine is equipped with fixed block (8) and movable block (9), allowing the workbench (2) to move only within a specified range, thereby preventing the workbench (2) from disengaging. The control mechanism of the non-standard labeling machine adopts a PLC control box (3) (Programmable Logic Controller), which can automatically control the switching of solenoid valves and simultaneously control the operation of the entire printing equipment. This ensures synchronization between the labeling mechanism (6) and the product handling, effectively guaranteeing the printing quality of the product.

Case 19: Non-standard Vertical Automatic Packaging Machine

I. Case Description

The non-standard vertical automatic packaging machine is suitable for packaging granular, short, and powdery materials such as puffed food, sugar, salt, and laundry detergent. The overall structure is shown in Fig. 2.27, mainly composed of control unit (1), film pulling mechanism (2), film feeding mechanism (3), film releasing mechanism (4), forming drive mechanism (5), cutting mechanism (6), forming mold (7), and frame (8).

This machine features a rational design, simple operation, and convenient use. The non-standard vertical automatic packaging machine controls the film pulling mechanism with a servo motor, pre-stretching the film with a frequency converter-controlled motor, and controls longitudinal and transverse sealing through the drive system to shape the packaging film. This not only improves work efficiency and saves space and labor but also enhances automation.

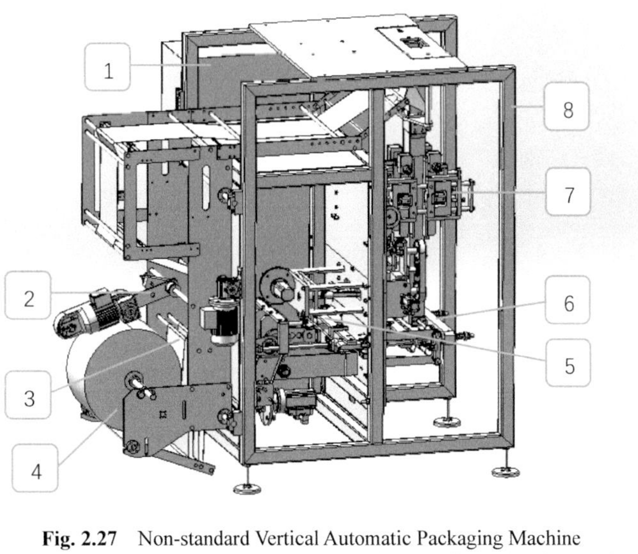

Fig. 2.27 Non-standard Vertical Automatic Packaging Machine
1-Control Unit 2-Film Pulling Mechanism 3-Film Feeding Mechanism 4-Film Releasing Mechanism 5-Forming Drive Mechanism 6-Cutting Mechanism 7-Forming Mold 8-Frame

II. Working Principle

During the operation of the non-standard vertical automatic packaging machine, the forming mold of the equipment automatically shapes the packaging film based on temperature, pressure, and time settings. The film feeding mechanism is primarily responsible for controlling the tension of the packaging film. The film releasing mechanism uses a motor to actively release the film and automatically controls the film releasing motor based on the length of the packaging film. When not in use, the brake under the film loading shaft automatically engages. The film loading mechanism is supported by double-sided bearings to ensure the alignment of the film loading shaft with the packaging film. The cutting mechanism is used to cut the sealed film. The control unit controls the automatic operation of the machine through a programmable controller, a touchscreen interface, servo motors, and frequency converters [47].

III. Main Structure Introduction

The film feeding mechanism and film guiding mechanism are shown in Fig. 2.28. The film feeding mechanism is driven by motors (1) and (2), ensuring the stable input of the packaging film into the forming mold. The film guiding mechanism (3) is equipped with multiple film guiding shafts to control the tension of the packaging film.

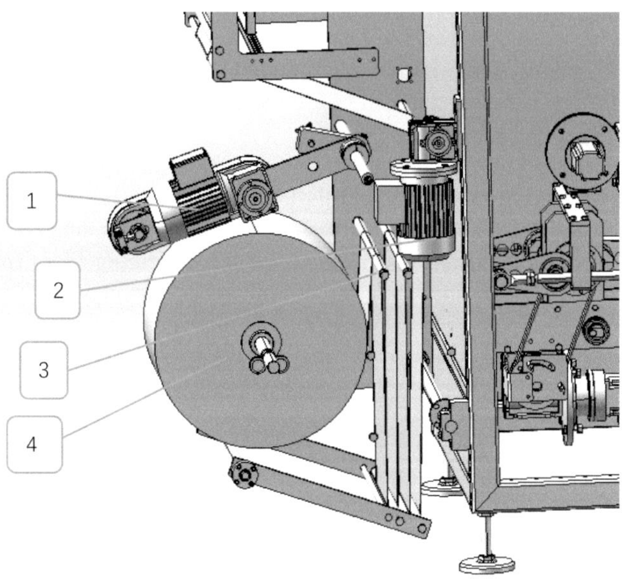

Fig. 2.28 Film Pulling Mechanism and Film Walking Mechanism
1, 2-Motors 3-Film Walking Shaft 4-Film Feeding Mechanism

Figure 2.29 shows the Forming Mechanism, mainly driven by the Forming Drive Mechanism. During operation, the heating elements (6) of the forming mechanism are preheated. Once they reach the appropriate temperature, the packaging film is fed from above the forming mechanism (1). The forming drive mechanism controls the sealing of the heating elements (6) on both sides, thus sealing the film. Simultaneously, the film sealed on the sides is wrapped around the product surface and transported to the cutting mechanism (3), where the packaging film is cut, completing the packaging process.

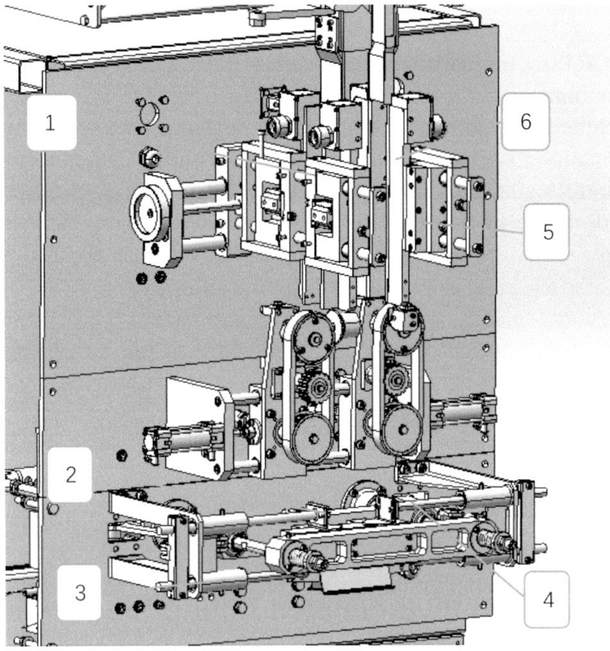

Fig. 2.29 Forming Mechanism
1 - Forming Mechanism 2 - Blade 3 - Cutting Mechanism 4 - Guide Rail 5 - Heating Mechanism 6 - Heat-sealing Strip

IV. Specific Parameters (see Table 2.8)

Table 2.8 Non-Standard Vertical Automatic Packaging Machine Parameters

Maximum Film Width:	600 mm
Bag Length:	80~350 mm
Maximum Film Roll Diameter:	320 mm
Packaging Speed:	10~50 bag/min
Measuring Range:	150~2000 ml
Film Thickness:	0.04~0.08 mm
Equipment Dimensions:	1815 mm(H) × 1970 mm(L) × 1350 mm(W)

V. Mechanical Design Highlights

(1) Advanced overall design with a rational structure and reliable performance;
(2) Equipped with dual synchronous belt film pulling mechanism, tension control by cylinders, automatic correction, and automatic alarm protection functions to minimize losses;
(3) This machine is compatible with the metering device, integrating bag making, filling, sealing, printing, and counting into one unit;
(4) The opening and closing methods for metering can be modified based on the material, allowing measurement of the packaged material quantity.

VI. Precautions

(1) Operators must follow the specified sequence of operations proficiently to avoid potential machine malfunctions;
(2) During the machine's operation, attention should be paid to the normality of machine sounds, temperature, remaining packaging film, and material supply. Operators should not leave the packaging machine unattended during production, and any issues discovered should prompt an immediate shutdown for inspection;
(3) When adjusting various components, remember the original dimensions and positions for restoring to the starting point in case of undesirable adjustments;
(4) New machines should undergo a run-in period at slow speeds for 100-150 hours;
(5) Avoid placing tools or foreign objects on the machine to prevent them from falling into the molds and causing damage;
(6) When pausing work, detach the sealing molds;
(7) After prolonged operation, use a dedicated copper brush to clean the surface of the molds, promoting sealing quality;
(8) Keep hands and objects away from moving parts during machine operation;
(9) Regularly check for any loose or disconnected parts on the machine.

Case 20: Highly Automated Filling Production Line

I. Case Description

Highly Automated Filling Production Line is used for mass production of products that require liquid filling, such as glass and plastic bottles for water injections, bottled biopharmaceuticals, etc. The overall structure is shown in Fig. 2.30, mainly composed of a robotic arm (1), automatic filling mechanism (2), conveyor mechanism (3), automatic capping mechanism (4), automatic sealing mechanism (5), and discharge mechanism (6). The control system of the highly automated filling production line adopts a PLC (Programmable Logic Controller), making the operation of the automated production line smoother, positioning more accurate, and operation more convenient [48]. This significantly improves the filling speed, bringing considerable productivity gains to the enterprise.

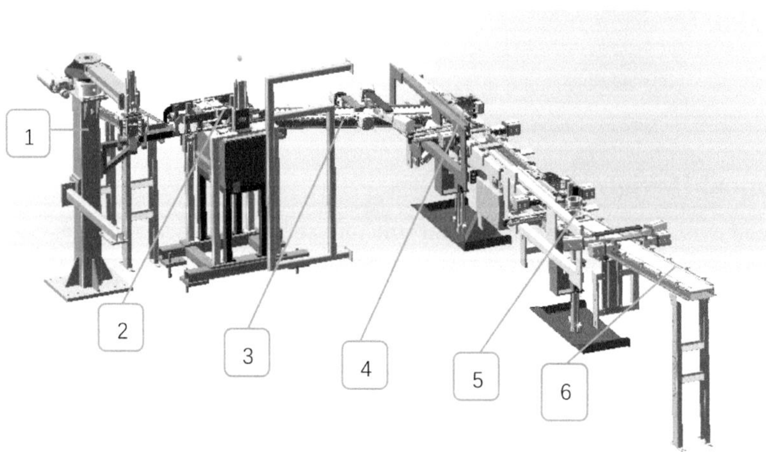

Fig. 2.30 Highly Automated Filling Production Line
1-Robotic Arm 2-Automatic Filling Mechanism 3-Conveyor Mechanism 4-Automatic Capping Mechanism
5-Automatic Sealing Mechanism 6-Discharge Mechanism

II. Working Principle

The highly automated filling production line operates by turning on the overall system power. The conveyor belt starts running, and the robotic arm moves an empty filling bottle to the production line. The pushing mechanism pushes the empty filling bottle onto the conveyor belt for transportation. When the empty filling bottle reaches the position of the automatic filling mechanism's sensor, the conveyor belt stops, and the automatic filling mechanism starts the filling process. When the preset quantity is reached, the filling stops, and the conveyor belt restarts until the next bottle to be filled is conveyed to the filling equipment or the equipment is stopped. The automatic filling mechanism can automatically count the filling process using sensors or manually reset the filling quantity to zero [49].

When the filled bottles reach the sensor position of the automatic capping mechanism, the sensor detects the filled bottle and activates the automatic capping mechanism, placing the cap on the filled bottle's neck. After capping, the filled bottles are conveyed to the sealing platform of the automatic sealing mechanism, where the bottles are sealed. After sealing, the products are transported to the packaging process through the conveyor belt.

III. Main Structure Introduction

The robotic arm, as shown in Fig. 2.31, is controlled by an electric motor (2) to rotate the mechanical arm (1) in the horizontal plane. The lifting and lowering of the robotic arm (3) is controlled by a cylinder (4).

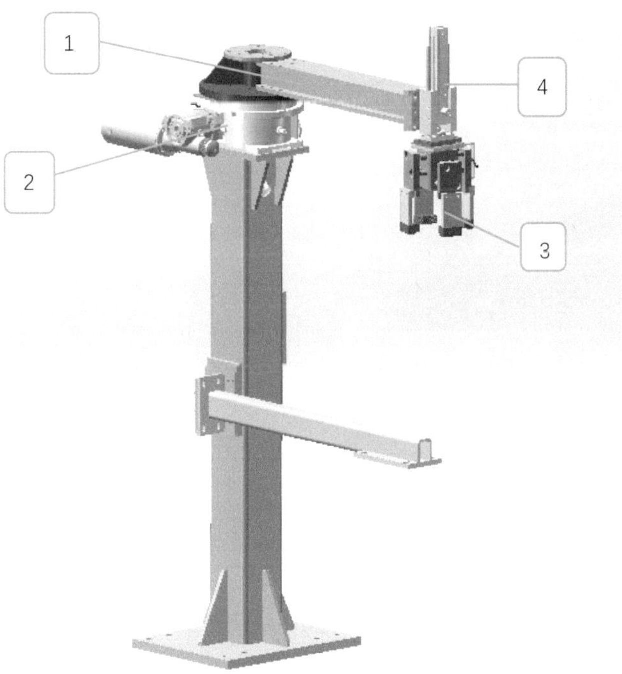

Fig. 2.31 Robotic Arm
1 - Mechanical Arm 2 - Electric Motor 3 - Robotic Hand 4 - Cylinder

Automatic Can Filling Mechanism as shown in Fig. 2.32 Empty cans are conveyed into the automatic can filling mechanism by the conveyor belt (2), where the mechanism fills the solution into the empty cans.

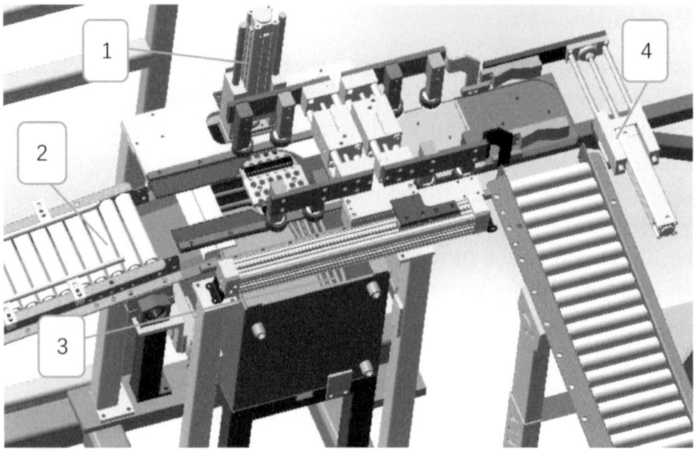

Fig. 2.32 Automatic Filling Mechanism
1 - Cylinder 2 - Conveyor Belt 3 - Infusion Tube 4 - Pushing Mechanism

Automatic Capping Mechanism as shown in Fig. 2.33 The bottles filled with solution are conveyed to the automatic capping mechanism (2) via conveyor belt (1). The pushing mechanism (4) pushes the bottles into the automatic capping mechanism (2), where the bottle caps are placed on the filled bottles.

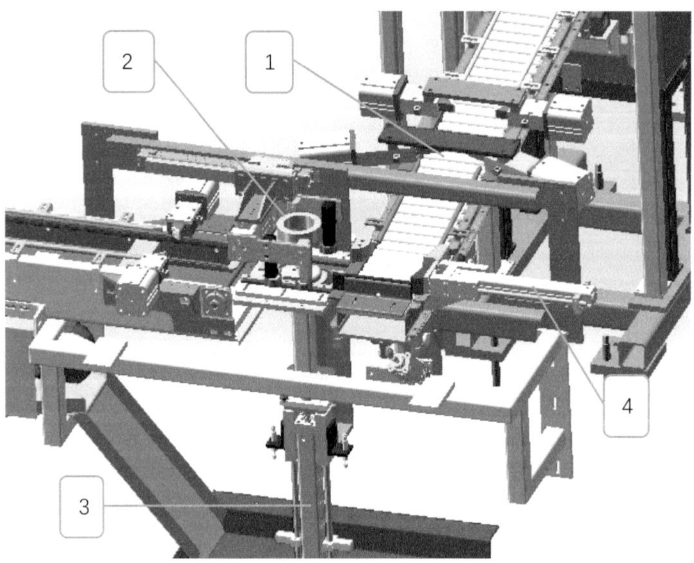

Fig. 2.33 Automatic Capping Mechanism
1 - Conveyor Belt 2 - Capping Mechanism 3 - Pneumatic Cylinder 4 - Pushing Mechanism

Automatic Sealing Mechanism as shown in Fig. 2.34, the filled bottles with caps are conveyed to the automatic sealing mechanism (1) via conveyor belt (3), where the bottles are secured and the caps are tightened.

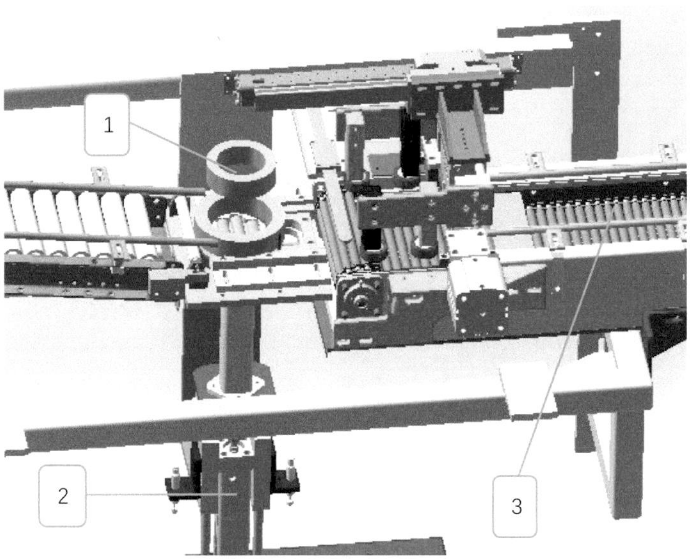

Fig. 2.34 Automatic Sealing Mechanism
1 - Sealing Mechanism 2 - Cylinder 3 - Conveyor Belt

IV. Specific Parameters (see Table 2.9)

Table 2.9 Highly Automated Filling Production Line Parameters

Operating Rate:	≥93%
Working Efficiency:	1200-1500 pcs/h
Environmental Temperature:	0~55°C
Permissible Humidity:	20%~95%
Equipment Dimensions:	1850 mm(H) × 5750 mm(L) × 3200 mm(W)

V. Mechanical Design Highlights

The electrical equipment control system of traditional canning production lines adopts the traditional relay contactor control method. In the process of use, the production efficiency is low. Human-machine interaction relies on the working mode of indicator lights, buttons, and audible alarms, which results in slow response, high failure rate, poor reliability, and the system's working status, fault handling, equipment monitoring, and maintenance can only be actively searched for faults based on experience. Moreover, in the production process, it is prone to secondary pollution, leading to low qualification rates and increased production costs. On the other hand, automated filling production lines are widely used in many fields. Their control mechanism adopts PLC (Programmable Logic Controller) control, making the automated production line run more smoothly, positioning more accurately, functions more comprehensively, and operations more convenient [49].

Case 21: Jelly Food Filling Machine

I. Case Description

Jelly food filling machine can be used for filling juice beverages, alcoholic beverages, purified water, oils, hair care products, skincare products, cosmetics, jelly, sauces, liquid alcohol, oral liquids, and more. The overall structure is shown in Fig. 2.35, mainly composed of lifting plate (1), placement trough (2), base (3), material storage tank (4), electrical control box (5), frame (6), hydraulic telescopic rod (7), and cam mechanism (8) [50]. The jelly food filling machine features stable operation, long service life, and high efficiency.

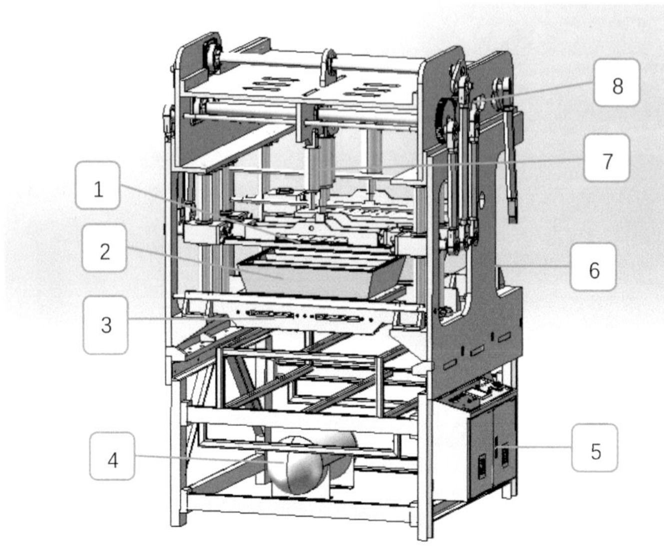

Fig. 2.35 Jelly Food Filling Machine
1-Lifting Plate 2-Placement Trough 3-Base 4-Material Storage Tank 5-Electrical Control Box 6-Frame 7-Hydraulic Telescopic Rod 8-Cam Mechanism

II. Working Principle

The jelly food filling machine works by first placing the tray containing empty bottles on the base. The cam mechanism drives the lifting plate downward, causing the filling head to move down and extend into the packaging shell of the jelly food for filling. While filling, as the jelly food enters the packaging shell, the filling tube rises under the action of the lifting plate. When it reaches the specified height, the filling stops. This process is repeated continuously to complete the batch filling of jelly food [51].

III. Main Structure Introduction

The cam drive mechanism, as shown in Fig. 2.36, is used to control the lifting of the lifting plate (2), completing the filling of jelly food.

The hydraulic expansion rod, as shown in Fig. 2.37, is connected to the bottom of the lifting platform (1). When the filling is not stopped in time, causing excessive jelly in the placement trough (2), the hydraulic expansion rod (3) drives the lifting platform (1) to move upward, allowing the jelly to flow back to prevent leakage of the packaging shell due to excessive jelly.

Placement Trough as shown in Fig. 2.38 is equipped with a large placement trough (2) and a small placement trough (1). According to the filling requirements, either a single placement trough (1) or placement trough (2) can be selected for operation, or filling operations can be conducted simultaneously.

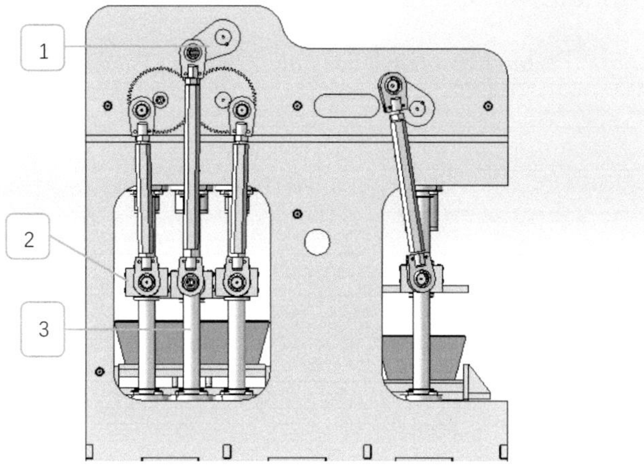

Fig. 2.36 Cam Drive Mechanism
1 - Cam Mechanism 2 - Lifting Plate 3 - Lifting Rod

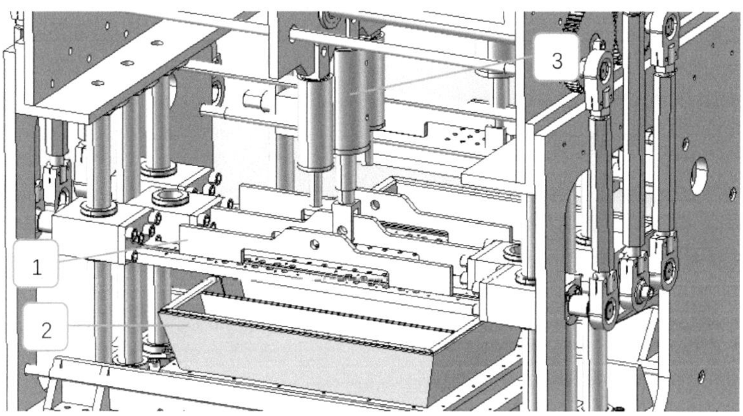

Fig. 2.37 Hydraulic Expansion Rod
1 - Lifting Platform 2 - Placement Trough 3 - Hydraulic Expansion Rod

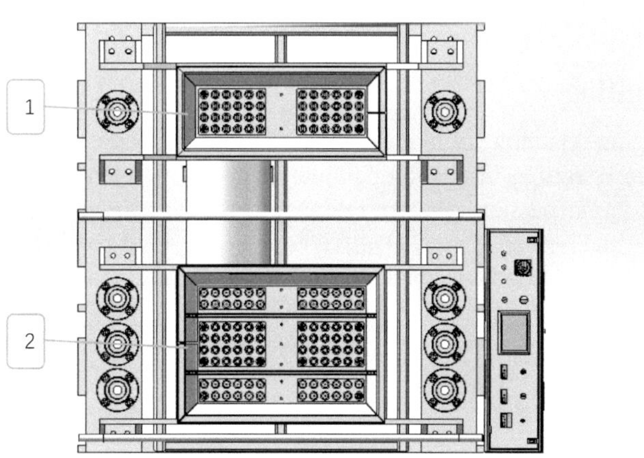

Fig. 2.38 Placement Trough
1 - Small Placement Trough 2 - Large Placement Trough

IV. Specific Parameters (see Table 2.10)

Table 2.10 Jelly Food Filling Machine Parameters

Filling Volume Range:	0~500 ml
Working Efficiency:	2700-4000 pcs/h
Filling Accuracy:	≤±1~3% (varies depending on the filling volume and liquid)
Air Pressure:	0.4-0.6 MPa
Ambient Temperature:	0~65°C
Allowed Humidity:	20%~95%
Motor Power:	0.75 kW
Equipment Dimensions:	2450 mm(H) × 1730 mm(L) × 1500 mm(W)

V. Mechanical Design Highlights

When the existing jelly filling machine fills jelly, it requires inserting the filling tube into the packaging shell of the jelly. As the filling progresses, the filling tube moves upward through the lifting mechanism. When the filling is about to be completed, the filling valve is closed, and the filling tube is positioned at the top of the jelly packaging shell. However, adopting this method of filling makes it difficult to manage the jelly's filling volume. Excess jelly can often seep out of the packaging shell due to a delayed closure of the filling valve, resulting in waste and increased manufacturing costs.

The jelly food filling machine, on the other hand, utilizes a hydraulic expansion rod to drive the lifting plate downward. The lifting plate, in turn, moves the filling can downward, creating a vacuum inside after vacuuming. Subsequently, the filling tube is inserted into the packaging shell of the jelly to initiate the filling process. As the jelly enters the packaging shell during filling, the filling tube moves upward under the influence of the lifting plate. When the jelly rises to a certain position, the filling stops. If the filling stops too late, causing an excess of jelly, the first electric expansion rod drives the piston upward, causing the jelly to recede to prevent excessive leakage from the packaging shell. This approach helps prevent wastage and ensures better control over the jelly filling process.

Case 22: Inspection Packaging Machine

I. Case Description

Inspection Packaging Machine is mainly used for the inspection and packaging of electronic components. It allows setting inspection parameters such as dimensions, surface defects, and dirt on the monitor of the inspection system to detect electronic components. The overall structure is shown in Fig. 2.39, and a detailed view is provided in Fig. 2.40. It consists of packaging mechanism (1), frame (2), fixed seat (3), first conveying mechanism (4), feeding mechanism (5), inspection mechanism (6), second conveying mechanism (7), and display screen (8). The detection and packaging machine offers advantages such as high inspection accuracy, fast detection and packaging speed, and high automation level.

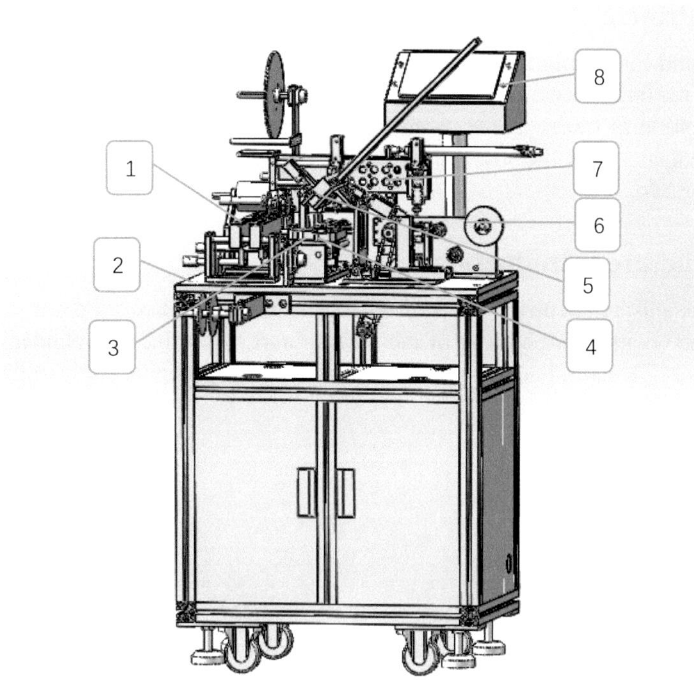

Fig 2.39 Inspection and Packaging Machine
1 - Packaging Mechanism 2 - Frame 3 - Fixed Seat 4 - First Conveying Mechanism 5 - Feeding Mechanism
6 - Inspection Mechanism 7 - Second Conveying Mechanism 8 - Display Screen

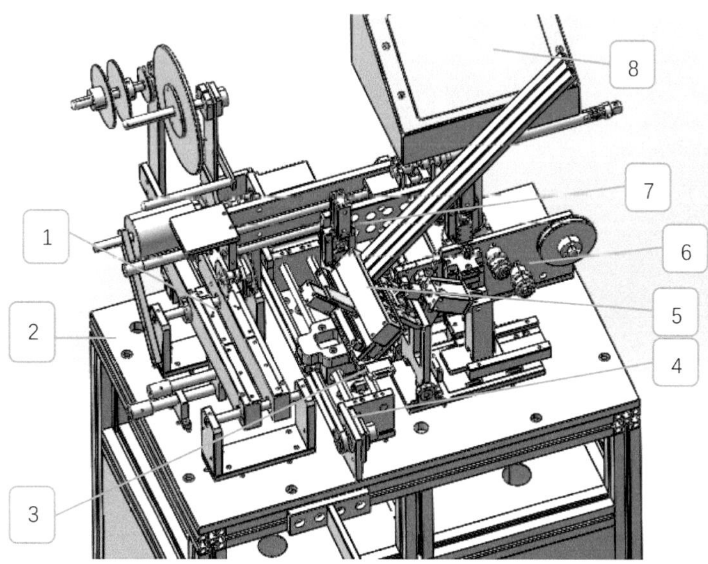

Fig. 2.40 Enlarged View of the Inspection and Packaging Machine
1 - Packaging Mechanism 2 - Frame 3 - Fixed Seat 4 - First Conveying Mechanism 5 - Feeding Mechanism
6 - Inspection Mechanism 7 - Second Conveying Mechanism 8 - Display Screen

II. Working Principle

The Inspection and Packaging Machine operates by introducing the electronic components to be tested from the feeding mechanism, placing them on the fixed seat. The conveying mechanisms work in coordination to transport the electronic components to the inspection mechanism. If the electronic components pass the inspection, they are conveyed to the packaging mechanism for further packaging.

III. Main Structure Introduction

The feeding mechanism is depicted in Fig. 2.41. Electronic components are introduced through the duct (1), with the opening and closing of the feeding port controlled by cylinder (2) to ensure the orderly placement of electronic components on the fixed seat. Cylinder (3) is utilized to control the angle of the feeding mechanism.

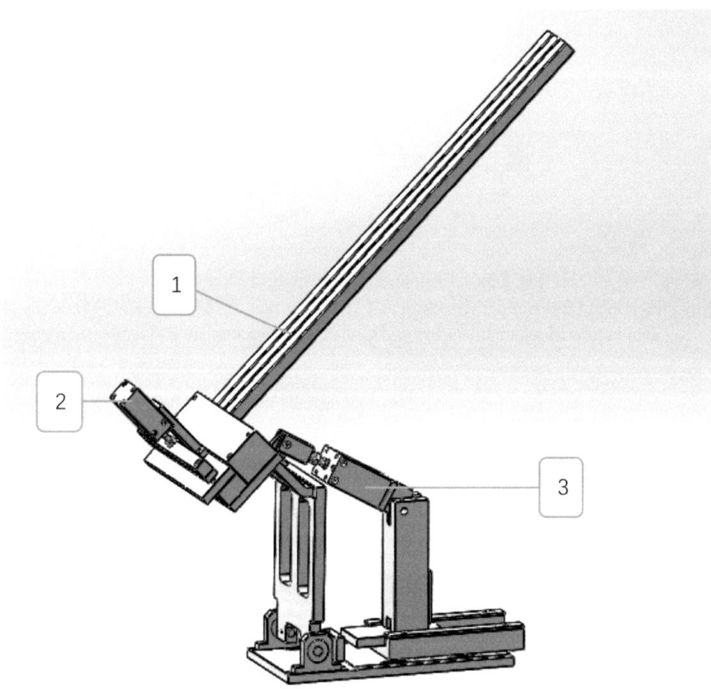

Fig. 2.41 Feeding Mechanism
1 - Duct 2, 3 - Cylinder

The first conveying mechanism and the second conveying mechanism are shown in Fig. 2.42. When the detection and packaging machine is in operation, the first conveying mechanism and the second conveying mechanism work together. The suction nozzle (5) on the second conveying mechanism sucks the electronic components and transports them to the inspection mechanism for testing. After testing, the suction nozzle (5) retracts the component and transports it to the fixed seat. Suction nozzle (4) synchronously displaces, extracting the electronic components that have already been tested and qualified on the fixed seat, and transports them to the packaging mechanism.

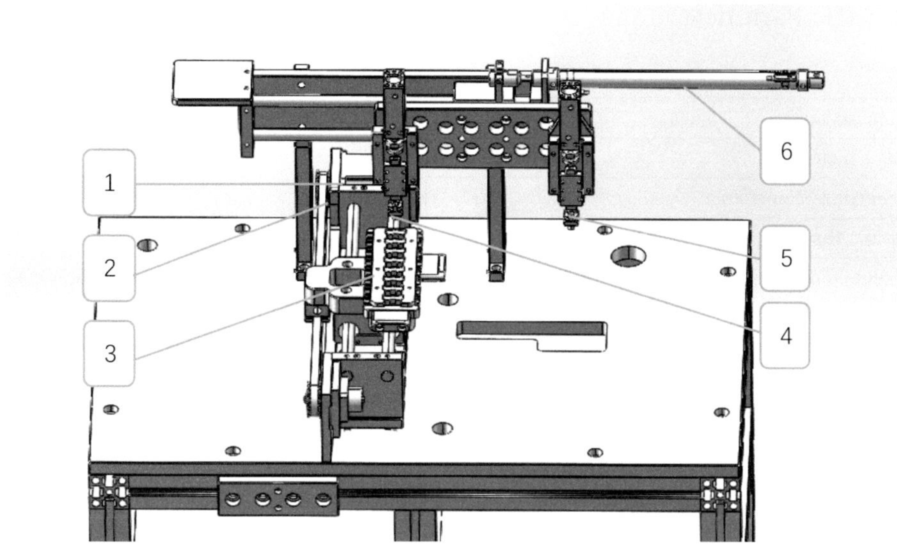

Fig. 2.42 Transport Mechanism
1 - Electric Motor 2 - Pulley 3 - Fixed Seat 4, 5 - Suction Nozzle 6 - Cylinder

Packaging Mechanism as shown in Fig. 2.43 is controlled by the rotating shaft driven by the electric motor (3). It continuously feeds the packaging material from the tray (1) into the packaging mechanism and seals the packaging material with the qualified electronic components. Blade (2) is used to trim the excess packaging material from the edges of the packaging box.

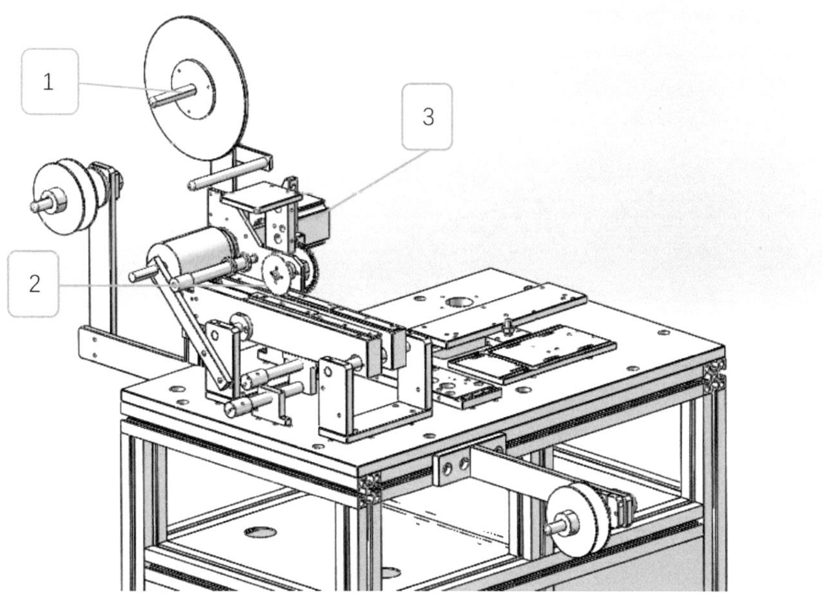

Fig. 2.43 Packaging Mechanism
1 - Tray 2 - Blade 3 - Electric Motor

IV. Specific Parameters (see Table 2.11)

Table 2.11 Detection Packaging Machine Parameters

Detection Accuracy:	≤0.01mm
Working Efficiency:	2000-2500 pcs/h
Environmental Temperature:	1~40°C
Permissible Humidity:	<75%
Equipment Dimensions:	1280 mm(H) × 650 mm(L) × 450 mm(W)

V. Mechanical Design Highlights

(1) Feed-side conveyor module: Feeding can be manually operated or connected to the front-end pinning machine.
(2) Fully automatic detection module, accurately controlling product quality.
(3) Defective product sorting module, classifying defective products detected in each group.
(4) Automatic tape splicing, suitable for both cold-sealing and heat-sealing tapes.

Case 23: Adhesive Tape Slitting Machine

I. Case Description

Adhesive Tape Slitting Machine is capable of automatically winding adhesive tape onto the adhesive tape reel. It is mainly used for existing medical adhesive tape rolls and industrial adhesive tape rolls. The overall structure is shown in Fig. 2.44, and a close-up view is shown in Fig. 2.45. It consists of tape disc (1), cutting mechanism (2), adhesive tape drum rotation mechanism (3), pressing mechanism (4), and frame (5), among other mechanisms [52]. The tape winding machine has the advantages of high automation, reliable structure, and significantly improved production efficiency.

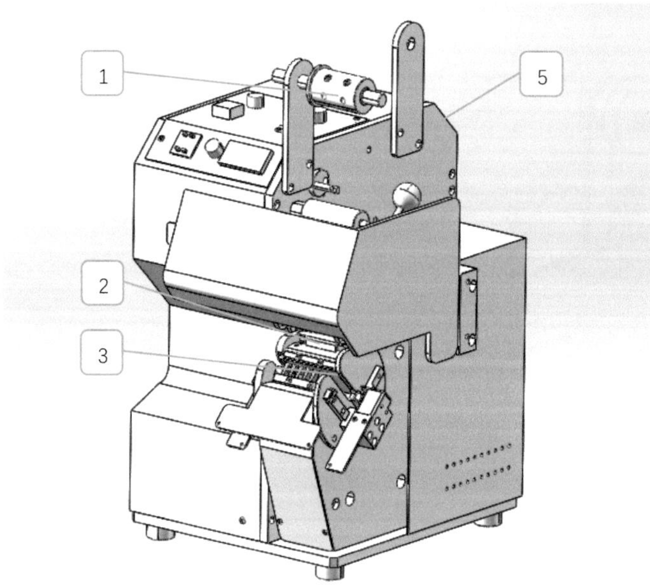

Fig. 2.44 Tape Winding Machine
1 - Tape Disc 2 - Cutting Mechanism 3 - Tape Roll Rotating Mechanism 5 - Frame

Fig. 2.45 Detailed View of Tape Winding Machine
1 - Tape Disc 2 - Cutting Mechanism 3 - Tape Roll Rotating Mechanism 4 - Pressing Mechanism 5 - Frame

II. Working Principle

The tape winding machine operates by placing the adhesive tape on the tape disc, positioning the tape roll on the tape roll rotating mechanism, and securing the tape from the tape disc onto the surface of the tape roll. The device is started, and the adhesive tape is clamped and fed to the rotating axis through the pressing mechanism, ensuring a secure wrap of the adhesive tape around the roll. When the predetermined length of adhesive tape is reached, the cutting mechanism automatically operates, cutting the tape. The tape roll rotating mechanism continues to rotate, winding the remaining tape onto the tape roll, completing the winding process. The operator then replaces the tape roll and repeats the winding operation.

III. Main Structure Introduction

The cutting mechanism and pressing mechanism are shown in Fig. 2.46. The adhesive tape is fed from the tape disc into the pressing mechanism (1). By adjusting the angle of the handle (3), the tension during the tape feeding process can be controlled, ensuring a secure wrap of the adhesive tape around the tape roll. When the predetermined length of tape is reached, the cutting mechanism (2) is lowered by the motor control, cutting the tape.

The roller rotating mechanism, as shown in Fig. 2.47, is driven and controlled by an electric motor (1). When the tape winding machine is operational, first place the tape roller in the groove (2), close the roller rotating mechanism, and start the electric motor (1). The roller rotating mechanism rotates by the gear transmission driven by the electric motor, thereby winding the tape onto the tape roller.

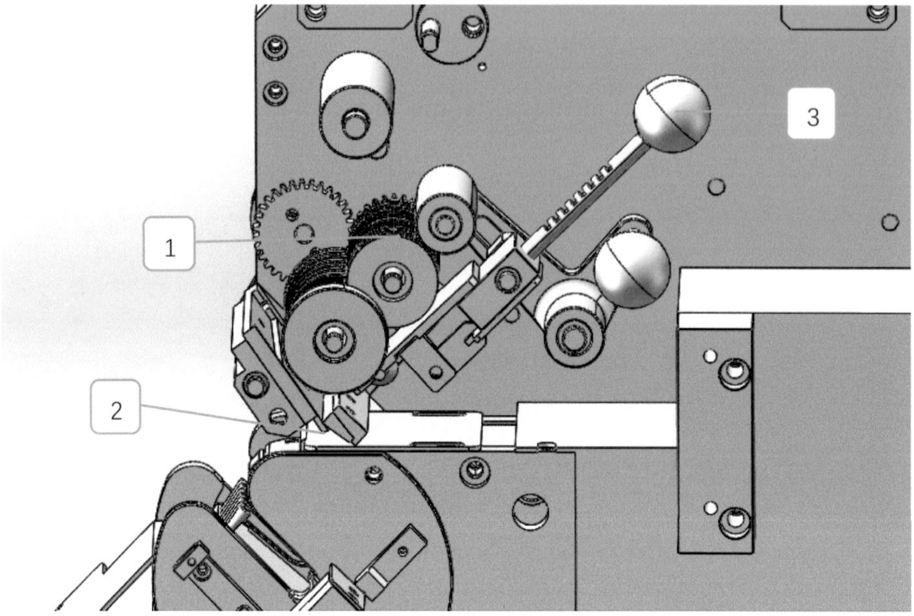

Fig. 2.46 Pressing Mechanism and Cutting Mechanism
1 - Pressing Mechanism 2 - Cutting Mechanism 3 - Handle

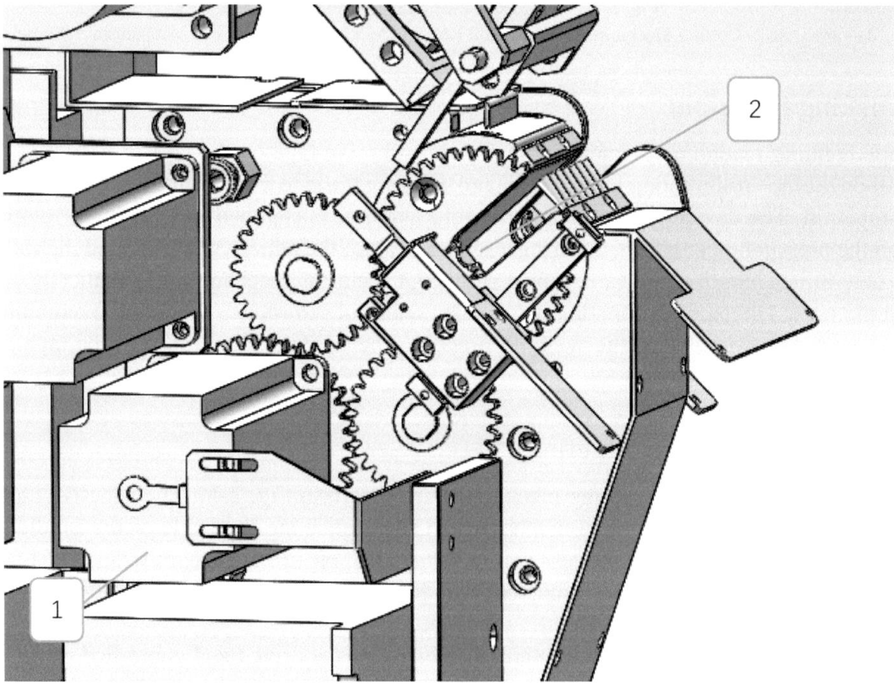

Fig. 2.47 Tape Roller Rotation Mechanism
1 - Electric Motor 2 - Groove

IV. Specific Parameters (see Table 2.12)

Table 2.12 Tape Rolling Machine Parameters

Tape Width:	0~60 mm
Belt Speed:	2.5 m/s
Conveying Capacity:	600 t/h
Winding Accuracy (Overlap):	±0.5~0.7 mm
Operating Temperature:	0~55°C
Allowable Humidity:	10%~95%
Equipment Dimensions:	460 mm(H) × 340 mm(L) × 275 mm(W)

V. Mechanical Design Highlights

The tape rolling machine can automatically wind the tape onto the tape roller, and the length of the tape is easy to control accurately. The high precision in pasting the tape onto the tape roller improves the yield of the tape roll. Additionally, the machine has a reliable structure and a high level of automation, significantly enhancing production efficiency. It is easy to achieve industrialization with this tape rolling machine.

VI. Precautions

(1) When replacing cutting components and the knife cutting mechanism, ensure that one has first turned off the sensors and power switch.
(2) When placing the tape, make sure to put it in smoothly and evenly.
(3) This machine is strictly prohibited for use with tapes other than those specified for cutting production.
(4) When the machine is in operation, it is strictly forbidden to place hands or other objects near the cutting edge.
(5) Ensure personal safety protection and equipment safety protection.

Case 24: Fully Automatic Packaging Machine

I. Case Description

The fully automatic packaging machine is suitable for packaging materials in the form of granules, tablets, liquids, powders, pastes, etc. It is mainly used in industries such as food, pharmaceuticals, chemicals, and material packaging for plant seeds. The overall structure is shown in Fig. 2.48, consisting mainly of a frame (1), film winding roller (2), synchronous belt (3), electric heating block (4), film supply roller (5), feeding mechanism (6), handling mechanism (7), and film guiding frame (8). The fully automatic packaging machine features high automation, high efficiency, adaptability to packaging films of different product specifications, and flexible and convenient use.

II. Working Principle

The automatic packaging machine operates by driving the film winding roller with the film winding motor, simultaneously rotating the synchronous belt with the conveyor motor to automatically transport the packaging film. The handling mechanism extracts the products to be packaged from the feeding mechanism and transports them onto the synchronous belt, synchronized with the

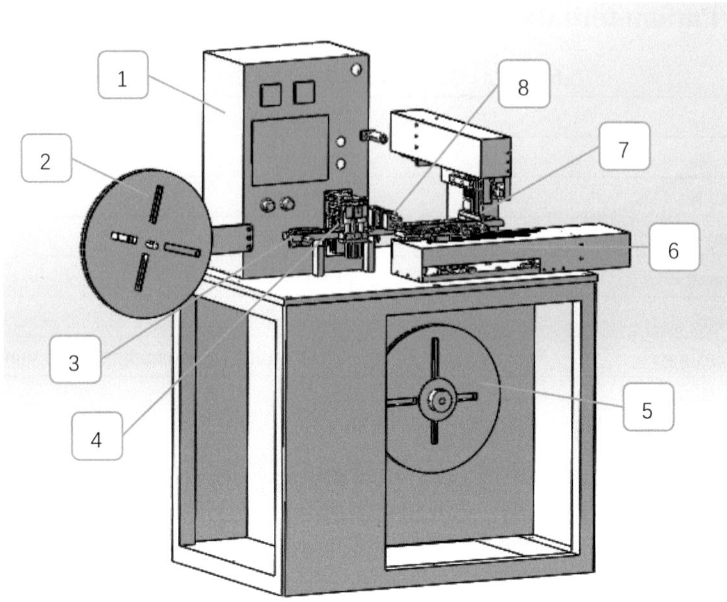

Fig. 2.48 Fully Automatic Packaging Machine
1 - Frame 2 - Film Winding Roller 3 - Synchronous Belt 4 - Electric Heating Block 5 - Film Supply Roller
6 - Feeding Mechanism 7 - Handling Mechanism 8 - Film Guiding Frame

movement of the packaging film. The automatically transported packaging film is clamped by the film guiding frame, lifting the sides of the film and continuously forwarding it. When the lifted sides of the packaging film pass through the electric heating block, the heat from the electric heating block automatically seals the sides of the packaging film. The sealed products and packaging film continue to be transported to the film winding roller. During use, the film guiding frame can be slid along its sliding track as needed to adjust the spacing of the film guiding frames on both sides of the frame, allowing flexibility and convenience to accommodate the different product specifications of packaging film.

III. Main Structure Introduction

The feeding mechanism is depicted in Fig. 2.49, where the products to be packaged are placed on the product rack (1). The product rack (1) is controlled for lateral movement by the servo motor (2).

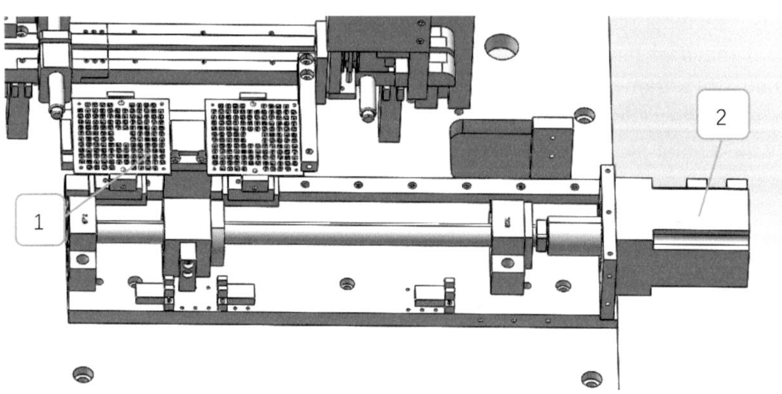

Fig. 2.49 Feeding Mechanism
1 - Product rack 2 - Servo motor

The conveying mechanism, as shown in Fig. 2.50, utilizes the gripping and lateral movement of the mechanical hand (2) to transport the products placed on the product rack to the synchronous belt. The vertical lifting movement of the mechanical hand (2) is controlled by the driven cylinder (3), while the lateral displacement is driven and controlled by the servo motor (1).

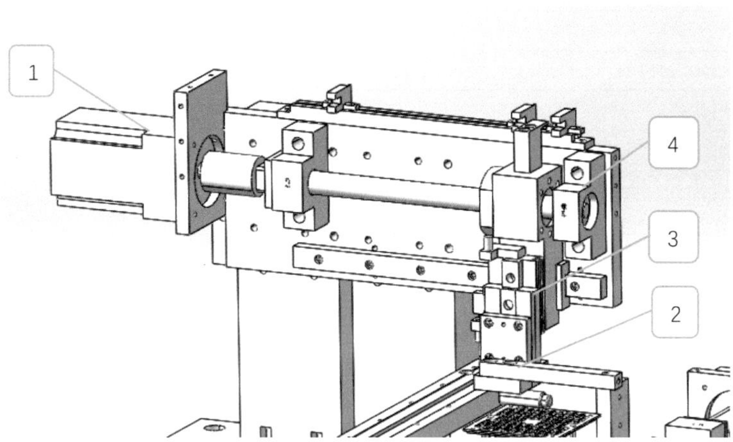

Fig. 2.50 Handling Mechanism
1 - Servo Motor 2 - Mechanical Hand 3 - Cylinder 4 - Stop Block

The electric heating block, as shown in Fig. 2.51, descends when the packaged film, held by the film guide frame (3), reaches the position of the electric heating block (1). Controlled by the cylinder (2), the electric heating block (1) automatically seals the two sides of the packaging film through heating action.

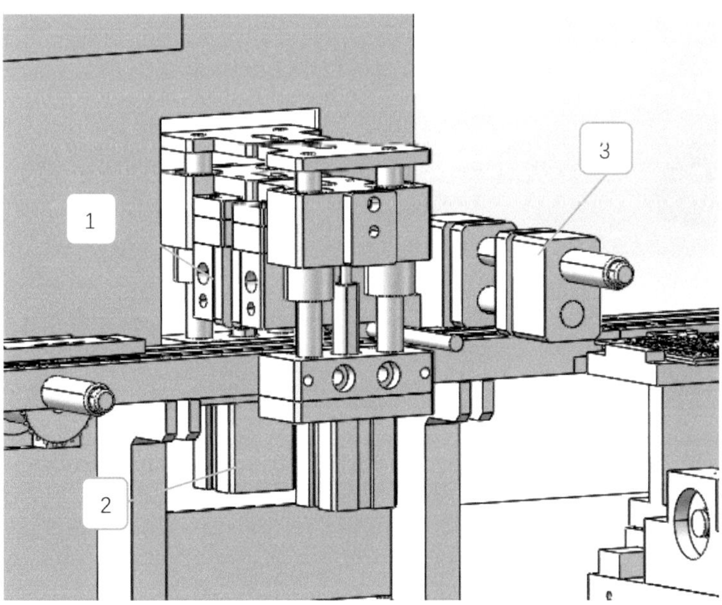

Fig. 2.51 Electric Heating Block
1 - Electric Heating Block 2 - Cylinder 3 - Film Guide Frame

IV. Specific Parameters (see Table 2.13)

Table 2.13 Automatic Packaging Machine Parameters

Working Efficiency:	1000-1500 pcs/h
Operation Rate:	≥90%
Yield Rate:	≥98%
Ambient Temperature:	0~65°C
Allowable Humidity:	10%~95%
Equipment Dimensions:	1145 mm(H) × 1300 mm(L) × 580 mm(W)

V. Mechanical Design Highlights

Automatic packaging machines can perform packaging and sealing automatically, achieving a higher level of automation and providing greater convenience. The guide frame is evenly distributed along the horizontal direction on both sides of the frame. The connection between the guide frame and the sliding track is a sliding connection, allowing the packaging film to smoothly pass through the guide frame. This facilitates the even sealing of both sides of the packaging film. The sliding connection also enables easy adjustment of the spacing between the guide frames on both sides of the frame, making the machine more flexible and convenient to use.

The electric heating block is controlled for vertical movement by a cylinder, allowing the packaging film passing through it to be automatically heated and sealed. This procedure reduces the need for manual involvement, which makes the operation more convenient.

Chapter 3

Automatic Assembly Machinery

Case 25: USB Fully Automatic Assembly Equipment

I. Case Description

The Universal Serial Bus (USB) is a serial bus standard and a specification for input/output interfaces. It is widely used in information and communication products such as personal computers and mobile devices, and has been extended to other fields including photography equipment, digital televisions (set-top boxes), gaming consoles, and more. The USB model is illustrated in Fig. 3.1.

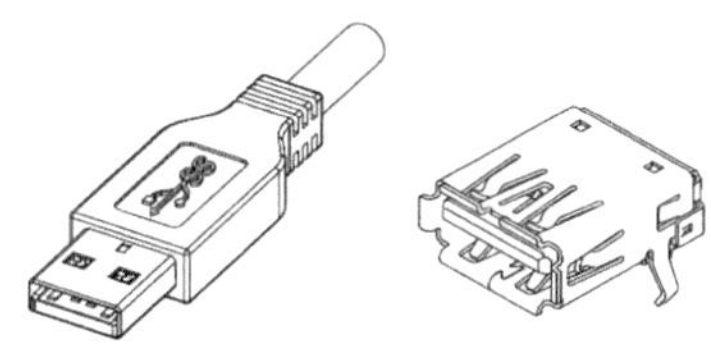

Fig. 3.1 USB Interface Model

The USB fully automatic assembly machine is designed for the automated plastic assembly of USB interfaces. The mechanical structure is illustrated in Fig. 3.2, and the top view is shown in Fig 3.3. It mainly consists of components such as the workbench (1), support frame (2), synchronous belt (3), rollers (4), electric motor (5), guide tube (6), sliding mechanism (7), first fixture (8), second fixture (9), precision assembly machine (10), station turntable (11), positioning hydraulic press (12), and preliminary assembly machine (13).

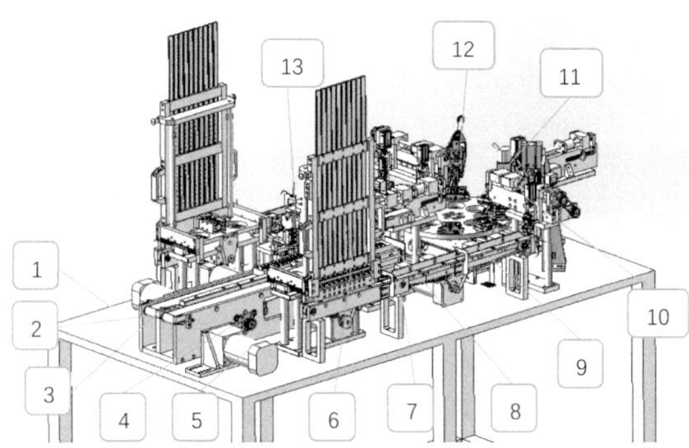

Fig. 3.2 USB Fully Automatic Assembly Machine

1 - Workbench 2 - Support frame 3 - Synchronous belt 4 - Rollers 5 - Electric motor 6 - Guide tube 7 - Sliding mechanism 8 - First fixture 9 - Second fixture 10 - Precision assembly machine 11 - Turntable 12 - Positioning hydraulic press 13 - Preliminary assembly machine

The sliding mechanism (7) includes an electric motor, sliding chain, slide rail, and slide table, with the stamping machine mounted on the slide table. The preliminary assembly machine (13) is equipped with a punching machine and a robotic arm. The station turntable (11) has eight stations, each equipped with the second fixture, allowing simultaneous precision assembly of eight USB housings.

The USB fully automatic assembly machine achieves high automation by coordinating multiple mechanisms, significantly saving labor, improving production efficiency and product yield, reducing scrap rates, and lowering production costs.

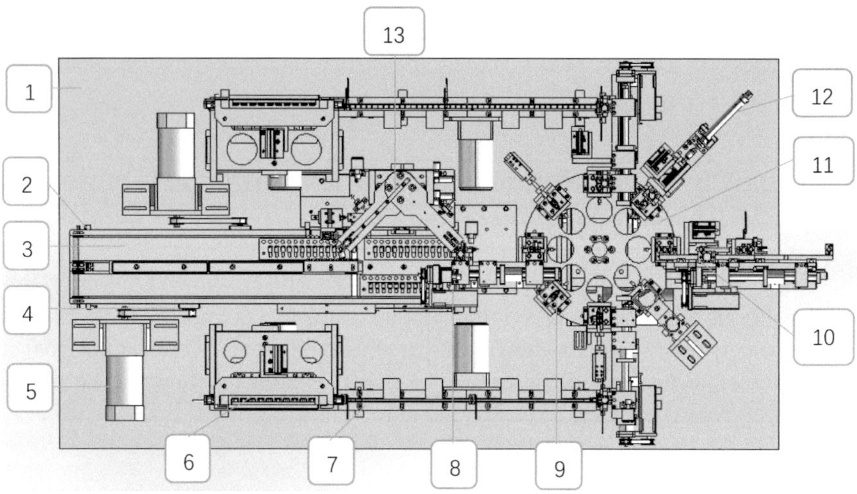

Fig. 3.3 Overhead Structure Diagram of USB Fully Automatic Assembly Machine
1 - Workbench 2 - Support frame 3 - Synchronous belt 4 - Rollers 5 - Electric motor 6 - Guide tube 7 - Sliding mechanism 8 - First fixture 9 - Second fixture 10 - Precision assembly machine 11 - Turntable 12 - Positioning hydraulic press 13 - Preliminary assembly machine

II. Working Principle

The equipment is set up in coordination with the electric motor and support frame. When it is necessary to perform plastic assembly on USB, the electric motor is started. The motor drives the rollers on top of it to rotate, thereby moving the synchronous belt horizontally on the support frame. As the synchronous belt is equipped with the first fixture, several USB shell blanks can be mounted at various clamping positions on the first fixture. The synchronous belt then slides to the assembly position. When the first fixture with the USB shell blanks slides to the position of the preliminary assembly machine, the first fixture stops below the preliminary assembly machine. The punching machine and robotic arm of the preliminary assembly machine punch holes in the USB shell blanks and preliminarily position the parts to be installed on the outer side walls of the USB shell. After the preliminary assembly, the synchronous belt conveys the preliminarily assembled USB shells to the turntable position. One of the precision assembly machines then secures the preliminarily assembled USB shells onto the second fixture. The rotary electric motor rotates the turntable, and through three precision assembly machines and a positioning hydraulic press, the installation of both internal and external components of the entire USB is completed.

III. Main Structure Introduction

The first fixture, as shown in Fig. 3.4, is mounted on the synchronous belt (2) and is primarily used for the preliminary positioning of the components to be installed on the USB shell and for bulk transportation of USB components. During operation, the first fixture is driven by the synchronous belt (2) through the preliminary assembly machine (1), achieving the preliminary positioning and installation of the components required for the USB shell.

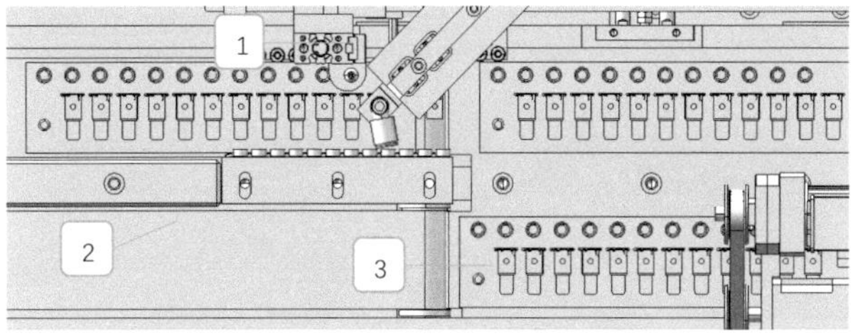

Fig. 3.4 The First Fixture
1 - Preliminary Assembly Machine 2 - Synchronous Belt 3 - First Fixture

The preliminary assembly machine, as shown in Fig. 3.5, is mainly composed of punching machine (1) and robotic arm (3), designed to accomplish the positioning and installation of the required components for the USB shell.

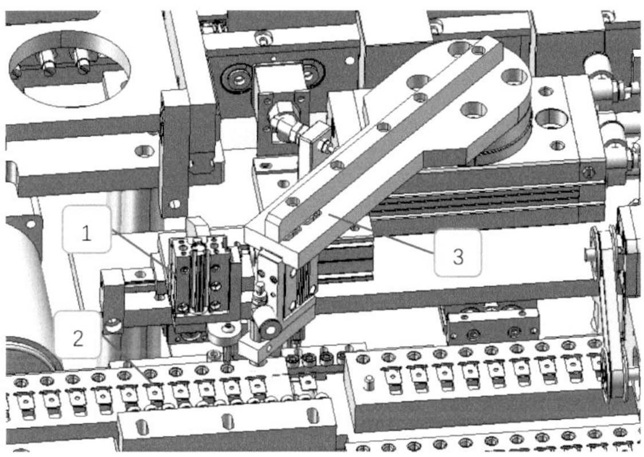

Fig. 3.5 Preliminary Assembly Machine
1 - Punching Machine 2 - First Fixture 3 - Robotic Arm

The Second Fixture as shown in Fig. 3.6 is mounted on the turntable (1), which applies pressure using springs (3) to secure the components of the USB. Additionally, there are eight stations on the turntable, each equipped with a second fixture (2), enabling simultaneous precision assembly of eight USB housings, resulting in high efficiency.

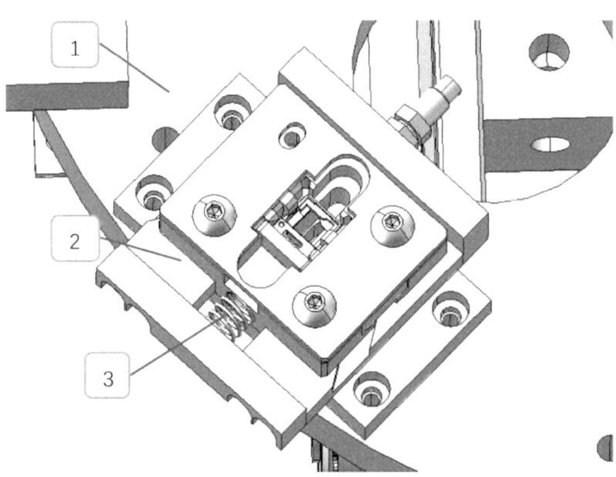

Fig. 3.6 Second Fixture
1 - Workstation Turntable 2 - Second Fixture 3 - Spring

Workstation Turntable as shown in Fig. 3.7 is driven by a positioning motor (4). The USB shell will pass through three precision assembly machines and a rotary hydraulic press to complete the installation of internal and external components of the entire USB.

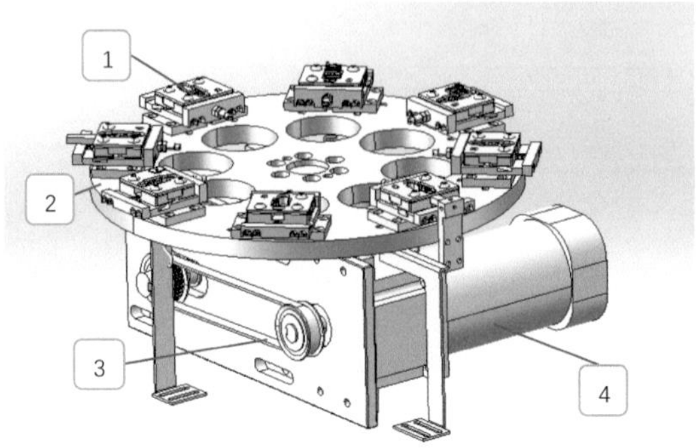

Fig. 3.7 Workstation Turntable
1 - Second Fixture 2 - Workstation Plate 3 - Belt Wheel 4 - Positioning Motor

Fine assembly machine as shown in Fig. 3.8, it is located on the upper side of the turntable along with the positioning hydraulic press, primarily used for accurately installing USB components transported from the conduit through manipulator (3).

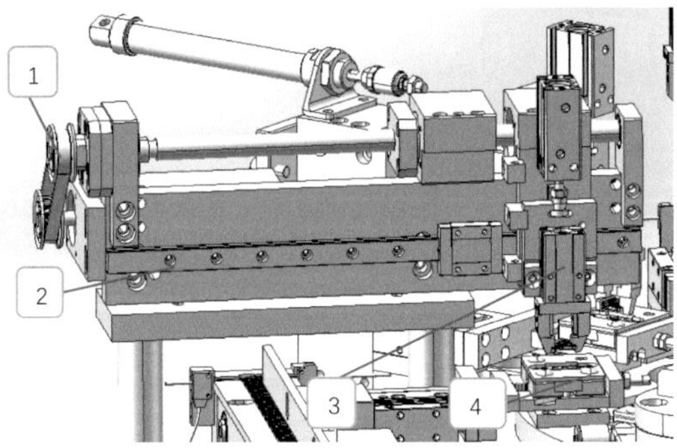

Fig. 3.8 Fine Assembly Machine
1 - Pulley 2 - Guide rail 3 - Manipulator 4 - Second clamp

IV. Specific Parameters (see Table 3.1)

Table 3.1 USB Fully Automatic Assembly Machine Parameters

Operating Pressure Range:	4-6 kg/cm^2
Variable Speed Range:	0.00-400.00 Hz
Working Efficiency:	2000-3000 pcs/h
Operating Temperature:	0~50°C
Permissible Humidity:	10%~95%
Equipment Dimensions:	1660 mm(H) × 1680 mm(L) × 885 mm(W)
Weight:	Approximately 150 kg
Equipment Footprint:	1680 × 885 mm = 1.49 m^2

V. Mechanical Design Highlights

The production process of USB interfaces typically involves multiple steps. Despite their small size, USB interfaces consist of numerous components and have intricate structures. Traditionally, USB interfaces were manufactured manually on assembly lines, requiring a dozen or more workers. This manual process led to low production efficiency and high defect rates.

The USB data interface automatic assembly machine is a complex integrated electromechanical device comprising mechanical and electrical systems. It features a sophisticated design and is composed of several components, each with specific functions. The production process is controlled through a configuration human-machine interface, which ensures ease of operation and maintenance. This automated system significantly reduces labor requirements and improves product quality.

VI. Other Functions

(1) The machine is equipped with emergency stop, automatic counting, fault alarm, and fault display and analysis functions.
(2) The unit adopts a modular structure. If any component of the equipment detects a lack of product or if the product is not in place, the machine will automatically issue an alarm and prompt, and the corresponding mechanism will stop automatically without affecting the mechanical operation of other components.

Case 26: Automotive Converter Assembly Equipment

I. Case Description

The automotive converter is a device that can convert 12V, 24V, 36V, or 48V DC power into 220V or 110V AC power, similar to household electricity, for general electrical appliances. It is a convenient power converter for vehicles, as shown in Fig. 3.9. The automotive converter assembly equipment is specifically designed for assembling automotive converters, as shown in the overall top view in Fig. 3.10. It mainly consists of electrical control box (1), rotating robot (2), feeding mechanism (3), synchronous belt (4), handling robot (5), punching mechanism (6), assembly mechanism (7), handling mechanism (8), and feeding mechanism (9) [53]. The assembly mechanism includes assembly support and positioning mechanism. The automotive converter assembly equipment can achieve integrated transportation and assembly, as well as automated collection of finished products, significantly improving production efficiency.

Fig. 3.9 Automotive Converter

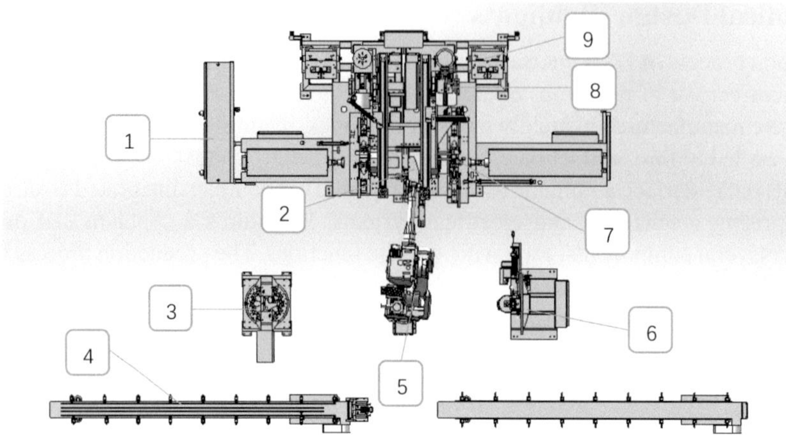

Fig. 3.10 Top View of Automotive Converter Assembly Equipment
1 - Electrical control box 2 - Rotary manipulator 3 - Feeding mechanism 4 - Synchronous belt 5 - Handling robot 6 - Punching mechanism 7 - Assembly mechanism 8 - Handling mechanism 9 - Feeding mechanism

II. Working Principle

The automotive converter assembly equipment operates as follows: the feeding mechanism provides the original components of the automotive converter, which are then transported to the assembly mechanism by the handling robot. Simultaneously, the handling mechanism transports the converter housing to the assembly mechanism. The assembly mechanism secures and assembles the various original components of the converter. After assembly, the rotary manipulator grabs the assembled converter and transports it to the punching mechanism. Once punching is completed, the converter is transported to the synchronous belt, where the fully assembled automotive converter is outputted.

III. Main Structure Introduction

The handling robot, as shown in Fig. 3.11, is driven by a servo motor (4). The handling robot has 5 degrees of freedom and is equipped with a rotating manipulator (1), which allows it to simultaneously grip two automotive converters or original components, significantly improving work efficiency.

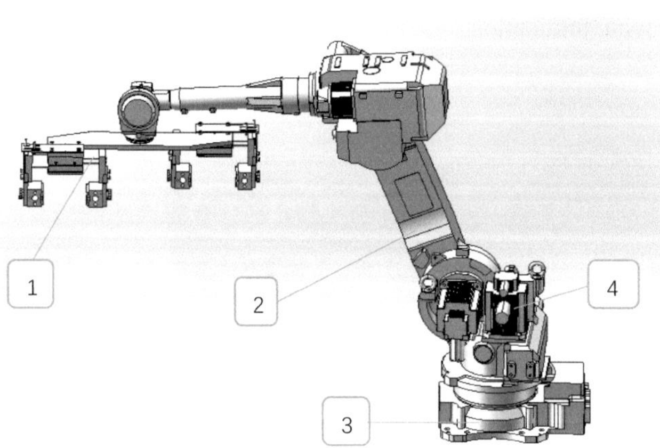

Fig. 3.11 Handling Robot
1 - Rotating Manipulator 2 - Mechanical Arm 3 - Base 4 - Servo Motor

The handling mechanism is shown in Fig. 3.12, controlled by the electric motor (2) to operate on the slide rail (3). The cylinders (1) and (7) respectively control the horizontal displacement and lifting of the handling robot (6). The handling mechanism (8) is mainly used to transport the components of the car converter from the feeding mechanism (5) to the assembly mechanism (4).

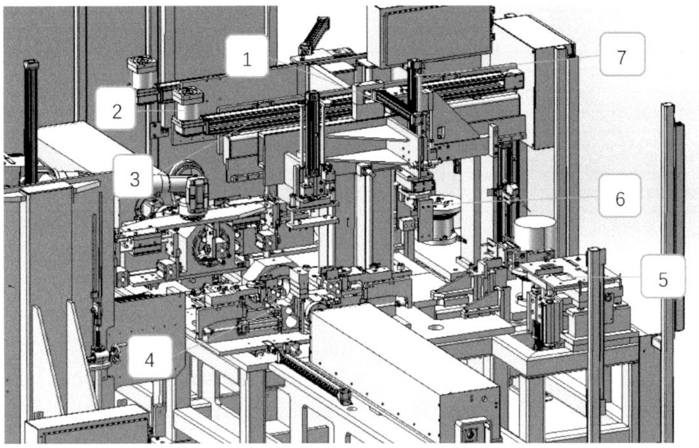

Fig. 3.12 Handling Mechanism
1, 7 - Cylinders 2 - Electric Motor 3 - Slide Rail 4 - Assembly Mechanism 5 - Feeding Mechanism 6 - Handling Robot

The assembly mechanism is shown in Fig. 3.13, primarily composed of a positioning ring (1), a fixture (2), and a locking mechanism (3). When the automotive converter assembly equipment is in operation, the handling robot and the handling mechanism sequentially feed the original components of the automotive converter into the assembly mechanism. The fixture (2) secures the original components and places them into the positioning ring (1). Subsequently, the locking mechanism (3) is activated to lock the various components of the automotive converter, completing the assembly process.

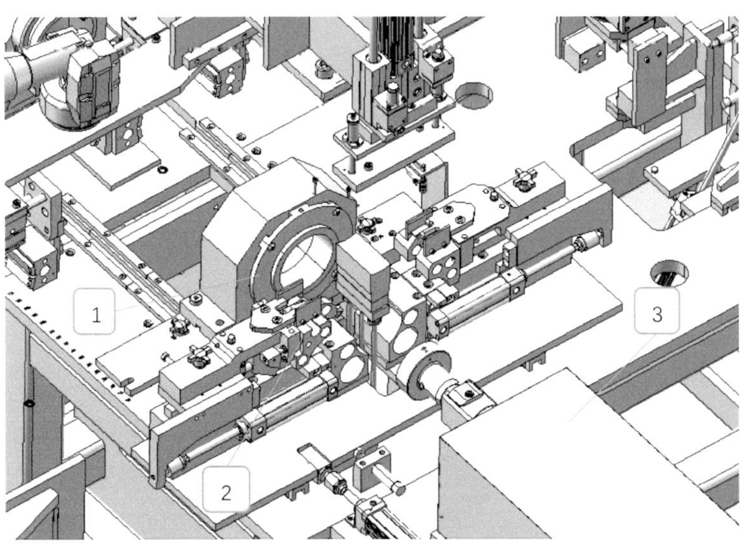

Fig. 3.13 Assembly Mechanism
1 - Positioning Ring 2 - Fixture 3 - Locking Mechanism

IV. Specific Parameters (see Table 3.2)

Table 3.2 Automotive Converter Assembly Equipment Parameters

Rated Power:	8.0 kW
Operating Efficiency:	1000-1500 pcs/h
Utilization Rate:	≥90%
Yield Rate:	≥98%
Environmental Temperature:	0~65°C
Permissible Humidity:	20%~95%
Equipment Dimensions:	2890 mm(H) × 9230 mm(L) × 5190 mm(W)

V. Mechanical Design Highlights

This equipment adopts a symmetrical assembly line for assembling automotive converters. The left and right assembly lines of the equipment can simultaneously assemble two automotive converters. The converters are then transported to the punching mechanism by a robotic arm equipped with a rotating manipulator. The robotic arm is designed with the ability to simultaneously grip two automotive converters or original components. The equipment features a sophisticated design, with each mechanism having clear responsibilities, significantly improving work efficiency.

Case 27: Traction Machine Assembly Line

I. Case Description

The traction machine, also known as the elevator main engine, is the power device of the elevator, responsible for conveying and transmitting power to make the elevator operate, as shown in Fig. 3.14. The traction machine assembly line is mainly used in the traction machine assembly workshop. The overall assembly line is shown in Fig. 3.15, composed of a flat belt, torque testing mechanism, turning mechanism, pressurization mechanism, feeding mechanism, assembly robot (6), assembly robot (7), and assembly robot (8). It can complete assembly, final assembly, transportation, and other production processes. With a high level of automation and reasonable structure, it can reduce the number of operators and their workload, thereby reducing the company's operating and production costs, while enhancing production efficiency and product quality [54].

Fig. 3.14 Traction Machine

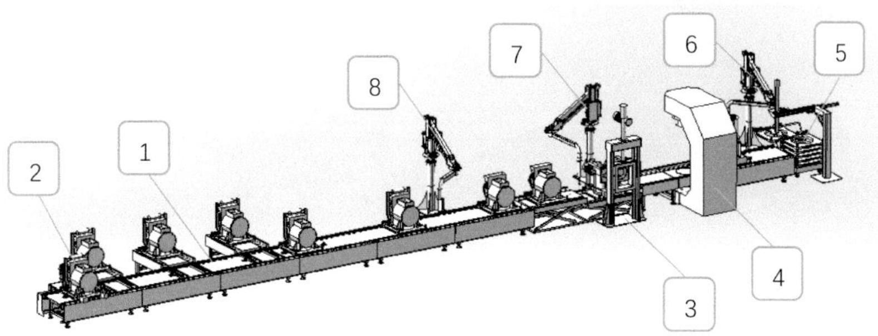

Fig. 3.15 Traction Machine Assembly Line Overall Assembly Diagram
1-Flat Belt 2-Torque Testing Mechanism 3-Positioning Mechanism 4-Pressurization Mechanism 5-Feeding Mechanism
6, 7, 8-Assembly Robots

II. Working Principle

The balance of the traction machine assembly line determines the equilibrium between personnel and machines. Therefore, a well-planned facility is needed to ensure that the total workload at a workstation is roughly equivalent to the total number of basic work units assigned to that workstation. The balance of the traction machine assembly line involves distributing all basic work units to various workstations, ensuring that each workstation is busy during the cycle time (the time interval between two consecutive products passing through the end of the assembly line) [55]. This allows each workstation to complete a significant amount of work, minimizing idle time. The balance of the traction machine assembly line directly affects the productivity of the manufacturing system.

III. Main Structure Introduction

Assembly manipulators are shown in Fig. 3.16, equipped with corresponding manipulators depending on the parts to be gripped. Assembly manipulator (1) is used to grip the stator of the traction machine and place it on the traction machine frame. Assembly manipulator (2) is used to rotate the traction machine in coordination with the turning mechanism. Assembly manipulator (3) is used to grip the guide wheel of the traction machine.

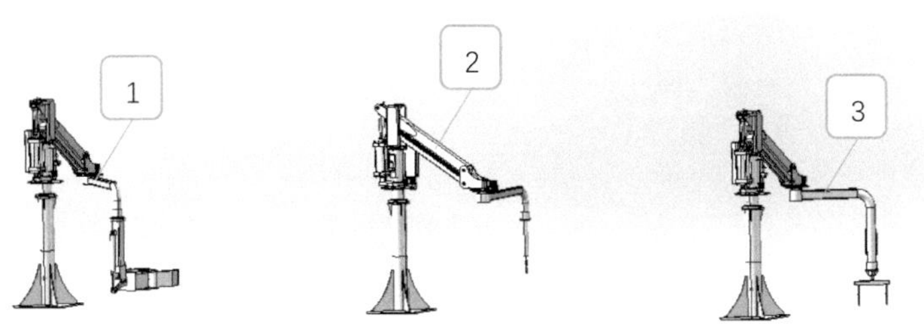

Fig. 3.16 Assembly Manipulators
1, 2, 3-Assembly Manipulators

The compression mechanism is shown in Fig. 3.17, used to press the traction machine stator into the traction machine frame.

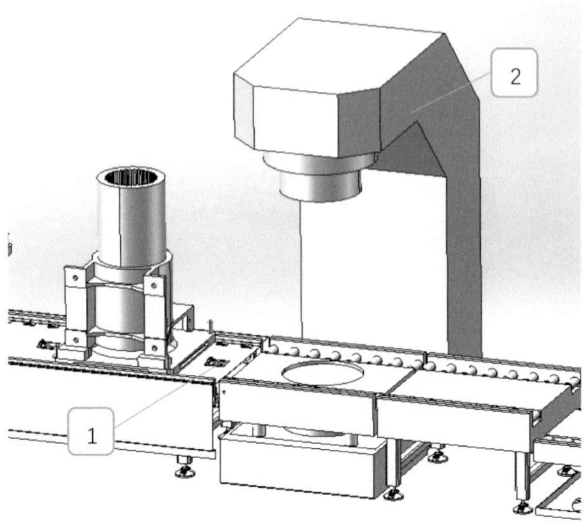

Fig. 3.17 Compression Mechanism
1 - Flat Belt 2 - Compression Mechanism

Turnover mechanism as shown in Fig. 3.18, cooperates with the assembly robot to turn the traction machine. When the turnover mechanism is operating, the flat belt (2) transports the traction machine to the underside of the clamping mechanism (1) of the turnover mechanism. The clamping mechanism clamps the traction machine. At this point, the assembly robot grabs the traction machine, the lifting platform (3) descends, and the servo motor (4) connected to the clamping mechanism (1) starts, driving the clamping mechanism (1) to rotate, thereby turning the traction machine group. After the turnover is completed, the lifting platform 3 rises, and the conveyor belt transports the traction machine.

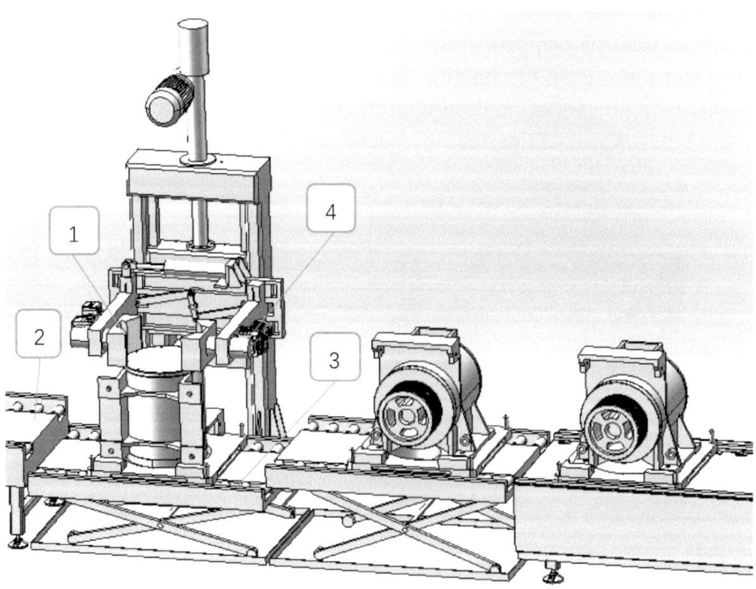

Fig. 3.18 Turning Mechanism
1 - Gripping Mechanism 2 - Flat Belt 3 - Lifting Mechanism 4 - Servo Motor

IV. Production Line Rhythm Setting

As shown in Table 3.3, the traction machine assembly line is divided into 7 stations based on the assembly process of the electric motor. These are:

Table 3.3 Traction Machine Assembly Line Cycle Time Setting Table

Process	*Process Name*	*Cycle Time*
1	Install Stator	300S
2	Stator, Rotor Press-Fit	720S
3	Stator, Rotor Assembly	300S
4	Traction Machine, Encoder, Brake Assembly	770S
5	Braking Torque Testing	500S
6	Traction Machine Testing	590S
7	Traction Machine Sorting	500S

V. Specific Parameters (see Table 3.4)

Table 3.4 Traction Machine Assembly Line Parameters

Working Efficiency:	50-80 pcs/h
Operating Rate:	≥93%
Yield Rate:	≥98%
Ambient Temperature:	0~60°C
Allowed Humidity:	20%~95%
Equipment Dimensions:	2500 mm(H) × 22190 mm(L) × 4100 mm(W)

VI. Mechanical Design Highlights

To assess congestion in each process, a professional assembly line simulation software was used to simulate the entire traction machine assembly line in terms of cycle time. The simulation generated a summary report outlining the maximum capacity and congestion issues at each workstation. To address congestion problems, corresponding processes can be configured as dual workstations. This modification enables the entire assembly line to efficiently complete the assembly of a traction machine in a shorter time frame [56]. Through the simulation design process, the traction machine assembly line achieved minimal congestion time, ensuring optimal performance in intelligent management.

Case 28: SIM Card Automatic Molding Machine

I. Case Description

The SIM card (Subscriber Identity Module), referred to as a user identification card. The SIM Card Automatic Molding Machine is mainly used for the automatic welding and molding of SIM cards. The overall structure is shown in Fig. 3.19, and an enlarged view is shown in Fig. 3.20. The SIM Card Automatic Molding Machine is mainly composed of a vibration conveyor (1), synchronous belt (2), hot melt device (3), pressurization mechanism (4), drive mechanism (5), vibration conveyor (6), discharge port (7), and feeding mechanism (8). The machine has a high level of automation, reducing manual operations and improving production efficiency.

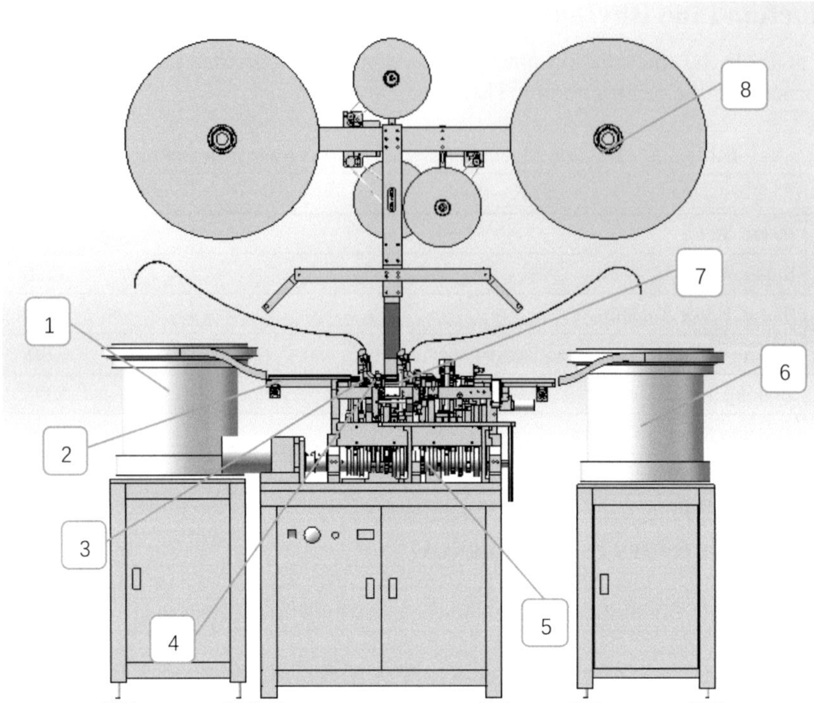

Fig. 3.19 SIM Card Automatic Molding Machine
1, 6 - Vibration Conveyor 2 - Synchronous Belt 3 - Hot Melt Device 4 - Pressurization Mechanism 5 - Drive Mechanism 7 - Discharge Port 8 - Feeding Mechanism

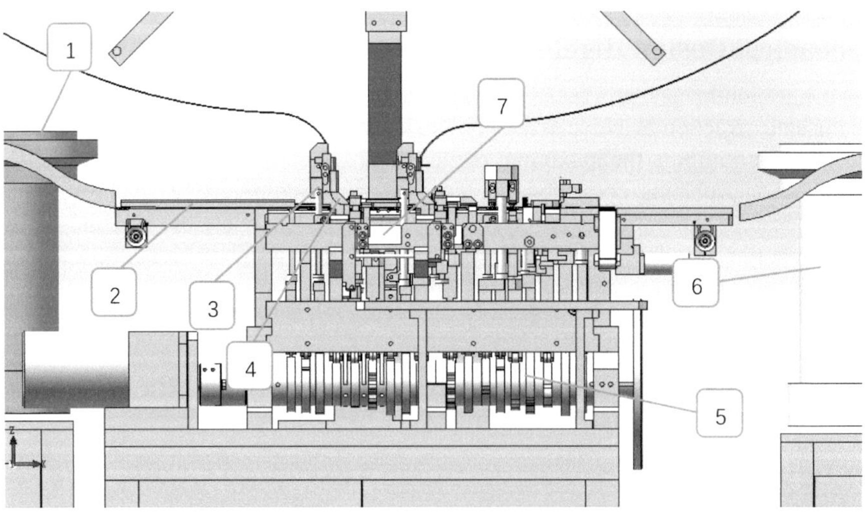

Fig. 3.20 Enlarged View of SIM Card Automatic Molding Machine
1, 6 - Vibration Conveyor 2 - Synchronous Belt 3 - Hot Melt Device 4 - Pressurization Mechanism 5 - Drive Mechanism 7 - Discharge Port

II. Working Principle

When the SIM Card Automatic Molding Machine is operational, the unprocessed SIM card cold mold is conveyed by the feeding vibration conveyor through the synchronous belt into the hot melt device. The welding joint is pressed down to imprint the card marks on the SIM card cold mold. Subsequently, it is transported by the synchronous belt to the pressurization mechanism, where the SIM card is laminated and cooled to solidify. At the same time, the feeding vibration conveyor introduces the card into the slot, combining the SIM card with the slot, and the molded SIM card is then output through the discharge port.

III. Main Structure Introduction

The feeding mechanism, as shown in Fig. 3.21, is driven by a servo motor (1) which rotates the feeding disk (3) through the pulley (2), sending the SIM card film and strip to the pressurization mechanism.

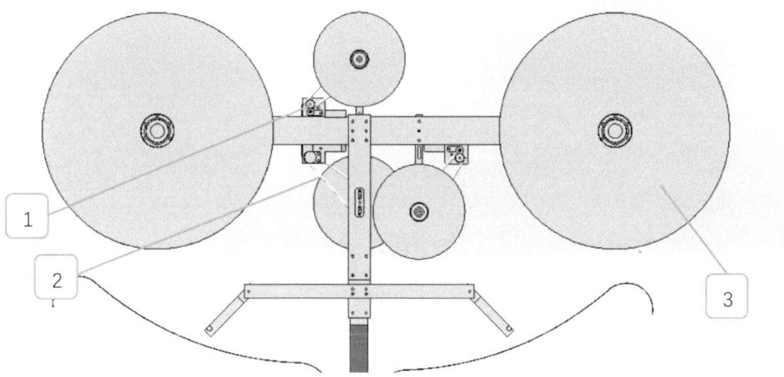

Fig. 3.21 Feeding Mechanism
1 - Servo Motor 2 - Pulley 3 - Feeding Disk

The hot melt device, as shown in Fig. 3.22, is controlled by the drive mechanism to press down, imprinting the SIM card cold mold.

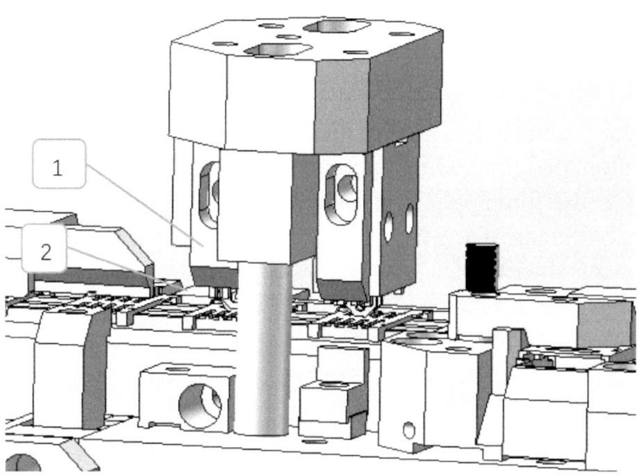

Fig. 3.22 Hot Melt Device
1 - Hot melt joint 2 - Synchronous belt

The drive mechanism is shown in Fig. 3.23, driven by servo motor (2) to rotate cam (3). Through the optimized design of the cam structure, the SIM card on the synchronous belt (1) is formed in one step according to the predetermined molding steps.

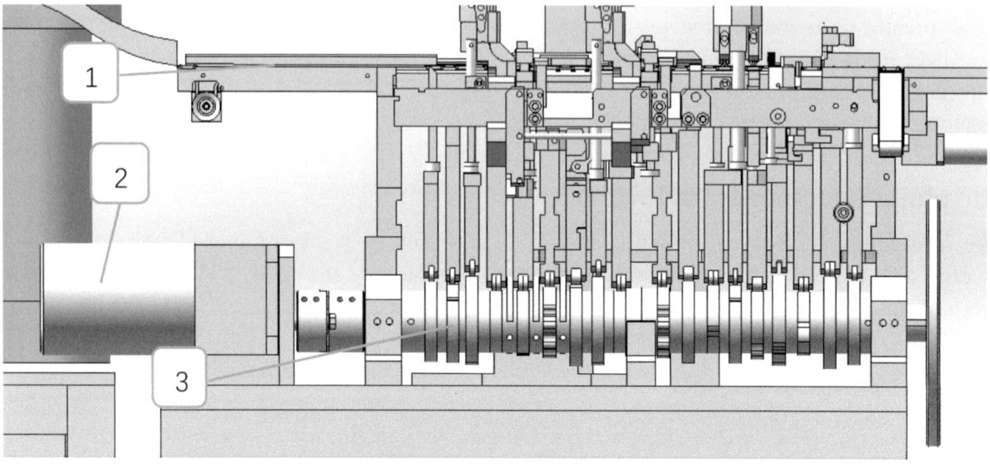

Fig. 3.23 Drive Mechanism
1 - Synchronous Belt 2 - Servo Motor 3 - Cam

IV. Specific Parameters (see Table 3.5)

Table 3.5 SIM Card Automatic Molding Machine

Docking Deviation:	≤0.03 mm
Temperature Deviation:	±5°C
Time Required to Reach Welding Temperature:	<20 min
Welding Temperature Range:	0~600°C
Operation Rate:	≥93%
Yield Rate:	≥98%
Equipment Dimensions:	2000 mm(H) × 1800 mm(L) × 1050 mm(W)

V. Mechanical Design Highlights

The processing steps for the SIM card slot are numerous, with a complex structure, mostly adopting a step-by-step semi-automated processing approach that involves manual loading and unloading of materials. Particularly in the film application process, it is necessary to apply film separately to the front and rear surfaces of the insert and the upper surface of the side block. This makes it impossible to achieve automatic and continuous film processing, leading to a significant consumption of production time. The SIM Card Automatic Molding Machine, designed with a servo motor and cam mechanism, enables integrated operations. The film application mechanism achieves interference-free, automated film application for three segments, greatly enhancing the efficiency of the film application process.

Case 29: E-Type Snap Ring Assembly Machine

I. Case Description

E-type snap rings, also known as E-type retaining rings, are a type of fastener used for installation in the grooves of machinery and equipment. They play a role in preventing axial movement of components on the shaft. This is illustrated in Fig. 3.24. The E-type snap ring assembly machine is designed to automatically assemble E-type snap rings into the grooves of product shafts. The overall structure is shown in Fig. 3.25, with an enlarged view in Fig. 3.26. It mainly consists of the snap ring pressing and assembly mechanism (1), left snap ring feeder (2), panel (3), base (4), pulley (5), support leg (6), feeding carrier (7), right snap ring feeder (8), and electrical control box (9). The E-type snap ring assembly machine features a compact footprint, high efficiency, robust durability, and flexible component connections.

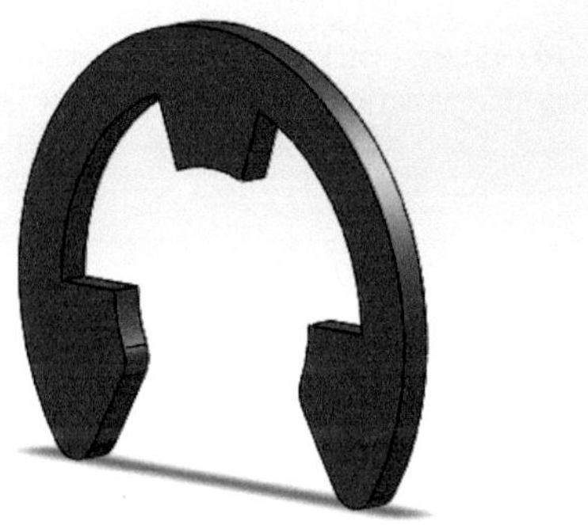

Fig. 3.24 E-type snap rings

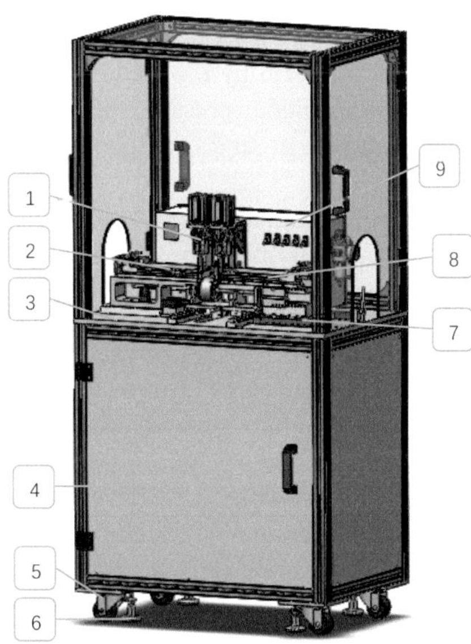

Fig. 3.25 E-type Retaining Ring Assembly Machine

1-Retaining Ring Pressing and Assembling Mechanism 2-Left-side Retaining Ring Feeder 3-Panel 4-Base 5-Pulley 6-Support Leg 7-Feeding Carrier 8-Right-side Retaining Ring Feeder 9-Control Box

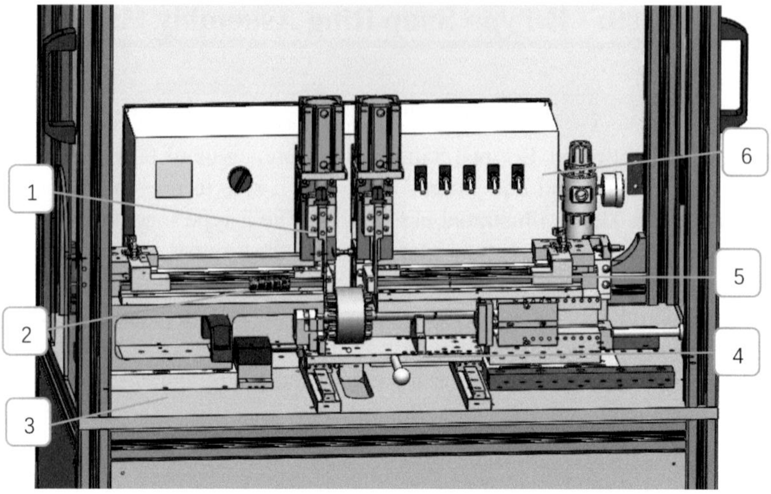

Fig. 3.26 E-type Snap Ring Assembly Machine Detailed View
1 - Snap Ring Pressing and Assembling Mechanism 2 - Left Snap Ring Feeder 3 - Panel 4 - Feeding Carrier 5 - Right Snap Ring Feeder 6 - Control Box

II. Working Principle

The automatic assembly machine is a mechanical device that combines several components of a product through tight fitting, locking, threaded connections, adhesion, riveting, etc., to obtain a finished product (or semi-finished product) with predetermined dimensional accuracy and functionality. Typically controlled by a Programmable Logic Controller (PLC), which receives various signal inputs and issues instructions to various execution mechanisms. The machine is equipped with various sensors and signal collectors to monitor the operation of each execution mechanism, and based on judgments, issue the next execution instructions [57].

The PLC of the E-type snap ring assembly machine can be adjusted as needed according to the actual situation. When the E-type snap ring assembly machine is in operation, the product to be assembled is loaded onto the work platform by the feeding carrier. The snap ring pressing and assembling mechanism moves downward under pressure, selects a snap ring from the snap ring feeder, and assembles the snap ring onto the product, completing the assembly process.

III. Main Structure Introduction

The feeding carrier, as shown in Fig. 3.27, places the product on the fixture (1) of the feeding carrier. By pushing the handle (2), the product is positioned under the snap ring pressing and assembling mechanism. When the E-type snap ring assembly machine is in operation, the left and right snap ring feeders continuously convey E-type snap rings towards the center. The snap ring pressing and assembling mechanism moves downward under pressure and selects a snap ring, which is then assembled onto the product.

The E-type card spring track is shown in Fig. 3.28. The E-type card springs to be assembled are stored on the left card spring feeding machine (3) and the right card spring feeding machine (4). When the E-type card spring assembly machine is in operation, it pushes the E-type card spring towards the center. The card spring pressing and assembling mechanism (2) is driven by cylinder (1) to press the E-type card spring down and assemble it onto the product.

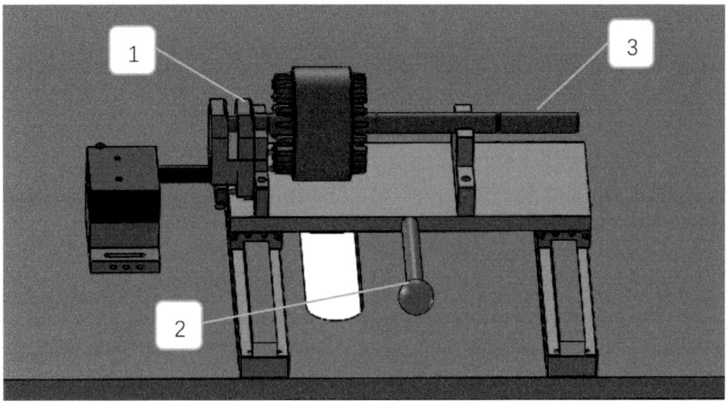

Fig. 3.27 Feeding Carrier
1 - Fixture 2 - Handle 3 - Rotating Shaft

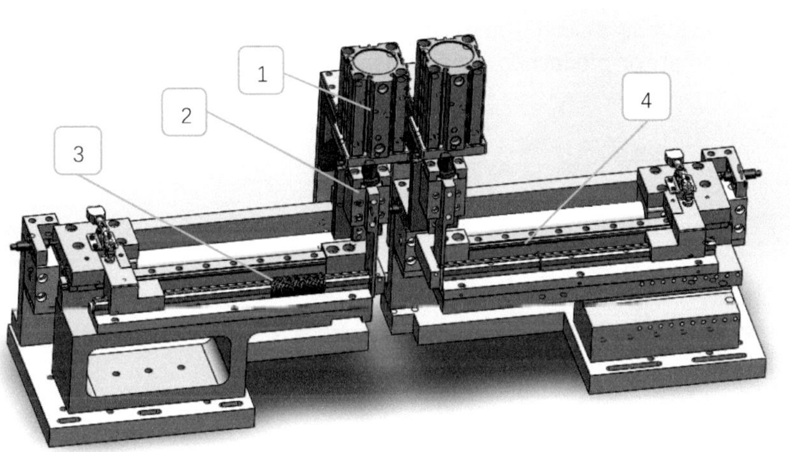

Fig. 3.28 Card Spring Feeding Track
1 - Cylinder 2 - Card Spring Pressing and Assembling Mechanism 3 - Left Card Spring Feeding Machine 4 - Right Card Spring Feeding Machine

IV. Specific Parameters (see Table 3.6)

Table 3.6 E-type Snap Ring Assembly Machine Parameters

Efficiency:	≥500 pc/h
Operation Rate:	≥93%
Qualification Rate:	≥98%
Environmental Temperature:	0~60°C
Permissible Humidity:	20%~95%
Equipment Dimensions:	1600 mm(H) × 750 mm(L) × 500 mm(W)

V. Mechanical Design Highlights

This equipment utilizes the left-side snap ring storage machine and right-side snap ring storage machine to store the E-type snap rings awaiting assembly. The control box can be adjusted as needed, and the overall structure is simple with flexible connections between components. It features a high level of automation, enhancing the efficiency of E-type snap ring assembly. Additionally, at the bottom of the equipment, there are wheels and support legs for easy mobility, improving space utilization and providing stability during operation.

Case 30: Automatic Button Assembly Machine

I. Case Description

The main function of the fully automatic button assembly machine is to use a rotary table to bring buttons together. Different stations assemble different button components. The overall structure is shown in Fig. 3.29, and a detailed view is provided in Fig. 3.30. It consists mainly of a display screen (1), vibration conveyor (2), frame (3), feeding mechanism (4), handling mechanism (5), rotary table (6), and keycap assembly mechanism (7). The fully automatic button assembly machine can significantly improve production efficiency and product quality while reducing production costs.

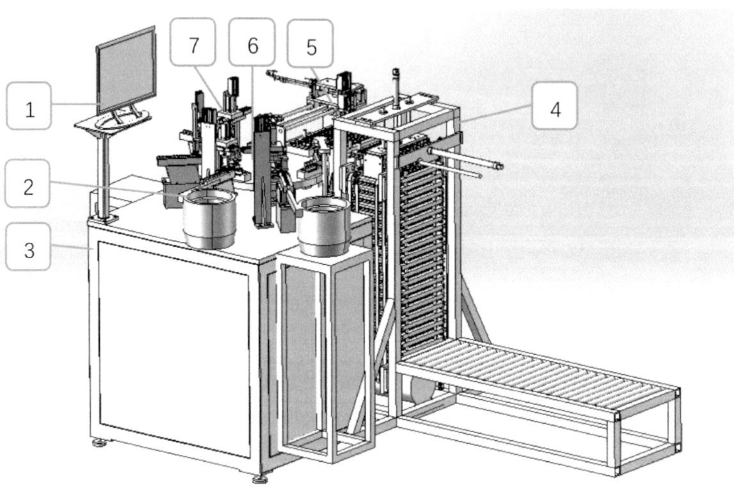

Fig. 3.29 Fully Automatic Button Assembly Machine
1-Display Screen 2-Vibration Conveyor 3-Frame 4-Feeding Mechanism 5-Handling Mechanism 6-Rotary Table 7-Keycap Assembly Mechanism

II. Working Principle

The button base, which requires the assembly of four buttons, is sequentially assembled by four keycap assembly mechanisms. When the fully automatic button assembly machine is in operation, the button base is input into the assembly machine through the feeding mechanism. The handling mechanism's fixture picks up the button base and transports it to the positioning mechanism on the rotary table. Meanwhile, the vibration conveyor supplies keycaps to the keycap assembly mechanism. As the rotary table rotates, the keycap assembly mechanism sequentially attaches keycaps to each button. After inspection, the machine outputs qualified products, completing the button assembly process.

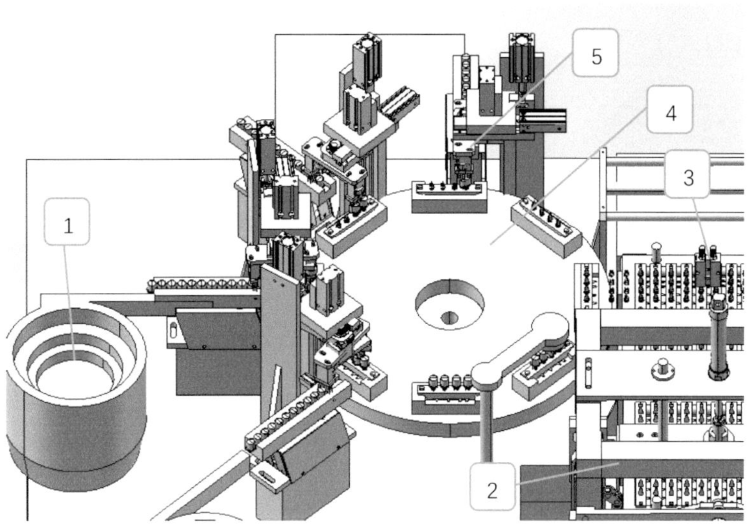

Fig. 3.30 Partial Enlarged View of Fully Automatic Button Assembly Machine
1 - Vibration Conveyor 2 - Feeding Mechanism 3 - Handling Mechanism 4 - Rotary Table 5 - Keycap Assembly Mechanism

III. Main Structure Introduction

Feeding mechanism is shown in Fig. 3.31. The button bases are placed in batches on the conveying platform (2). The servo motor (3) of the feeding mechanism drives the pulley (4) to rotate, continuously lifting the conveying platform (2). The pushing mechanism (1) is driven by a cylinder, continually pushing the conveying platform (2) under the handling mechanism (5).

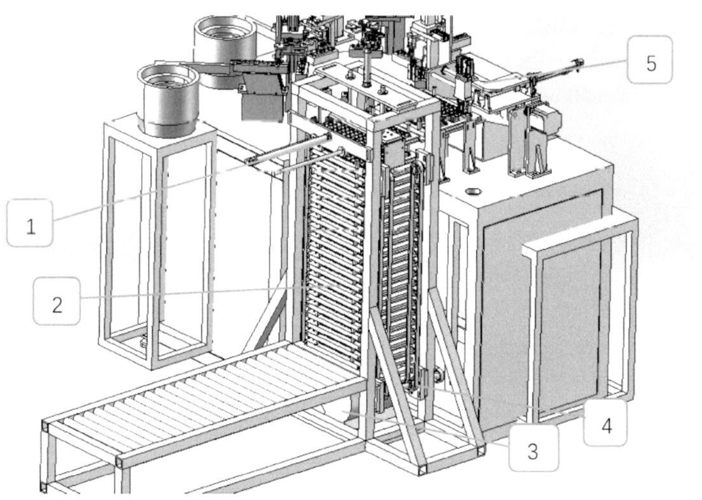

Fig. 3.31 Feeding Mechanism
1 - Pushing Mechanism 2 - Conveying Platform 3 - Servo Motor 4 - Pulley 5 - Handling Mechanism

Handling Mechanism as shown in Fig. 3.32 is controlled by servo motor (4) to control the lateral displacement of the fixture (3) in the horizontal plane, and by cylinder (5) to control the vertical displacement of the fixture (3) in the horizontal plane. The opening and closing of the fixture (3) are controlled by cylinder (2), continuously transporting the button base from the conveying platform (1) to the turntable (6).

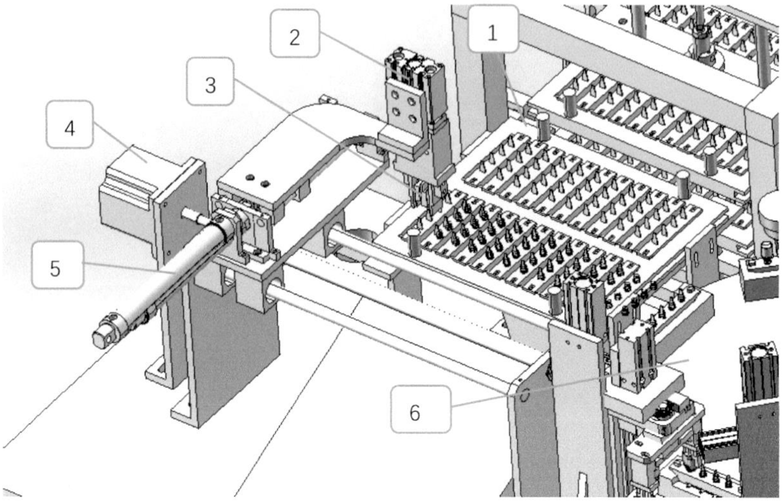

Fig. 3.32 Handling Mechanism
1 - Conveying platform 2, 5 - Cylinder 3 - Fixture 4 - Servo motor 6 - Turntable

Button Cap Assembly Mechanism as shown in Fig. 3.33, this equipment is equipped with a total of 4 button cap assembly mechanisms, which sequentially place button caps on each button. When the button fully automatic assembly machine is operational, the vibrating conveyor (2) provides button caps to the button cap assembly mechanisms. The fixture (3) of the button cap assembly mechanism grips the button cap and places it on the button base. The button cap assembly mechanism is driven by cylinders (1) and (5). Cylinder (1), driven by a gear mechanism, controls the rotation of the fixture (3) in the horizontal plane, while cylinder (5) controls the vertical movement of the fixture (3), thus enhancing operational efficiency.

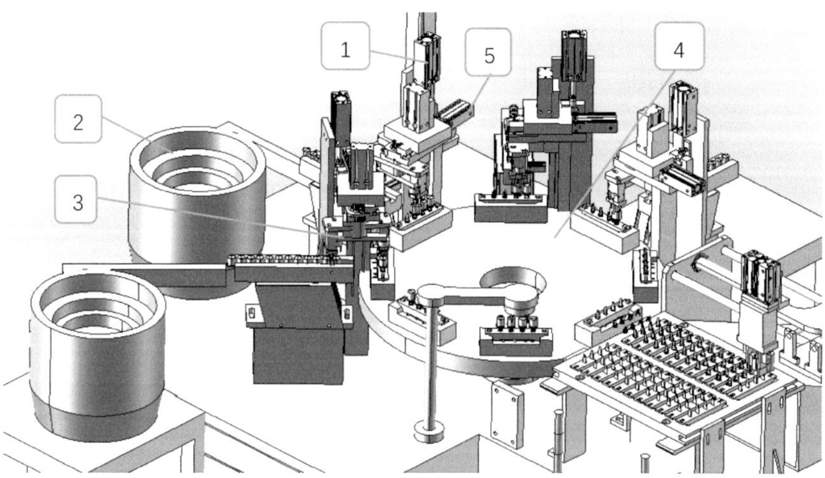

Fig. 3.33 Button Cap Assembly Mechanism
1, 5 - Cylinders 2 - Vibrating Conveyor 3 - Fixture 4 - Turntable

IV. Specific Parameters (see Table 3.7)

Table 3.7 Button Fully Automatic Assembly Machine Parameters

Positioning Accuracy:	±0.01 mm
Working Efficiency:	1500-2000 pcs/h
Yield Rate:	≥98%
Operating Temperature:	0~50°C
Allowable Humidity:	20%~95%
Drive Mode:	PLC-controlled electric motor and pneumatic components
Feeding Method:	Vibrating conveyor automatic arrangement
Equipment Dimensions:	1790 mm(H) × 1750 mm(L) × 2510 mm(W)

V. Mechanical Design Highlights

The fixture (3) of the key cap assembly mechanism of the fully automatic button assembly machine can simultaneously grip two key caps, as shown in Fig. 3.34. The fixture (3) is equipped with a rotating shaft for rotation coordination. The cylinder (1) pushes the rack (2), which in turn drives the rotating shaft to rotate, thereby causing the fixture (3) to rotate around the rotating shaft. By combining the cylinder with the gear mechanism, this equipment greatly simplifies the key cap gripping process, thereby improving the speed of key cap assembly.

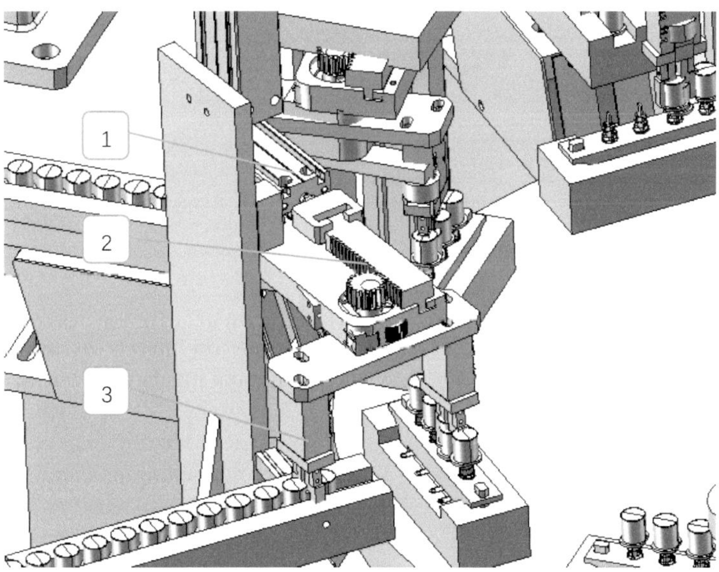

Fig. 3.34 Fixture
1 - Cylinder 2 - Rack 3 - Fixture

Case 31: Semi-finished Product Screw Locking Machine

I. Case Description

The semi-finished product screw locking machine is suitable for efficiently locking multiple identical or different specification screws on the same workpiece simultaneously. The overall structural diagram is shown in Fig. 3.35, consisting mainly of the frame (1), feeding system (2), screw locking mechanism (3), turntable (4), discharge box (5), handling mechanism (6), and display screen (7). This machine offers flexibility in operation, high efficiency, precise torque, and ensures the quality of screw tightening [58].

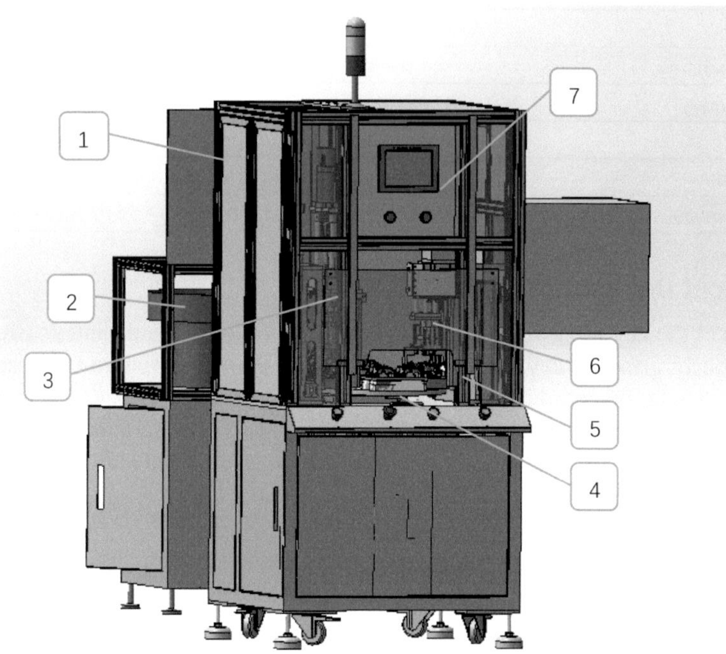

Fig. 3.35 Semi-finished product screw locking machine
1 - Frame 2 - Feeding system 3 - Screw locking mechanism 4 - Turntable 5 - Discharge box 6 - Handling mechanism 7 - Display screen

II. Working Principle

The semi-finished product screw locking machine operates by having the handling mechanism transport the product that needs screw locking to the positioning mechanism on the turntable. As the turntable rotates, the product is conveyed to the screw locking mechanism, which locks the screw in place. The turntable continues to rotate, and the completed product is outputted. The screws for locking are provided by the feeding system. When the screw locking machine finishes locking a screw, the incoming material sensing system sends a signal to the signal receiver. The signal receiver then sends instructions to the feeding sensor, which starts conveying screws from the hopper. Screws that pass the incoming material detector's inspection are directed into the nail-feeding tube. Using compressed air, these screws are then sent from the nail-feeding tube to the nozzle, where they wait for the next screw locking operation [59]. This process is cyclic, as the incoming material sensor continues to send signals to the signal receiver each time a screw is locked in place.

III. Main Structure Introduction

The turntable, as shown in Fig. 3.36, is equipped with four stations, each featuring a positioning mechanism (2) designed to secure the product awaiting screw locking. This prevents accidental dislodging of the product during the operation.

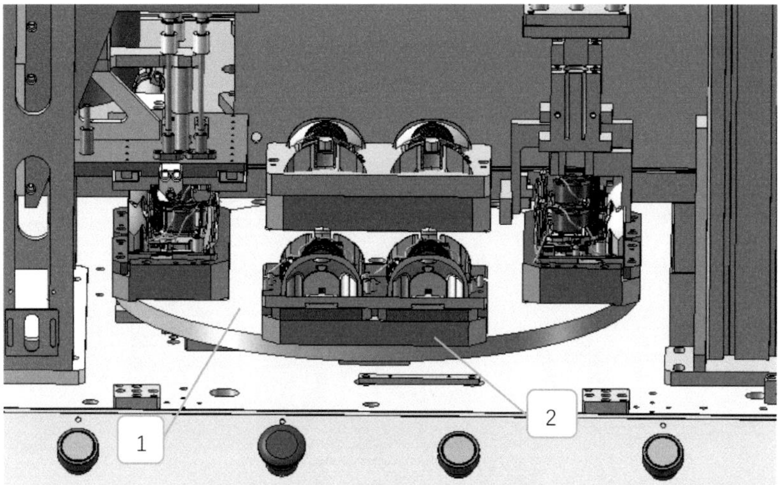

Fig. 3.36 Turntable
1 - Turntable 2 - Positioning Mechanism

The handling mechanism is shown in Fig. 3.37. It is controlled by an electric motor (6) for the horizontal displacement of the clamp (1), an air cylinder (5) for the vertical displacement, and another air cylinder (4) for the opening and closing of the clamp (1). This mechanism is used to transport the products needing screw locking to the positioning mechanism of the turntable (3) and then transport the products with completed locking to the discharge box (2).

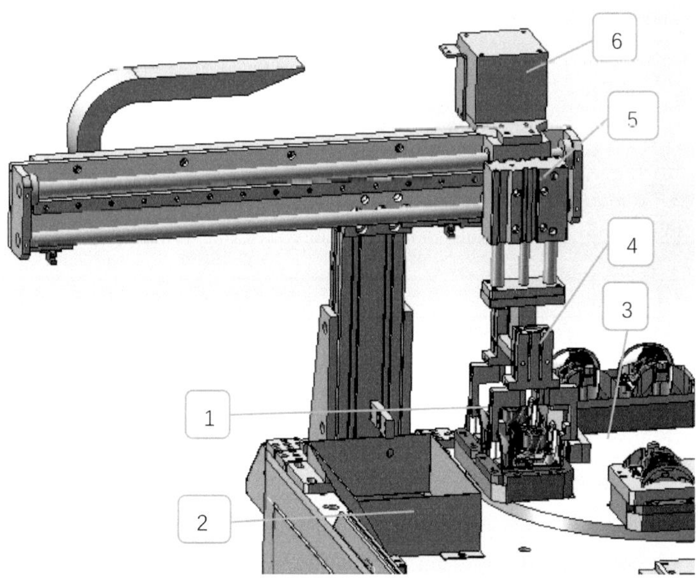

Fig. 3.37 Handling Mechanism
1 - Clamp 2 - Discharge Box 3 - Turntable 4, 5 - Air Cylinder 6 - Servo Motor

The screw locking mechanism, as shown in Fig. 3.38, receives qualified screws from the screw feeding pipe (1). At this point, cylinder (2) is activated, controlling the downward pressure of the screw locking mechanism. The screws are then conveyed by the screw clamp (4) to the screw hole of the product, where they await locking. Subsequently, the product is moved to the electric screwdriver for final tightening.

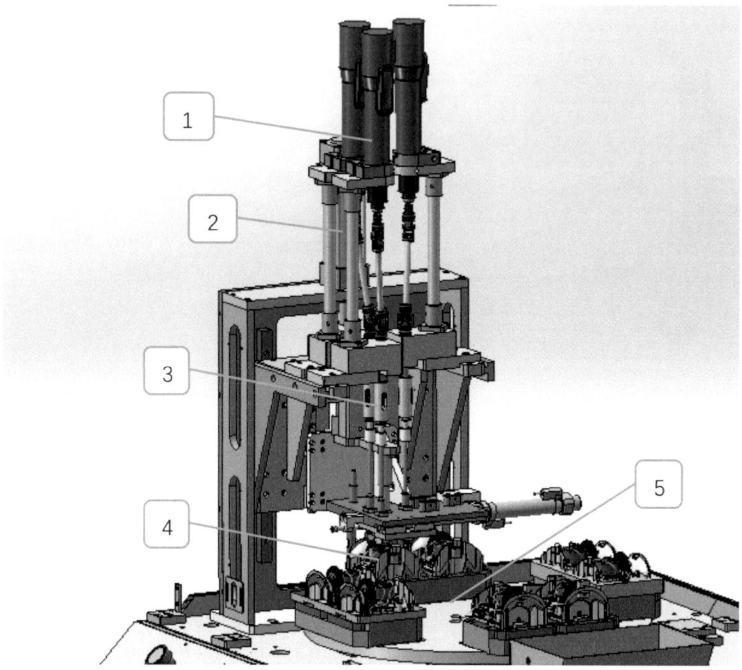

Fig. 3.38 Screw Locking Mechanism
1 - Screw Feeding Pipe 2 - Cylinder 3 - Screw Detection Mechanism 4 - Screw Clamp 5 - Turntable

IV. Specific Parameters (see Table 3.8)

Table 3.8 Semi-finished Screw Locking Machine Parameters

Maximum Torque:	150 kg.cm
Air Pressure Source:	5.5 (±0.5) kg/cm^2
Operating Temperature:	–10°C~40°C
Environmental Humidity:	5%-90%
Screw Feeding Mechanism:	Vibrating Arm Drum Type
Suitable Screw Specifications:	M2.0-8.0 mL 4.0-25.0 mm D3.0-10 mm L > 1.1-1.5D
Equipment Dimensions:	1900 mm(H) × 1100 mm (L) × 1270 mm(W)

V. Mechanical Design Highlights

(1) Efficient and high-speed screw placement, capable of simultaneously placing 4-6 screws, suitable for applications requiring the locking of multiple screws.
(2) Multi-functional integrated machine, flexible integration of various peripheral auxiliary operations, such as ultrasonic welding, date code printing, part assembly, etc., truly versatile with multiple functions in one machine [60].
(3) Automated design with automatic feeding, automatic unloading and retrieval, and automatic detection and judgment for sorting defective products.

(4) Intelligent detection, reducing inspection steps by catching floating locks and slipping teeth during the locking process.
(5) High-quality design for continuous and uninterrupted operation, significantly reducing average downtime, effectively enhancing productivity.
(6) Precision mechanical design with innovative mechanisms for easy adjustment, saving time, and ensuring stable and reliable operation.
(7) Advanced feeding system to reduce the probability of material jamming, greatly improving overall efficiency.
(8) Excellent operability with user-friendly design, allowing operators to start working without training. Different authorization levels are designed for personnel to avoid unnecessary mistakes.
(9) Compact structural layout to effectively save floor space and facilitate maintenance.

Case 32: Automatic Assembly Equipment for Magnets

I. Case Description

The automatic assembly equipment for magnets is designed to replace manual feeding and assembly, enhancing the efficiency of magnet assembly operations. The overall structure is depicted in Fig. 3.39, comprising dispensing mechanism (1), frame (2), ratchet indexing disk (3), magnet assembly equipment (4), positioning mechanism (5), and electrical control box (6). This automated system significantly improves the production yield of products and reduces manufacturing costs.

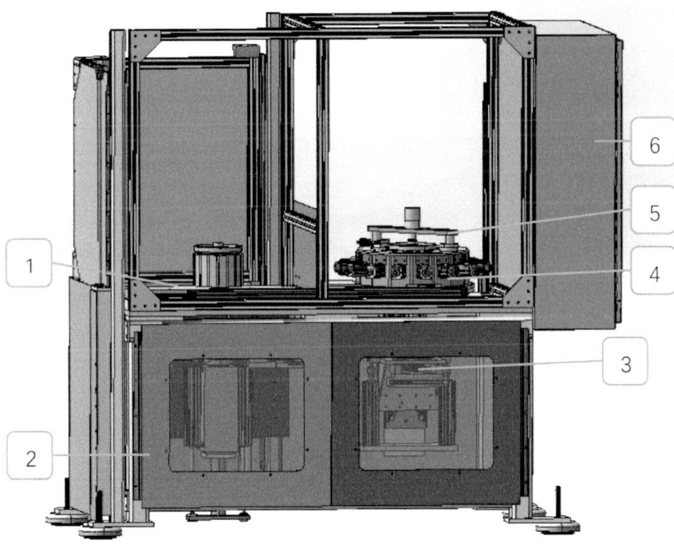

Fig. 3.39 Automatic Assembly Equipment for Magnets
1-Dispensing Mechanism 2-Frame 3-Ratchet Indexing Disk 4-Magnet Assembly Equipment 5-Positioning Mechanism 6-Electrical Control Box

II. Working Principle

The magnetic automation assembly equipment operates by dispensing adhesive on the product through the dispensing mechanism. After the adhesive application, the product is transported to the magnet assembly equipment. The ratchet indexing disk positions the product and transports the iron pieces to the specified location. The assembly process for magnets is simultaneously carried out at 16 workstations, completing the magnet assembly process.

III. Main Structure Introduction

The dispensing mechanism, as shown in Fig. 3.40, is controlled by cylinder (2) to fix the product with the base (5). Cylinder (4) drives and controls the hook (3), rotating the base (5) by flipping. The dispensing mechanism (1) dispenses adhesive on the surface of the product.

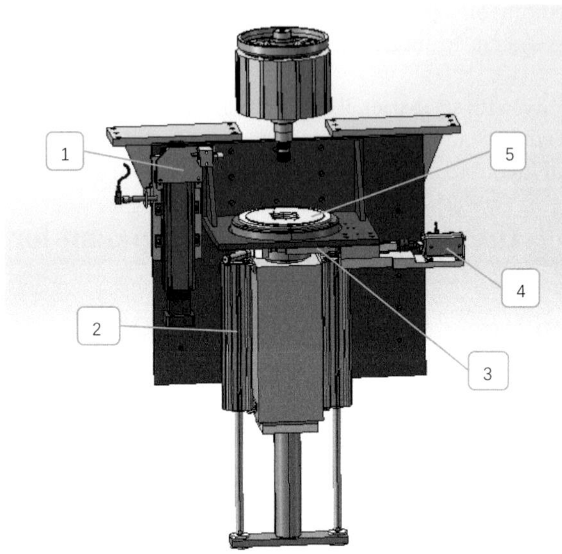

Fig. 3.40 Dispensing Mechanism
1 - Dispensing Mechanism 2, 4 - Cylinders 3 - Hook 5 - Base

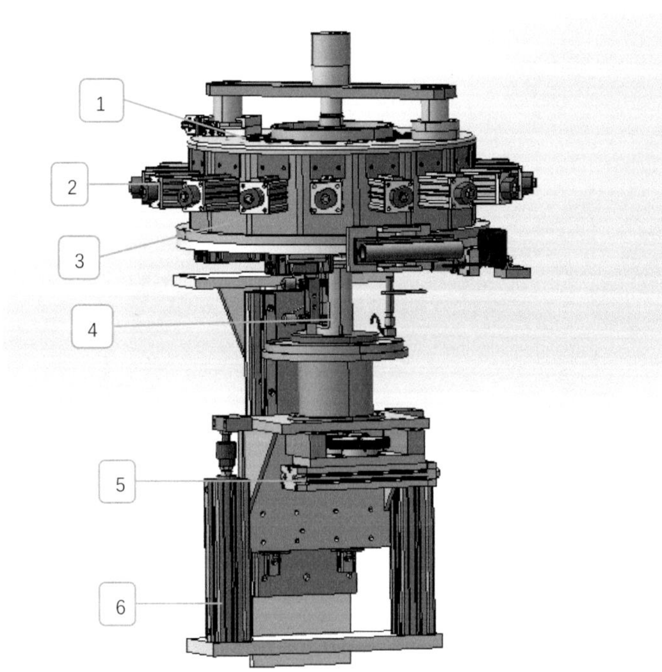

Fig. 3.41 Magnetic Assembly Equipment
1 - Cover Plate 2, 6 - Cylinder 3 - Ratchet Indexing Disc 4 - Spindle 5 - Servo Motor

The magnetic assembly equipment, as shown in Fig. 3.41, operates by placing the product with completed dispensing work at the center of the ratchet indexing disc (3). The cylinder (6) drives the cover plate (1) to secure the product, and the servo motor (5) controls the rotation of the spindle (4). This aligns the magnetic position on the product surface with the magnet. The cylinder (2) then drives the attachment of the magnet to the product surface.

IV. Specific Parameters (see Table 3.9)

Table 3.9 Magnetic Automation Assembly Equipment Parameters

Working Efficiency:	300-500 pcs/h
Operating Rate:	≥95%
Yield Rate:	≥98%
Ambient Temperature:	0~45°C
Allowed Humidity:	30%~95%
Equipment Dimensions:	1800 mm(H) × 2000 mm(L) × 800 mm(W)

V. Mechanical Design Highlights

The current magnetic automatic assembly machines in the market not only have a complex structure but also possess limited functionality. Existing machines rely on a combination of manual and fixture-based production for the gluing process. This results in high labor intensity, low assembly precision, low production capacity, and a large footprint. Compared to traditional manual gluing, this magnetic automatic assembly machine significantly reduces labor requirements and space utilization. It saves half the space compared to traditional gluing methods and allows real-time monitoring of various production parameters, facilitating problem tracking during the production process.

Case 33: Electric Motor Stator Filling Machine

I. Case Description

The electric motor stator filling machine is designed for the automatic assembly of electric motor stators. The overall structure is illustrated in Fig. 3.42, consisting mainly of the stator filling

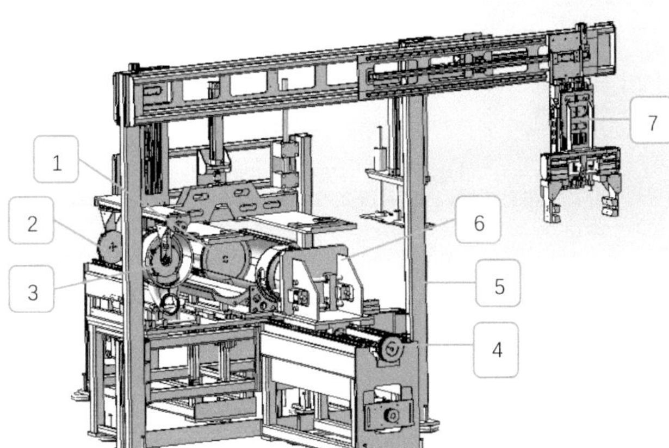

Fig. 3.42 Electric Motor Stator Filling Machine
1 - Stator Filling Mechanism 2 - Mounting Ring 3 - Stator Adjustment Mechanism 4 - Material Feeding Mechanism 5 - Frame 6 - Fixed Base 7 - Large Manipulator

mechanism (1), mounting ring (2), stator adjustment mechanism (3), material feeding mechanism (4), frame (5), fixed base (6), and large manipulator (7) [61]. This equipment effectively reduces safety risks associated with manual labor, and it possesses advantages such as a simple structure, easy maintenance, low processing costs, and high production efficiency.

II. Working Principle

The electric motor stator filling machine operates by having the large manipulator grab the electric motor casing, making a horizontal movement, and placing the motor casing on the fixed base with a columnar structure. The stator is secured on the stator adjustment mechanism, which adjusts the position and angle of the stator through lateral displacement and rotation. Subsequently, the material feeding mechanism propels the fixed base, positioning and assembling the stator with the electric motor casing, completing the filling process for the electric motor stator [62].

III. Main Structure Introduction

The large manipulator, as shown in Fig. 3.43, is used to grab the electric motor casing. During operation, the control of the manipulator is positioned above the shelf. Cylinder (5) initiates a vertical descent of the manipulator to the height of the stator. Simultaneously, cylinder (3) activates, causing the manipulator to open and grasp the electric motor casing. Cylinder (5) then controls the upward movement of the manipulator. The electric motor (6) starts, directing the manipulator to move horizontally and transport the electric motor casing to the fixed base.

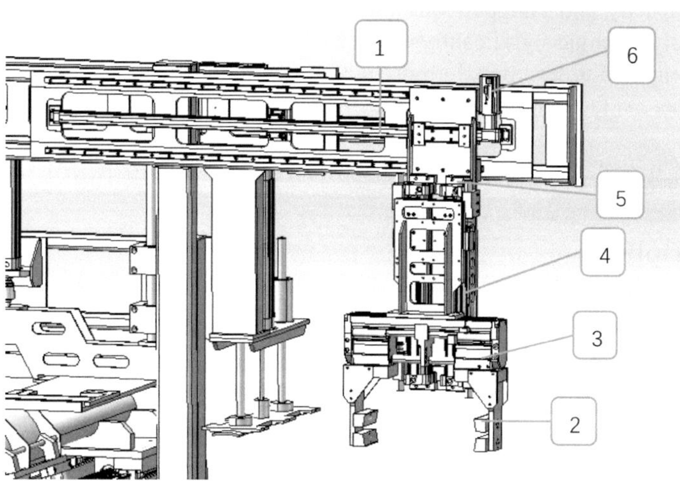

Fig. 3.43 Large Manipulator
1, 4 - Guide Rails 2 - Spacer 3, 5 - Cylinders 6 - Electric Motor

The stator filling mechanism is shown in Fig. 3.44. During operation, the motor housing is placed within the cylindrical filling mechanism. The housing adjustment mechanism (6) adjusts the angle of the motor housing. At this point, cylinders (2), (3), (4), and (5) are activated, pushing plates (1) and (7) tightly together to secure the motor housing in place, ensuring alignment with the stator assembly angle. Simultaneously, the stator adjustment mechanism transports the stator to the stator filling mechanism, and the pusher mechanism pushes the fixture base laterally to insert the stator into the housing.

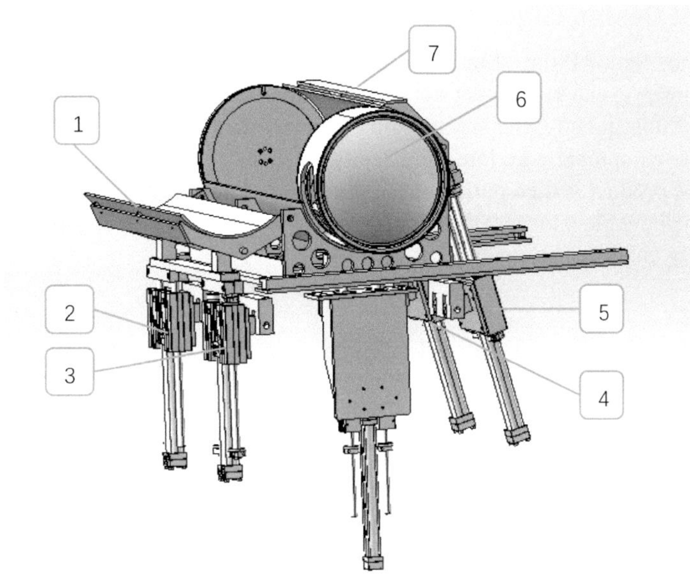

Fig. 3.44 Stator Filling Mechanism
1, 7 - Cover plates 2, 3, 4, 5 - Cylinders 6 - Housing adjustment mechanism

The feeding mechanism is shown in Fig. 3.45, driven by motors (2) and (3), which push the motor housing to tightly fit with the stator.

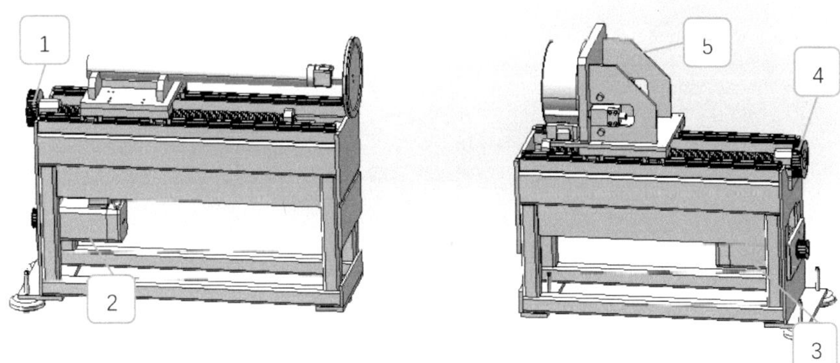

Fig. 3.45 Feeding Mechanism
1, 4 - Pulley 2, 3 - Motors 5 - Fixed Base

IV. Specific Parameters (see Table 3.10)

Table 3.10 Parameters of the Electric Motor Stator Filling Machine

Travel Accuracy:	0.05 mm
Pressure Control Accuracy:	0.1%
Operating Temperature:	Room temperature
Rated Pressure:	30 N
Maximum Downward Speed:	100 mm/s
Equipment Dimensions:	2430 mm(H) × 3600 mm(L) × 3720 mm(W)

V. Mechanical Design Highlights

The Electric Motor Stator Filling Machine operates by utilizing the pushing mechanism to bring the stator core closer to the equipment core. This allows the stator core to nest at one end of the equipment core. Subsequently, the stator filling mechanism compresses, forcing the copper wire wound around the equipment core into the stator core. This completes the processing of the stator core. The finished product is then pulled out of the equipment core by a connecting rope, which, in turn, drives the pushing plate to eject the stator core that was nested outside the equipment core. This process effectively enhances processing efficiency.

Case 34: Automatic Harpoon Assembly Machine

I. Case Description

The VGA connector (Video Graphics Array connector) is commonly used in computers for graphics cards, monitors, and other devices. It is employed to transmit analog signals. The Automatic Harpoon Assembly Machine is designed for installing harpoons on VGA connectors. The overall structure is shown in Fig. 3.46 and mainly consists of the assembly robot (1), feeding mechanism (2), transfer guide (3), feeding mechanism (4), base (5), and riveting pressure mechanism (6). The automatic harpoon assembly machine is characterized by its simplicity, durability, stable and efficient production process, effectively improving work efficiency, and reducing production costs.

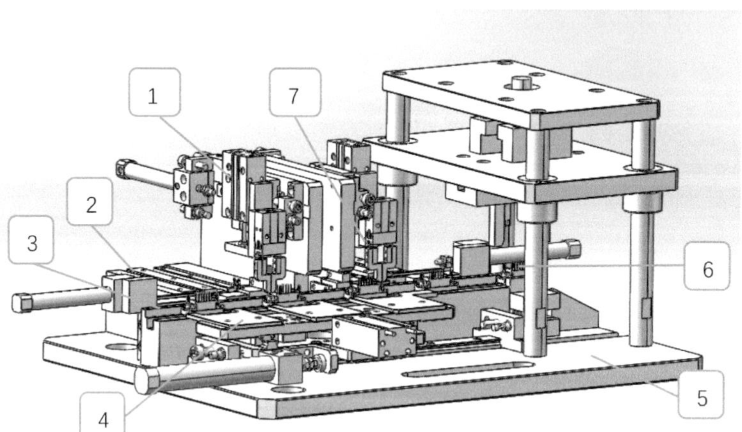

Fig. 3.46 Automatic Harpoon Assembly Machine
1, 7 - Assembly Robot 2, 4 - Feeding Mechanism 3 - Transfer Guide 5 - Base 6 - Riveting Pressure Mechanism

II. Working Principle

The Automatic Harpoon Assembly Machine operates by feeding the semi-finished product from the left side of the transfer guide. The feeding mechanism organizes and transports the semi-finished product, introducing the harpoon and rivet under the action of the cylinder. After the assembly is complete, the assembly robot grabs the harpoon assembly and transports it to the left side of the connector. Similarly, the assembly robot places the harpoon assembly on the right side of the connector. The synchronous belt operates continuously, transporting the semi-finished product with the assembled harpoon to the riveting pressure mechanism. The riveting pressure mechanism presses and secures the harpoon, and after the pressing is complete, the assembled component is output, completing the installation of the harpoon on the connector.

III. Main Structure Introduction

The feeding mechanism, as shown in Fig. 3.47, is primarily composed of three clamping plates, two cylinders, and a stopper block. The semi-finished product enters from the left side of the transfer guide, and the cylinders control the horizontal and vertical positions of the clamping plates (cylinder (2) controls the horizontal position, and cylinder (3) controls the vertical position), sequentially transporting the semi-finished product to the assembly, pressing, and other processes.

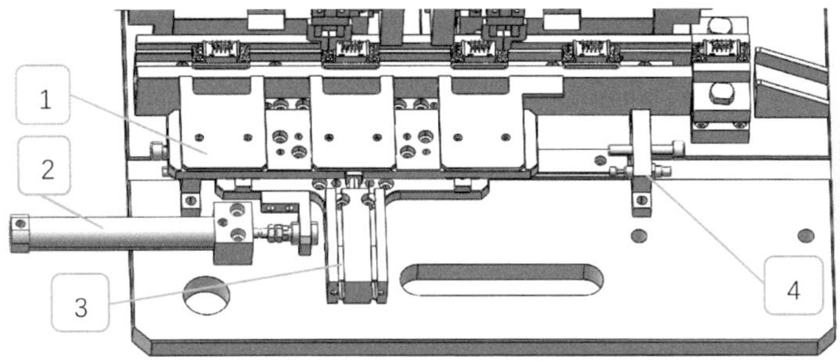

Fig. 3.47 Feeding Mechanism
1 - Clamping plate 2, 3 - Cylinder 4 - Stopper block

Figure 3.48 shows the Feeding Mechanism and Assembly Robot. The SUB harpoon components are respectively fed in from guides (2) and (3). Cylinder (4) pushes the parts together at the end of guide (2). The assembled harpoon is then transferred by robot (1) and placed in the left slot of the sub-assembly.

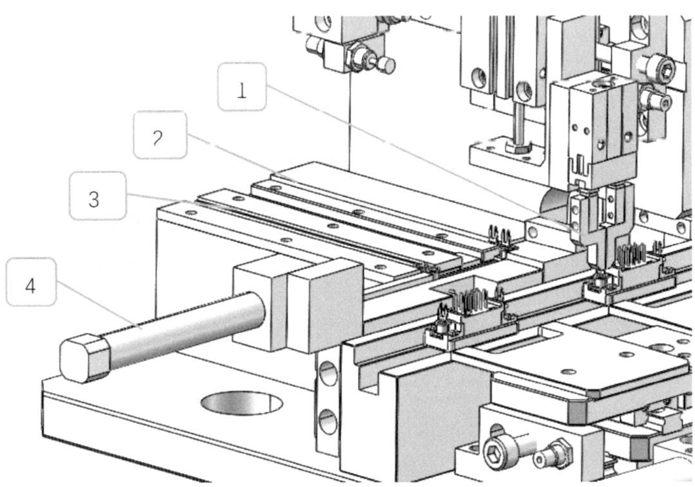

Fig. 3.48 Feeding Mechanism and Assembly Robot
1 - Assembly Robot 2, 3 - Guides 4 - Cylinder

IV. Specific Parameters (see Table 3.11)

Table 3.11 Electric Motor Stator Filling Machine Parameters

Ambient Temperature:	0~60°C
Allowable Humidity:	20%~95%
Operating Pressure Range:	4-6 kg/cm^2
Working Efficiency:	2500-3000 pcs/min
Equipment Dimensions:	280 mm(H) × 380 mm(L) × 300 mm(W)

V. Mechanical Design Highlights

The automatic fish spear installation machine, through the set clamping blocks and clamping slots, effectively joins various boards together. This method efficiently avoids the slow screw installation and non-dismountable issues found in traditional techniques. Additionally, it occupies less space, saving both financial and labor resources. It exhibits excellent creativity and practicality.

Case 35: Stator Encapsulation Machine

I. Case Description

Stator Encapsulation Machine is mainly used for encapsulating the stator of an electric motor, which is composed of the stator, rotor, and other accessories. The overall structure is illustrated in Fig. 3.49 and includes components such as the frame (1), upper die (2), platen (3), lower die (4), top rod column (5), and electrical control box (6). The Stator Encapsulation Machine features a compact structure and high efficiency. Additionally, it can adapt to different sizes of electric motors by utilizing corresponding molds for operations.

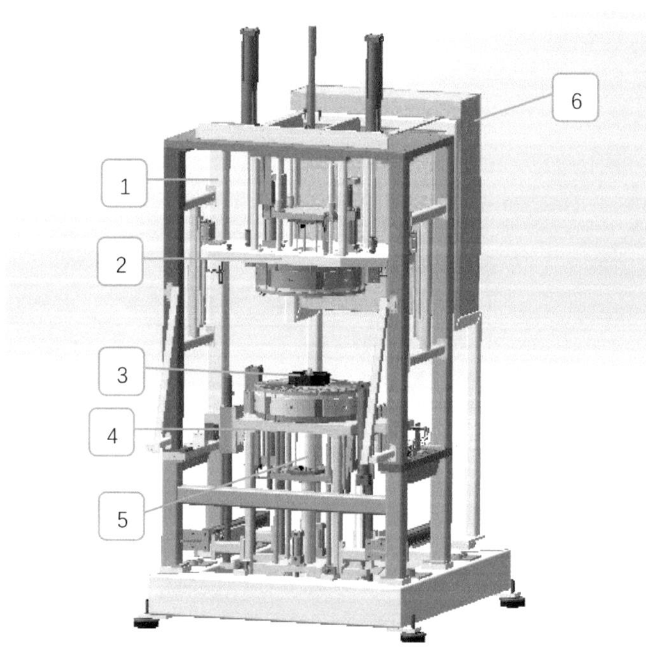

Fig. 3.49 Stator Encapsulation Machine
1-Frame 2-Upper Die 3-Platen 4-Lower Die 5-Top Rod Column 6-Electrical Control Box

II. Working Principle

When the stator encapsulation machine is in operation, the operator places the stator assembly on the turntable. The coils are positioned and evenly arranged on the lower die through the positioning column and the lower die platen. The lead-out end and the connection end of the coils are respectively placed in the lead-out end slot and the connection end slot. Each segment is placed in the corresponding slot. The upper die is then combined with the lower die, fixing the coil winding within the encapsulation mold. High molecular materials are injected into the encapsulation mold through the injection port. After the coil winding is securely encapsulated, the stator disc is lifted out using the top rod column through the top rod column hole. The connection ends of the multi-phase coil windings are welded, forming the stator of the electric motor.

III. Main Structure Introduction

The upper die structure is shown in Fig. 3.50, driven by cylinders (1) and (5). The upper die is equipped with a pressure plate (3), driven by cylinders (2) and (4), and works together with the lower die to secure the motor stator.

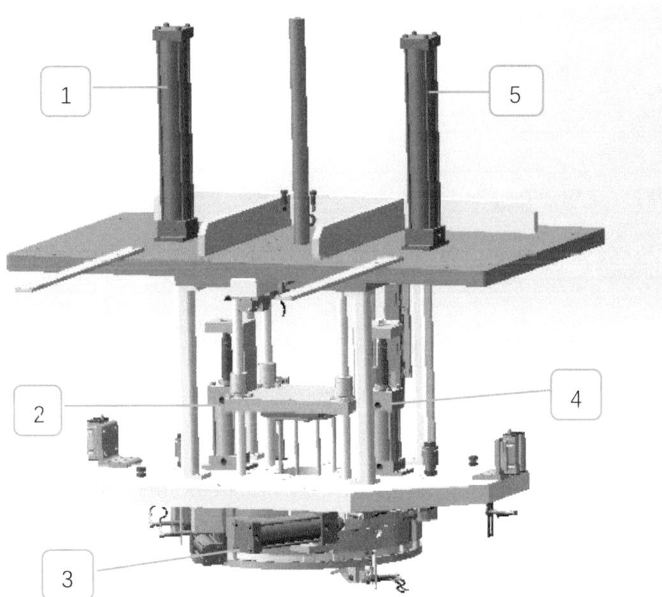

Fig. 3.50 Upper Die
1, 2, 4, 5 - Cylinders 3 - Pressure Plate

The lower die structure is shown in Fig. 3.51, and its vertical movement is driven by cylinders (1) and (7). Above the base plate (3), there is a positioning column (8) for locating the coil, and platform (2) is used to secure the stator. Cylinders (4) and (6) control the vertical movement of the base plate (3). When the stator encapsulation machine is in operation, the upper die and lower die are closed, and the coil winding is fixed within the encapsulation mold. High polymer material is injected into the mold through the injection port. After the coil winding is solidified, the stator disc is ejected using plunger (5).

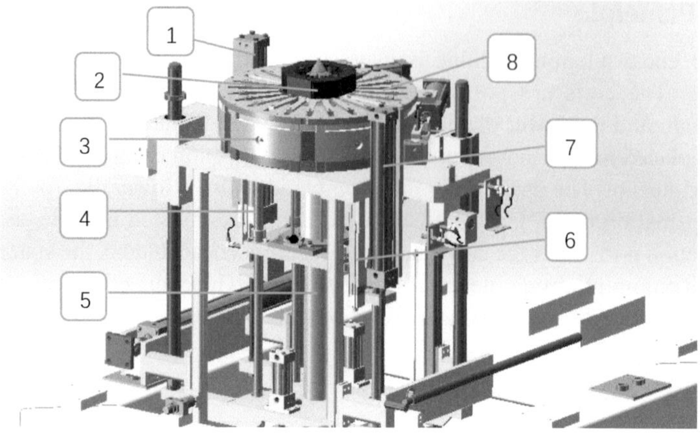

Fig. 3.51 Lower die structure
1, 4, 6, 7 - Cylinders 2 - Platform 3 - Base plate 5 - Plunger 8 - Positioning column

IV. Specific Parameters (see Table 3.12)

Table 3.12 Stator Encapsulation Machine Parameters

Yield Rate:	≥98%
Utilization Rate:	≥93%
Allowed Humidity:	25%~95%
Weight:	320 kg
Equipment Dimensions:	2400 mm(H) × 1340 mm(L) × 1700 mm(W)

V. Mechanical Design Highlights

(1) No preforming of the coil is required during use; instead, it undergoes one-time molding in the pressing process, making the process simpler.
(2) Automatically generates reinforcing ribs, heat dissipation holes, and a surface covering layer on the stator of the electric motor. This ensures that the coil is less prone to being pulled out during motor use and provides effective heat dissipation, allowing for higher power in large electric motors.
(3) The internal and external positioning columns of the encapsulation mold play a role in positioning the coil, ensuring the dimensional accuracy of the coil and maintaining consistency in batch production.

Case 36: Copper Head Assembly Machine

I. Case Description

The Copper Head Assembly Machine is used for the rapid and precise assembly of copper heads. The overall structure is shown in Fig. 3.52, with a detailed view in Fig. 3.53. It mainly consists of the discharge mechanism (1), vibratory conveyor (2), frame (3), assembly robot (4), turntable (5), base mold (6), and compression molding device (7). The Copper Head Assembly Machine offers advantages such as low cost, user-friendly operation, and high efficiency.

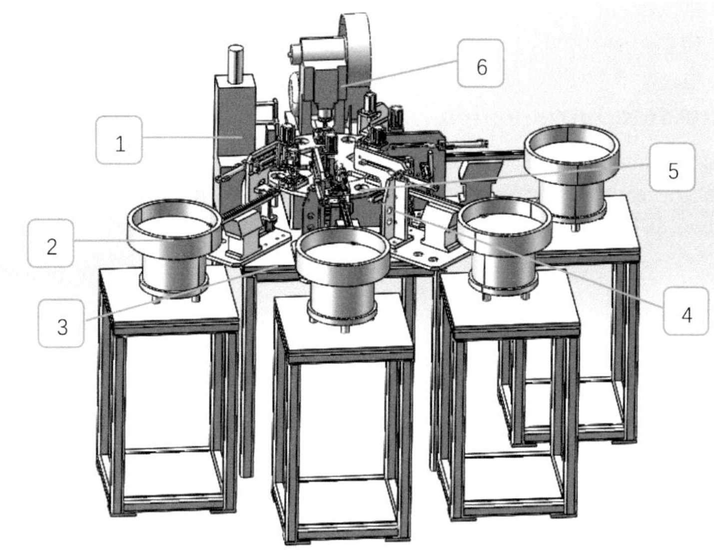

Fig. 3.52 Copper Head Assembly Machine
1 - Discharge Mechanism 3 - Frame 4 - Assembly Robot 5 - Turntable 6 - Bottom Mold

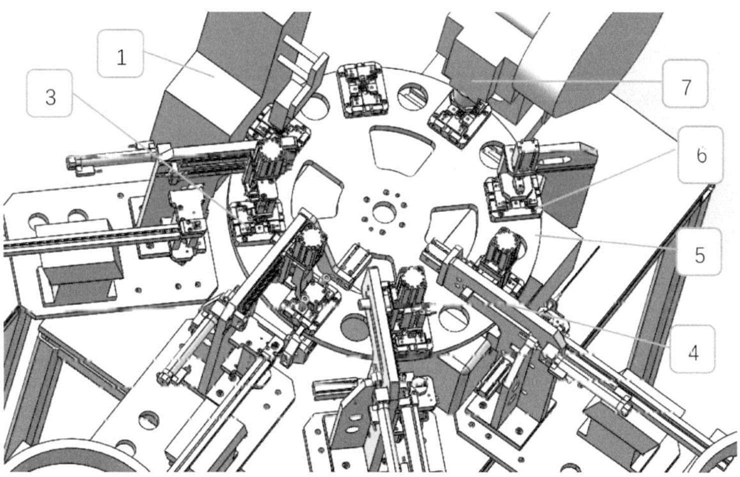

Fig. 3.53 Close-up View of Copper Head Assembly Machine
1 - Discharge Mechanism 3 - Frame 4 - Assembly Robot 5 - Turntable 6 - Bottom Mold
7 - Pressure Molding Device

II. Working Principle

The copper head assembly machine operates by arranging and conveying the individual components of the 3M copper head in an orderly manner using the vibrating conveyor. These components are then transported to the assembly robot located below. The assembly robot, positioned at four different workstations, assembles the various parts of the 3M copper head according to the bottom mold. Once the assembly is complete, the turntable transports the assembled components to the pressure molding device. The pressure device applies pressure to fix the outer cover and inner core of the assembled component, thereby completing the automatic assembly of the 3M copper head.

The assembled 3M copper head is then grabbed from the bottom mold by the discharge mechanism and transported to the next step in the process.

III. Main Structure Introduction

The bottom mold is shown in Fig. 3.54. The equipment has a total of 8 workstations, and the bottom mold is installed on each of the 8 positions on the turntable. It is used to sequentially place the various components of the 3M copper head.

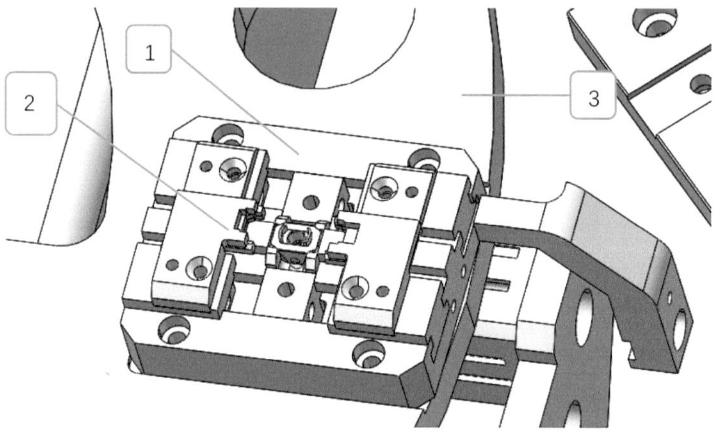

Fig. 3.54 Bottom Mold
1-Bottom Mold 2-Fixture 3-Turntable

The assembly robot, as shown in Fig. 3.55, is controlled by cylinder (1) for vertical displacement, cylinder (2) for opening and closing, and cylinder (3) for horizontal displacement. During operation, the assembly robot grips the parts from the discharge track of the vibration conveyor and accurately places them in the center of the bottom mold (6).

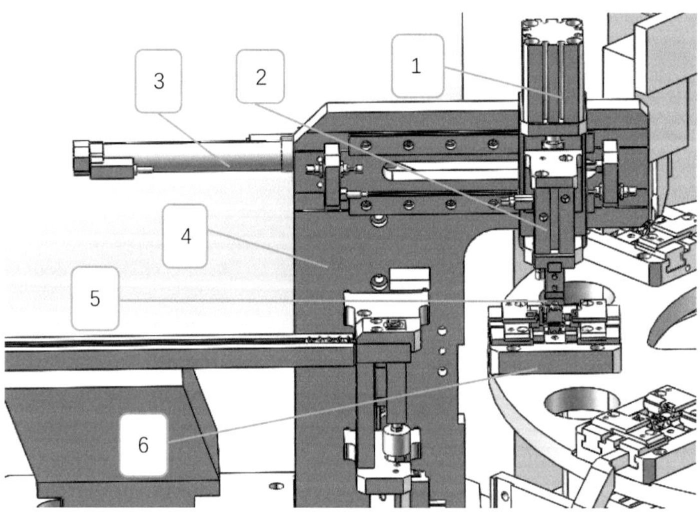

Fig. 3.55 Assembly Robot
1, 2, 3-Cylinders 4-Assembly Robot Bracket 5-Robot Arm 6-Bottom Mold

The pressure molding device, as shown in Fig. 3.56, is driven by an electric motor (2) to lower the pressing head (1), applying pressure to fix the external patch and inner core of the assembled component, thereby completing the automatic assembly operation of the 3M copper head.

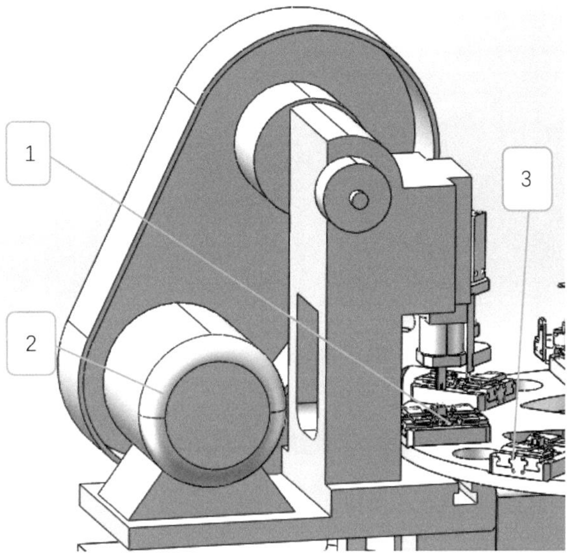

Fig. 3.56 Pressure Molding Device
1 - Pressure Head 2 - Electric Motor 3 - Bottom Mold

IV. Specific Parameters (see Table 3.13)

Table 3.13 Copper Head Assembly Machine Parameters

Yield Rate:	≥95%
Repeatability:	0.05 mm
Environmental Requirements:	Temperature: 1~50°C
Allowable Humidity:	<75%
Equipment Dimensions:	1220 mm(H) × 1730 mm(L) × 1400 mm(W)

V. Mechanical Design Highlights

Above the base plate of the copper head assembly machine, there is a rotating disc connected to the motor. The rotating disc is linked to the feeding mechanism, which, in turn, connects to the feeding slide table. The feeding slide table is connected to the vibratory conveyor. On the rotating disc, there are operational buttons, and four feeding mechanisms are installed. The copper head assembly machine enables fast and precise positioning and assembly of 3M copper heads, offering low cost, ease of use, and time-saving features.

Case 37: Double-Tower Non-Standard Assembly Machine

I. Case Description

The overall structure of the Double-Tower Non-Standard Assembly Machine is shown in Fig. 3.57, and the top view is illustrated in Fig. 3.58. It mainly consists of the tower robot (1), electrical control box (2), frame (3), automatic assembly mechanism (4), turntable (5), feeding mechanism (6), component detection mechanism (7), and other components. The Double-Tower Non-Standard Assembly Machine features stable operation, high yield, and high efficiency.

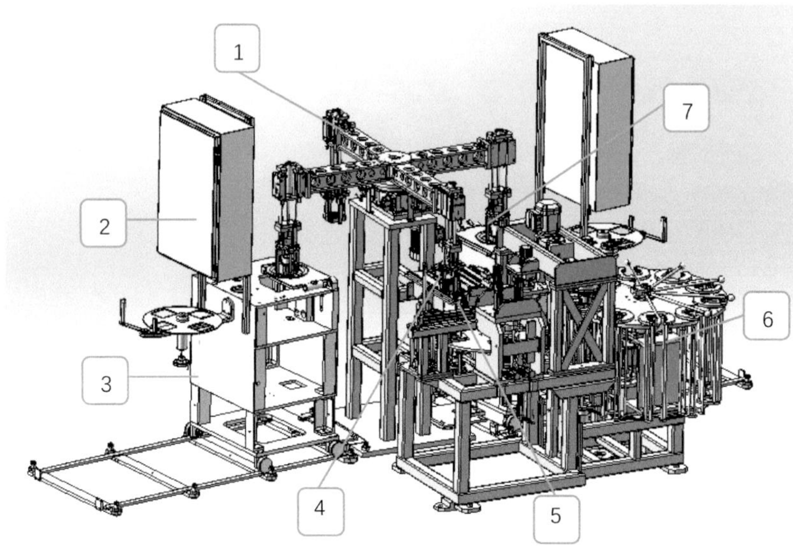

Fig. 3.57 Double-Tower Non-Standard Assembly Machine
1 - Tower Robot 2 - Electrical Control Box 3 - Frame 4 - Automatic Assembly Mechanism 5 - Turntable 6 - Feeding Mechanism 7- Component Detection Mechanism

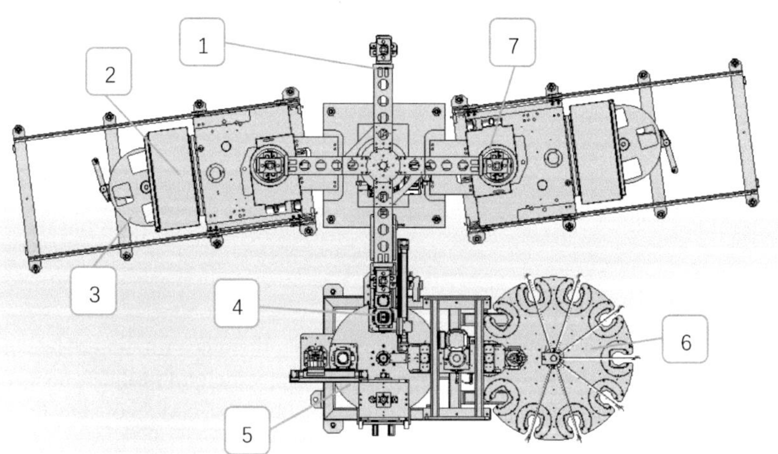

Fig. 3.58 Top View of Double-Tower Non-Standard Assembly Machine
1 - Tower Robot 2 - Electrical Control Box 3 - Frame 4 - Automatic Assembly Mechanism 5 - Turntable 6 - Feeding Mechanism 7 - Component Detection Mechanism

II. Working Principle

When the double-tower non-standard assembly machine is operating, the parts to be assembled are loaded into the matching feeding mechanism. The feeding robot transports the product's core to the automatic assembly turntable, while the tower robot places the inspected and qualified product parts into the assembly turntable. The turntable has a total of 4 workstations, each equipped with a bottom mold. The automatic assembly mechanism achieves the automatic assembly of products by controlling servo motors and cylinders. During the assembly process, lubricating oil can be sprayed into the pump body to enhance the durability of the produced products.

III. Main Structure Introduction

The feeding mechanism is shown in Fig. 3.59. The parts to be assembled are placed on the shelf (3). The shelf (3) can rotate around the central axis (2). During operation, the feeding robot (1) takes the parts and transports them to the bottom mold on the turntable.

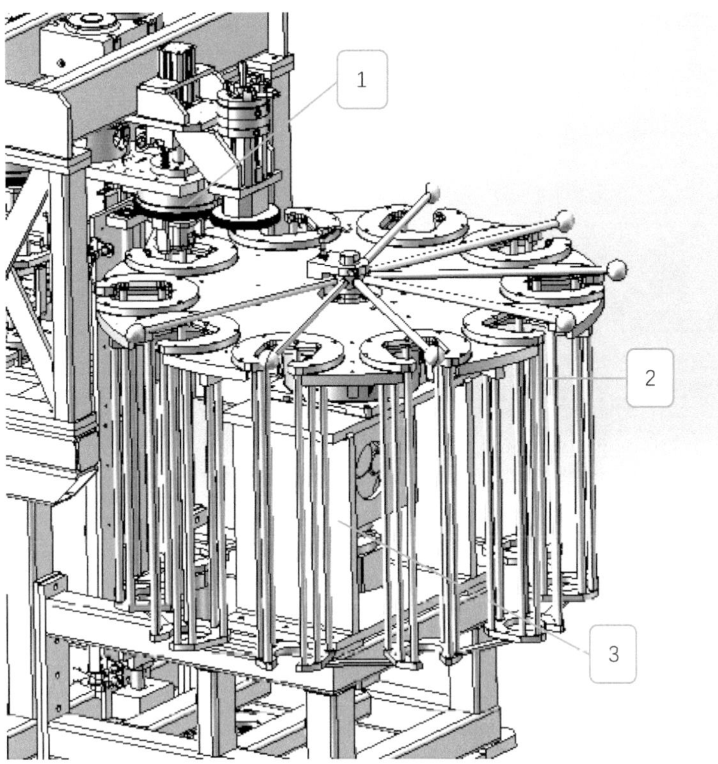

Fig. 3.59 Feeding Mechanism
1 - Feeding Robot 2 - Central Axis 3 - Shelf

The tower-type robot, as shown in Fig. 3.60, mainly consists of four robotic arms (1) and a central motor (4). It sequentially and in batches transfers products from the automatic assembly mechanism to the component inspection mechanism through rotation.

The automatic assembly mechanism and the turntable are shown in Fig. 3.61. The turntable (4) is driven by the electric motor (5) to rotate clockwise around the central axis. The turntable (4) is equipped with 4 stations, each with a base mold (3) for positioning product components. The automatic assembly mechanism achieves the automatic assembly of products through cylinders and servo motors.

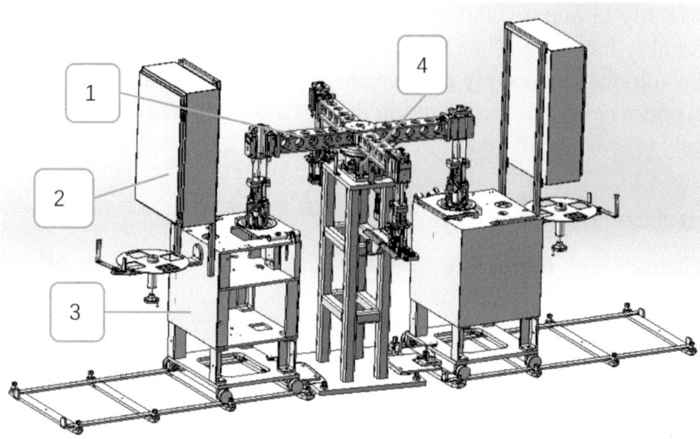

Fig. 3.60 Tower-style Robotic Arm
1 - Robotic Arm 2 - Electrical Control Box 3 - Frame 4 - Central Electric Motor

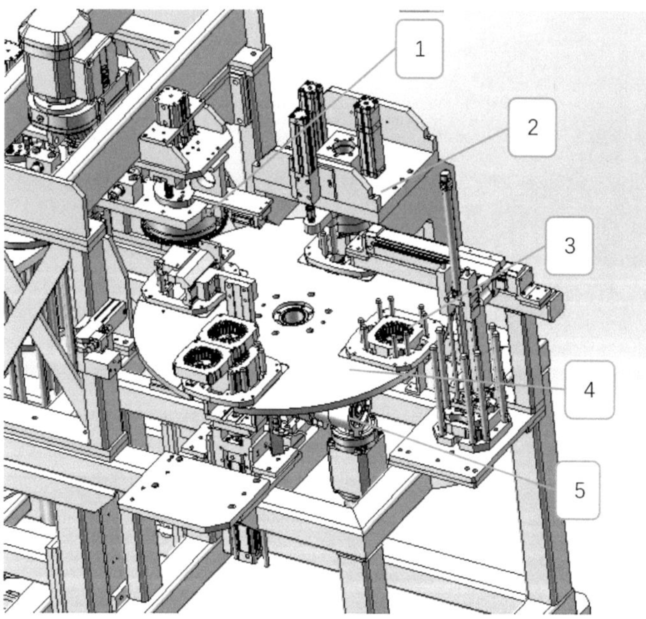

Fig. 3.61 Automatic Assembly Mechanism and Turntable
1 - Turntable Motor 2 - Turntable 3 - Base Mold 4 - Automatic Assembly Mechanism 5 - Feeding Robot Arm

IV. Specific Parameters (see Table 3.14)

Table 3.14 Double Rotary Tower Non-standard Assembly Machine Parameters

Operating Temperature:	0~45°C
Permissible Humidity:	25%~95%
Operational Efficiency:	≥90%
Yield Rate:	≥95%
Equipment Dimensions:	2160 mm(H) × 4700 mm(L) × 2500 mm(W)

V. Mechanical Design Highlights

The reversing mechanism in existing assembly machines is highly customized, resulting in poor versatility. Additionally, the adjustment of height and installation angles of the reversing mechanism in the current double rotary tower assembly machines is challenging. The double rotary tower non-standard assembly machine employs a servo motor to drive the adjustment positioning block, facilitating quick and precise product reversal. This mechanism offers fast response speed and high accuracy. Furthermore, the detection device is capable of accurately identifying the presence of products and detecting whether stacking has occurred.

Case 38: Pipeline Assembly Machine

I. Case Description

A pipeline is a frequently used medium transport structure in manufacturing operations, generally composed of connectors and fittings. The connectors and fittings are shown in Fig. 3.62. To ensure that the medium in the pipeline does not overflow, the fittings must fit tightly with the connectors without any gaps. This equipment is a pipeline assembly machine, as shown in Fig. 3.63 for the overall structure and in Fig. 3.64 for a detailed view. It mainly consists of a pipe pressing device (1), a frame (2), a movable work table (3), and an automatic assembly mechanism (4). The pipeline assembly machine completes the assembly of the pipeline in one step, greatly improving work efficiency and ensuring the safety and consistency of product production.

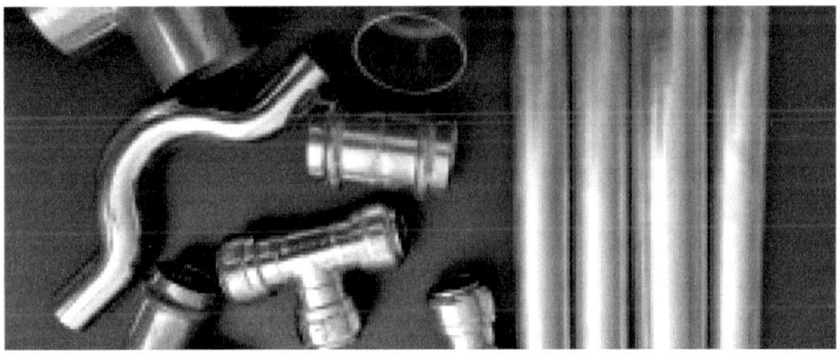

Fig. 3.62 Pipeline

II. Working Principle

The pipeline assembly machine is primarily designed to complete tasks such as pressing the ports and fitting heads of pipe components. During operation, the product is first placed in the limit slot of the connecting part in the pipe pressing device. The cylinder drives the limiting support block to secure the pipe component. Subsequently, the cylinder controls the movement of the workbench to dock with the pipe component and fix the interface. At this point, the automatic assembly mechanism's cylinder is activated, lowering and tightening the assembly head onto the connecting part and the pipe component. Once the pipeline assembly is complete, the pipe pressing device and the mobile workbench release the pipe component, allowing for the retrieval of the assembled product.

III. Main Structure Introduction

The pipe pressing device and the automatic assembly mechanism are depicted in Fig. 3.65. During the operation of the pipeline assembly machine, the product is placed in the limit slot (3), where

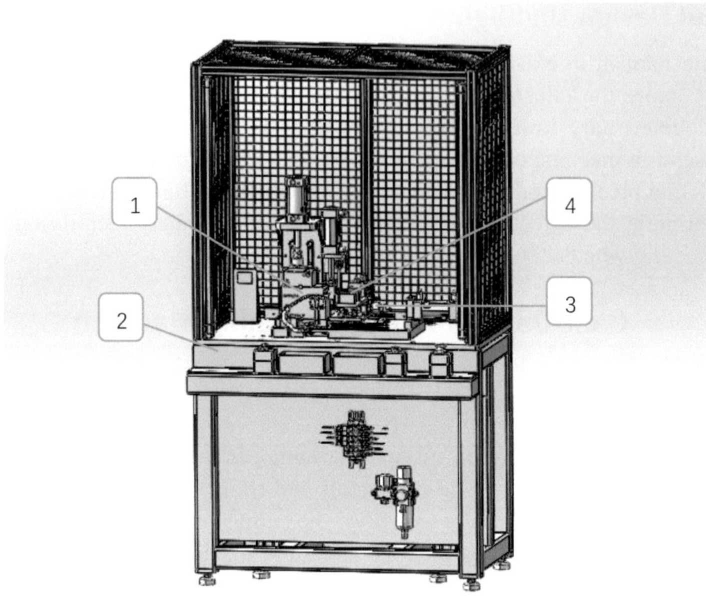

Fig. 3.63 Pipeline Assembly Machine
1 - Pipe Pressing Device 2 - Frame 3 - Mobile Workbench 4 - Automatic Assembly Mechanism

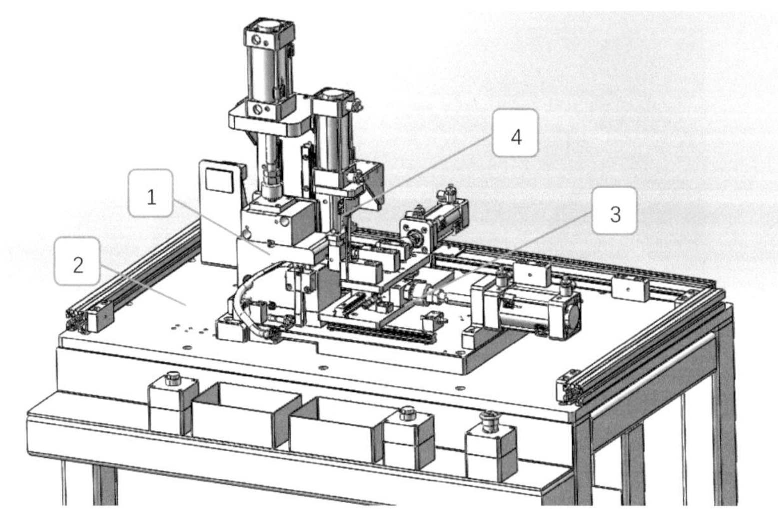

Fig. 3.64 Pipeline Assembly Machine-Detailed View
1 - Pipe Pressing Device 2 - Frame 3 - Mobile Workbench 4 - Automatic Assembly Mechanism

cylinder (1) drives the limiting support block (2) downward to secure the pipe component. The automatic assembly mechanism, driven by cylinder (5), controls the downward movement of the assembly head (4) to tighten the connecting part and the pipe component.

The mobile workbench is depicted in Fig. 3.66, driven by cylinders (1) and (4) to control the docking of the workbench with the pipe fittings and secure the pipe fitting interface.

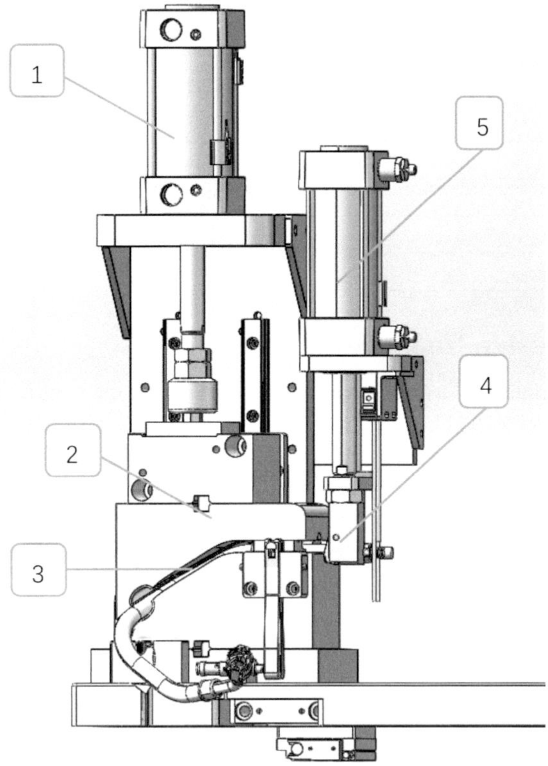

Fig. 3.65 Illustrates the pipe pressing device and the automatic assembly mechanism.
1, 5 - Cylinder 2 - Limiting support block 3 - Limiting slot 4 - Assembly head

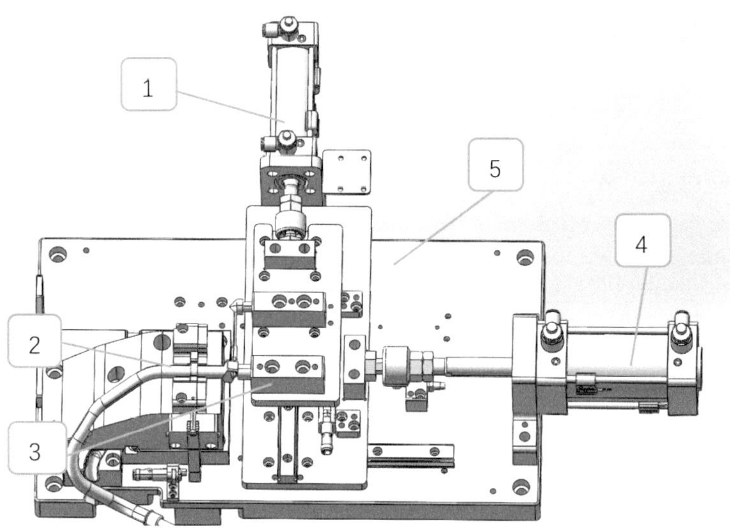

Fig. 3.66 Mobile Workbench
1, 4 - Cylinders 2 - Limiting Slot 3 - Docking Head 5 - Workbench

IV. Specific Parameters (see Table 3.15)

Table 3.15 Pipe Assembly Machine Parameters

Air Pressure:	>0.5 MPa
Efficiency:	15-20 pcs/min
Yield Rate:	≥98%
Operating Temperature:	0~55°C
Allowable Humidity:	1%~90%
Equipment Dimensions:	810 mm(H) × 1950 mm(L) × 1090 mm(W)

V. Mechanical Design Highlights

The pipe assembly machine avoids inaccuracies in judgment results caused by the subjective awareness of the operators. This ensures accurate assessment of whether the pipe fittings and connectors are tightly fitted, thereby improving the quality of pipe assembly and reducing the rate of defective products. This not only guarantees the normal progress of production operations but also saves maintenance costs for the pipes. Simultaneously, the connection component's positioning groove restricts the movement of the connection component during assembly, preventing displacement and ensuring the correct dimensions of the assembled pipe. This, in turn, enhances the precision of pipe assembly.

VI. Precautions

1. When connecting pipes, ensure that there is sufficient allowance for deformation to prevent the pipes from experiencing tensile stress.
2. Avoid subjecting the pipes to lateral forces during connection, as excessive lateral force can lead to poor sealing.
3. Connect the pipes properly in one attempt to avoid multiple disassembly operations, as repeated disassembly can also degrade the sealing performance.

Case 39: Machine Case Assembly Machine

I. Case Description

Machine Case Assembly Machine is designed for the automated assembly of stator casings. The overall structure is shown in Fig. 3.67, mainly consisting of the pushing mechanism (1), angle adjustment mechanism (2), conveying mechanism (3), casing assembly mechanism (4), robotic arm (5), control box (6), and frame (7). The casing assembly machine has the advantages of high precision, a high degree of automation, and high assembly efficiency, greatly reducing labor costs.

II. Working Principle

The machine case assembly machine operates by conveying the components of the casing to be assembled through the conveying mechanism to the area below the robotic arm. The robotic arm is controlled by the control box to grasp the casing components and transport them to the casing assembly mechanism. The base of the casing assembly mechanism detects the presence of casing components and checks the angle of the casing. The detected angle information is then sent to the control box, which, based on the received angle information, controls the angle adjustment mechanism to adjust the angle of the casing components until it meets the required angle. At this

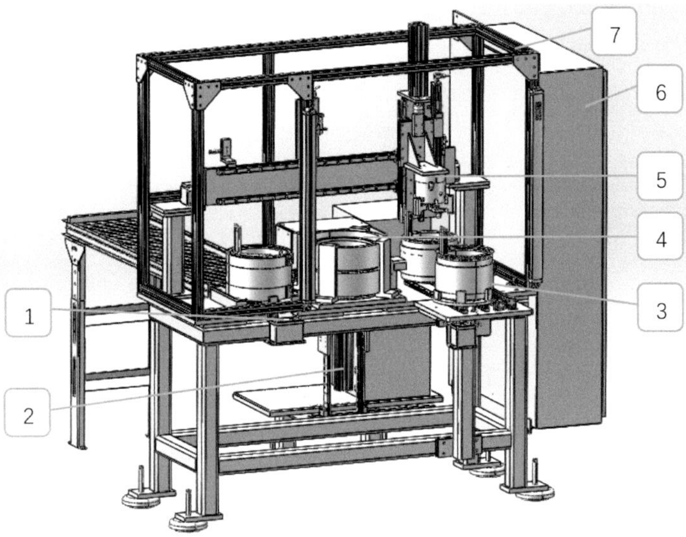

Fig. 3.67 Machine Case Assembly Machine
1-Pushing Mechanism 2-Angle Adjustment Mechanism 3-Conveying Mechanism 4-Casing Assembly Mechanism 5-Robotic Arm 6-Control Box 7-Frame

point, the casing is assembled, and after assembly, the robotic arm picks up the assembled casing and places it on the output track. The pushing mechanism then pushes out the assembled casing.

III. Main Structure Introduction

The conveying mechanism, as shown in Fig. 3.68, works by placing the casing to be assembled on the casing base (2). It is driven by a servo motor (3), which transports the casing base (2), along with the casing to be assembled, along the guide rail (1) to the area below the robotic arm.

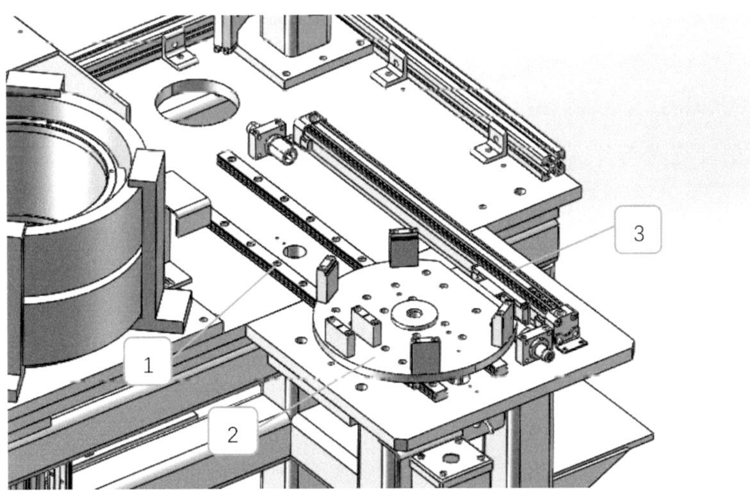

Fig. 3.68 Conveying Mechanism
1-Guide Rail 2-Casing Base 3-Servo Motor

The mechanical arm, as shown in Fig. 3.69, is controlled by a servo motor (4) to move horizontally along the guide rail (2). The vertical displacement of the mechanical arm (3) is controlled by a cylinder (1). The mechanical arm (3) secures and grips the casing component using internal expansion.

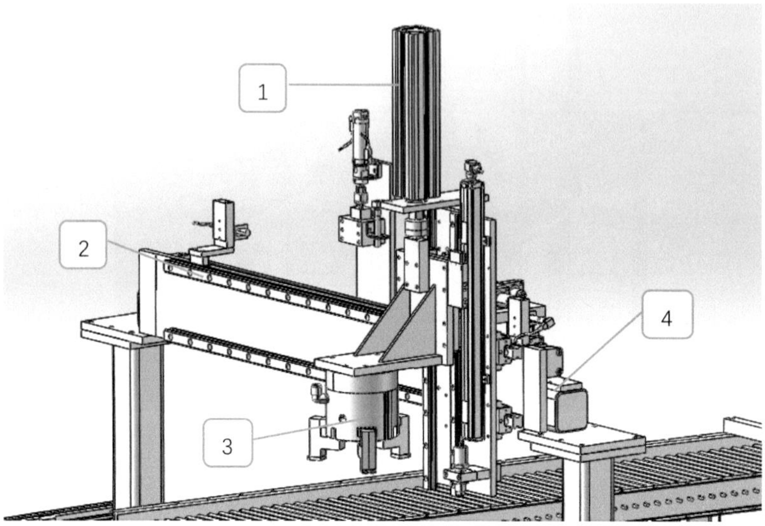

Fig. 3.69 The Mechanical Arm
1 - Cylinder 2 - Guide rail 3 - Mechanical arm 4 - Servo motor

The shell assembly mechanism is shown in Fig. 3.70. When the shell assembly mechanism (1) is in operation, the mechanical arm carries the shell component to the top of the shell assembly mechanism and slowly lowers it. The angle adjustment mechanism (2) takes over the shell component and simultaneously corrects its position and angle. After the correction is complete, the shell component is manually assembled.

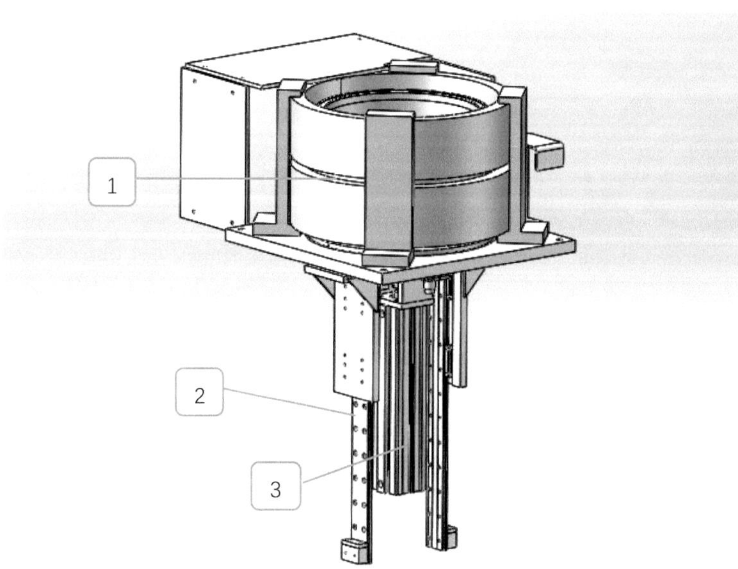

Fig. 3.70 Machine Casing Assembly Mechanism
1 - Casing Assembly Mechanism 2 - Angle Adjustment Mechanism 3 - Cylinder

The pushing mechanism is shown in Fig. 3.71, controlled and driven by a servo motor (2). It continuously pushes the assembled casing onto the output track (1) and forwards it to the next process.

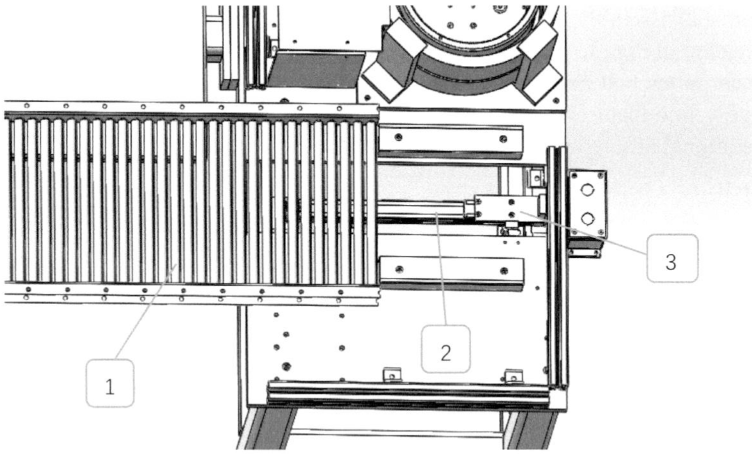

Fig. 3.71 Pushing Mechanism
1 - Output Track 2 - Servo Motor 3 - Pushing Mechanism

IV. Specific Parameters (see Table 3.16)

Table 3.16 Machine Shell Assembly Machine Parameters

Working Pressure:	0.45~0.65 MPa
Working Efficiency:	20-30 pcs/h
Operation Rate:	≥93%
Yield Rate:	≥98%
Environmental Temperature:	0~55°C
Allowable Humidity:	25%~95%
Equipment Dimensions:	2000 mm(H) × 2300 mm(L) × 2030 mm(W)

V. Mechanical Design Highlights and Precautions

(1) The shell assembly machine is capable of detecting the position and angle of the shell to ensure compliance, and adjusts them through the controller by controlling the angle adjustment mechanism.

(2) Operators at the workstation should wear gloves to prevent injection parts from getting contaminated with oil stains, sweat, etc.

(3) Before assembly, quality inspection of the motor shell should be conducted. The surface should not have obvious scratches, cracks, or deformations, and the surface coating should not have bubbles, cracks, or peeling.

(4) The general assembly procedure for panels, shells, and other components is to start with lighter components before heavier ones, lower components before higher ones, and inner components before outer ones.

Case 40: Robot Automated Assembly Line

I. Case Description

The robot automated assembly line is designed for the automated assembly of products. The overall structure is depicted in Fig. 3.72 and comprises essential components such as a five-axis robot (1), frame (2), synchronous belt (3), positioning mechanism (4), and automatic assembly mechanism (5). The assembly line features symmetrical arrangements of two sets of equipment, significantly enhancing operational efficiency. Its high level of automation effectively reduces labor costs.

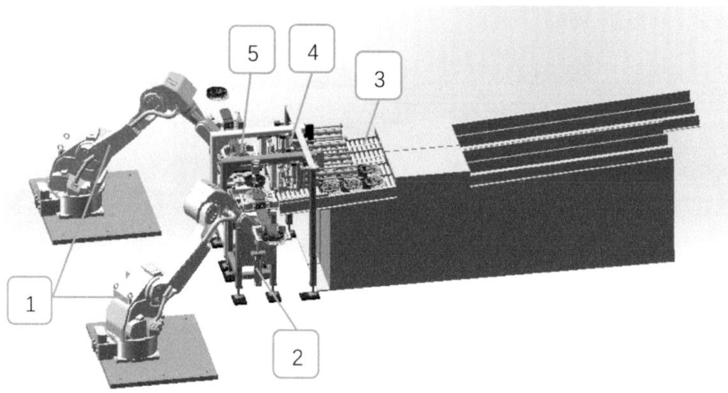

Fig. 3.72 Robot Automated Assembly Line
1 - Five-axis robot 2 - Frame 3 - Synchronous belt 4 - Positioning mechanism 5 - Automatic assembly mechanism

II. Working Principle

The robot automatic assembly line operates by transporting the components to be assembled by the synchronous belt to the positioning mechanism. The positioning mechanism adjusts the angle of the product, aligning the components for seamless assembly. The product, with the adjusted angle, is then conveyed to the automatic assembly mechanism, where the assembly process is completed automatically. After assembly, the five-axis robot transfers the finished product to the shelf, completing the automation assembly operation. The gripping part of the five-axis robot is equipped with two fixtures, allowing it to handle two products simultaneously, significantly improving the efficiency of the equipment and enhancing the safety of assembly personnel.

III. Main Structure Introduction

The five-axis robot, as shown in Fig. 3.73, is driven and controlled by a servo motor (1). Additionally, two fixtures are installed on the mechanical hand (2) of the five-axis robot, allowing it to handle two products simultaneously. This significantly enhances the efficiency of the equipment.

The positioning mechanism is shown in Fig. 3.74. When the components to be assembled are transported by the synchronous belt (3) to the positioning mechanism, while the automatic assembly mechanism is in operation, the fixed column (2) is pressed down by cylinder (1) to fix the components to be assembled on the synchronous belt (3). This prevents the product from continuing to be transported to the automatic assembly mechanism and adjusts the angle of the product, ensuring that the components of the product match each other.

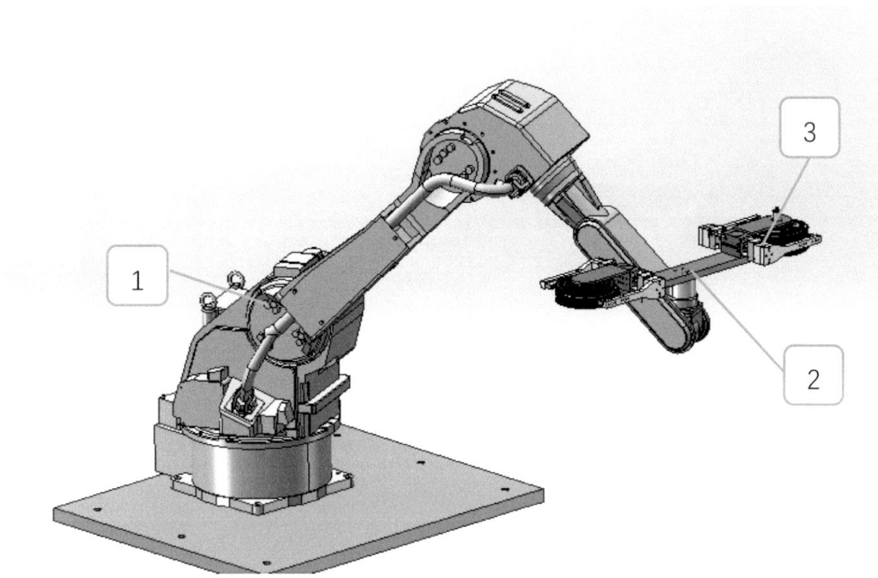

Fig. 3.73 Five-axis Robot
1 - Servo Motor 2 - Mechanical Hand 3 - Fixture

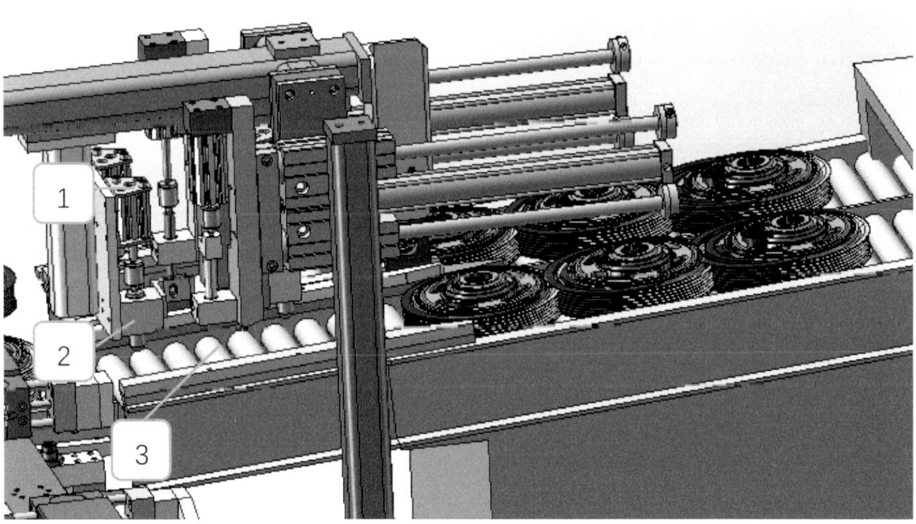

Fig. 3.74 Positioning Mechanism
1 - Cylinder 2 - Positioning Column 3 - Synchronous Belt

The automatic assembly mechanism is shown in Fig. 3.75. The product, with its angle adjusted, is transported to the automatic assembly mechanism via the synchronous belt. The components are fixed using the base (2), and then the servo motor (3) is activated to assemble the product components. Once assembly is complete, the product is carried to the shelf by the five-axis robot.

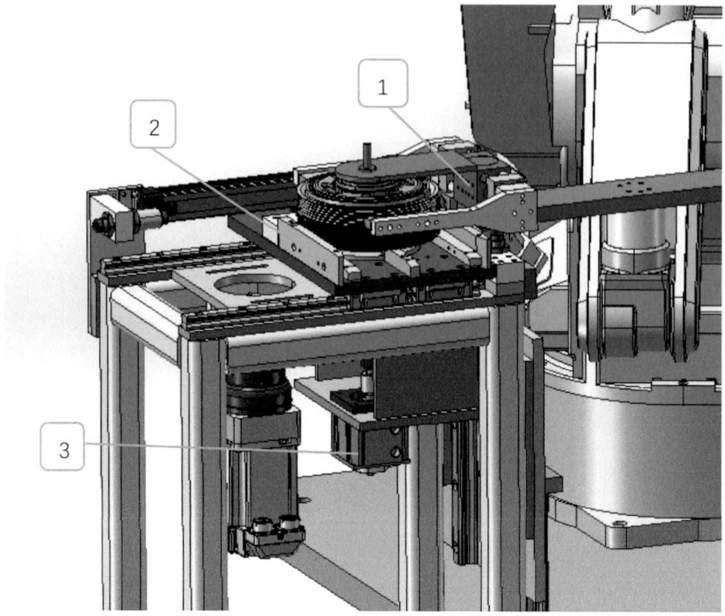

Fig. 3.75 Automatic Assembly Mechanism
1 - Fixture 2 - Base 3 - Servo Motor

IV. Specific Parameters (see Table 3.17)

Table 3.17 Parameters of Robot Automatic Assembly Line

Assembly Accuracy:	0.01 mm
Working Efficiency:	800-1000 pcs/h
Utilization Rate:	≥93%
Yield Rate:	≥98%
Environmental Temperature:	0~65°C
Permissible Humidity:	1%~95%
Equipment Dimensions:	1250 mm(H) × 4850 mm(L) × 3580 mm(W)

V. Mechanical Design Highlights

(1) The five-axis robot is equipped with two grippers at the grabbing location, allowing it to simultaneously handle two products, greatly improving the efficiency of the equipment and enhancing the safety of assembly personnel's work.
(2) The use of highly sensitive and high-precision sensor detection ensures the accuracy and stability of the detection, with minimal dynamic errors and no cumulative errors, resulting in high precision [63].
(3) Digital automatic control is employed, providing high speed, quick dynamic response, and minimal motion inertia.
(4) The structure is compact and stable, with high machine rigidity, large load-bearing capacity, and a relatively small working space.
(5) High-precision fixtures can be designed and installed according to production needs, offering high flexibility and strong adaptability.

Case 41: Fully Automatic Relay Assembly Machine

I. Case Description

Relays, as shown in Fig. 3.76, are electrical control devices used to produce predetermined step changes in the controlled quantity in an electrical output circuit when the input quantity meets certain specified conditions. Typically employed in automated control circuits, relays essentially serve as "automatic switches" where small currents control the operation of larger currents. They play essential roles in circuits for automatic regulation, safety protection, and circuit switching. The fully automatic relay assembly machine is an automated device designed for assembling relays. Its overall structure, as depicted in Fig. 3.77 and the top view in Fig. 3.78, consists primarily of a vibrating conveyor (1), a synchronous belt (2), the first indexing mechanism (3), the first feeding mechanism (4), the second feeding mechanism (5), the feeding end mechanism (6), the second assembly mechanism (7), the second indexing mechanism (8), the second handling robot (9), the first handling robot (10), the first assembly mechanism (11), and the frame structure (12), among others. The fully automatic relay assembly machine offers advantages such as high production efficiency and high product qualification rate.

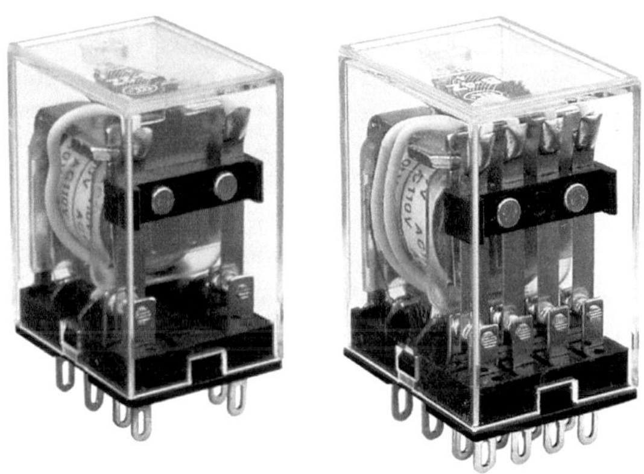

Fig. 3.76 Relays

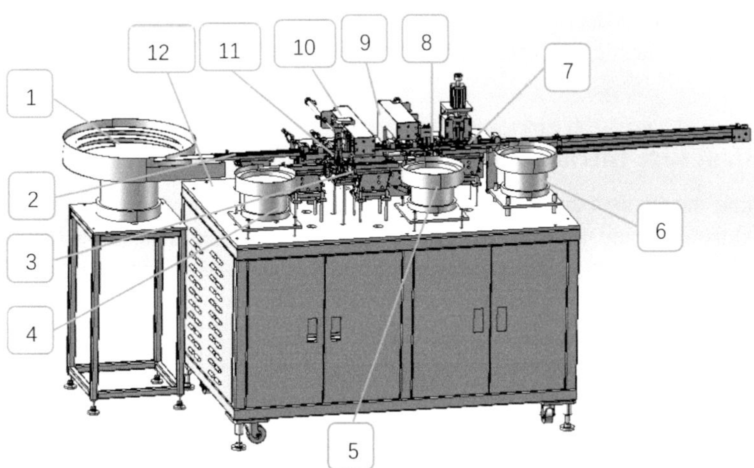

Fig. 3.77 Fully Automatic Relay Assembly Machine

1 - Vibratory Conveyor 2 - Synchronous Belt 3 - First Rotational Mechanism 4 - First Feeding Mechanism 5 - Second Feeding Mechanism 6 - Feeding End Mechanism 7 - Second Assembly Mechanism 8 - Second Rotational Mechanism 9 - Second Handling Robot 10 - First Handling Robot 11 - First Assembly Mechanism 12 - Frame

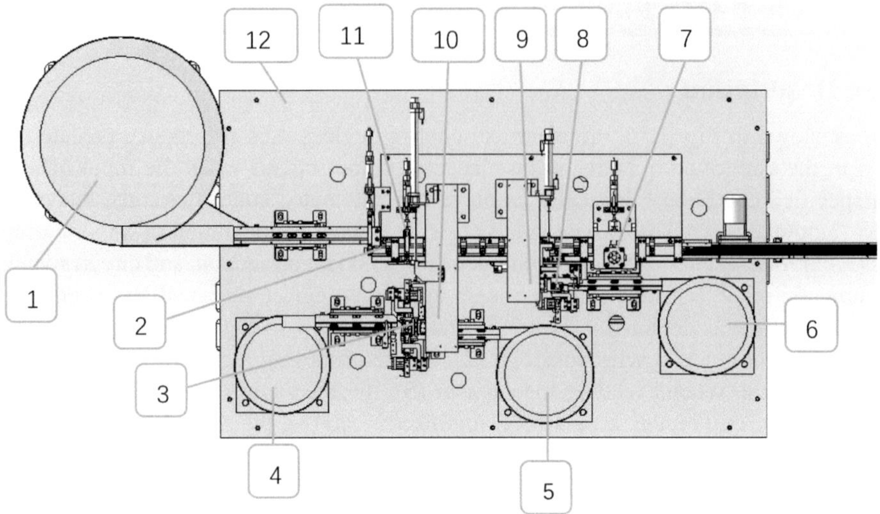

Fig. 3.78 Top View of Fully Automatic Relay Assembly Machine
1 - Vibratory Conveyor 2 - Synchronous Belt 3 - First Rotational Mechanism 4 - First Feeding Mechanism 5 - Second Feeding Mechanism 6 - Feeding End Mechanism 7 - Second Assembly Mechanism 8 - Second Rotational Mechanism 9 - Second Handling Robot 10 - First Handling Robot 11 - First Assembly Mechanism 12 - Frame

II. Working Principle

The fully automatic relay assembly machine operates as follows: The relay housings are conveyed into the synchronous belt by the vibratory conveyor. The first and second feeding mechanisms transport the first and second contact springs to the first rotational mechanism. The pneumatic cylinder of the first rotational mechanism is activated to align and connect the two contact springs. The assembled component is then transferred to the synchronous belt, where it is picked up by the first handling robot and placed on the relay housing. The first assembly mechanism completes the assembly of the component with the relay housing.

Simultaneously, the feeding end mechanism transports terminals to the second rotational mechanism. The second handling robot picks up the terminals and transfers them to the synchronous belt where the relay housing is located. The second assembly mechanism completes the assembly of the relay.

III. Main Structure Introduction

The first feeding mechanism and the second feeding mechanism are shown in Fig. 3.79. The first feeding mechanism (2) and the second feeding mechanism (3) transport the first contact spring and the second contact spring, respectively, to the first rotational mechanism (1). The pneumatic cylinder of the first rotational mechanism is activated to align and connect the contact springs of the two relays.

The first rotational mechanism, as shown in Fig. 3.80, is driven by cylinder (3) to push the contact pieces, allowing the contact pieces of two relays to connect with each other.

The first conveying manipulator, as shown in Fig. 3.81, is driven and controlled by cylinder (1) to horizontally display fixture (2), transporting the aligned contact pieces to the synchronous belt where the relay casing is located. The assembly of the component with the casing is completed by the first assembly mechanism.

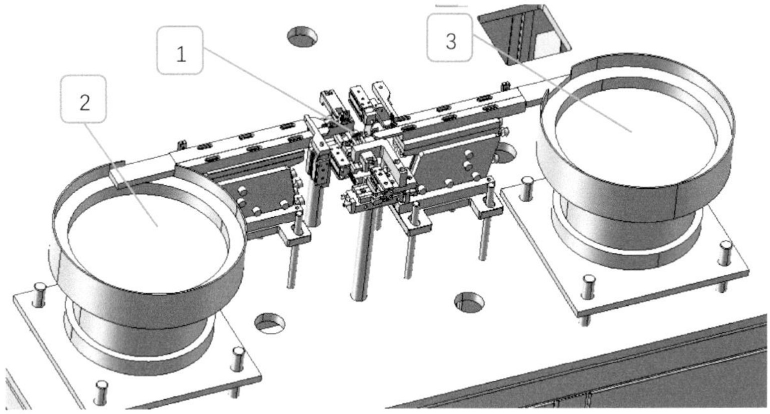

Fig. 3.79 Feeding Mechanism
1 - First Rotational Mechanism 2 - First Feeding Mechanism 3 - Second Feeding Mechanism

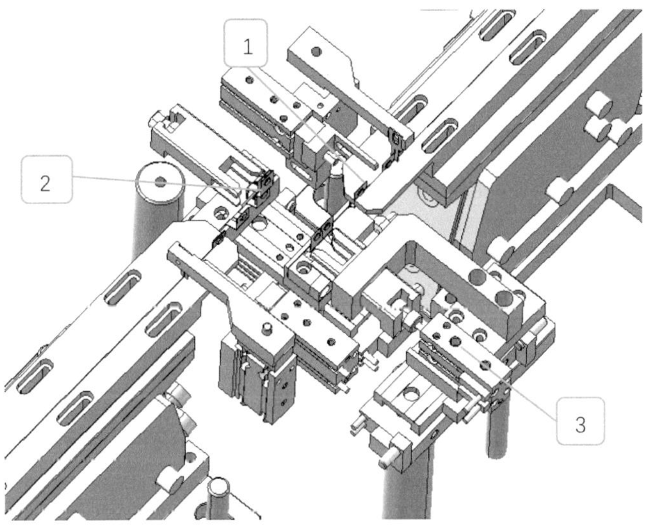

Fig. 3.80 First Rotational Mechanism
1 - Second Contact Piece 2 - First Contact Piece 3 - Cylinder

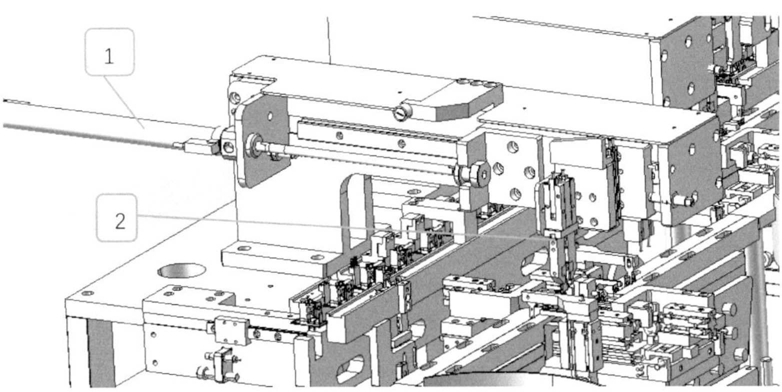

Fig. 3.81 First Transfer Manipulator
1 - Cylinder 2 - Fixture

The first assembly mechanism is shown in Fig. 3.82. It is driven by cylinder (1) to assemble the relay casing on the synchronous belt (3) with the piece transported by the first transfer manipulator.

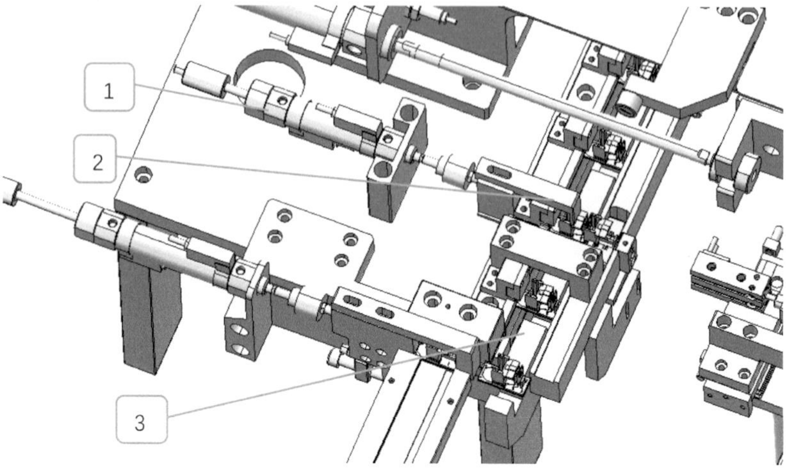

Fig. 3.82 Assembly Mechanism
1 - Cylinder 2 - Assembly Mold 3 - Synchronous Belt

The second assembly mechanism, as shown in Fig. 3.83, is driven by Cylinder (1) to lower Mold (2), assembling the relay housing on the synchronous belt (3) with the terminals carried over by the second handling robot.

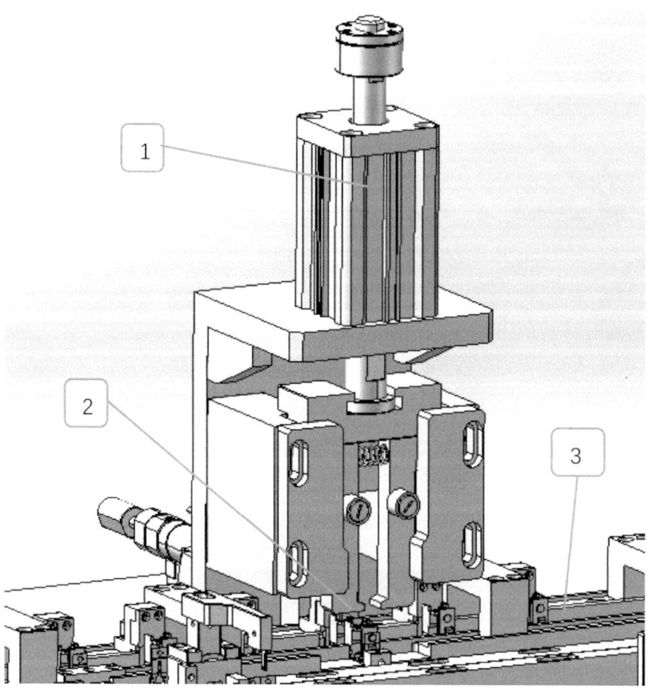

Fig. 3.83 Second Assembly Mechanism
1 - Cylinder 2 - Mold 3 - Synchronous Belt

IV. Specific Parameters (see Table 3.18)

Table 3.18 Automatic Relay Assembly Machine Parameters

Power:	150 KVA
Working Efficiency:	1800-2500 pcs/h
Operation Rate:	≥93%
Qualified Rate:	≥98%
Environmental Temperature:	0~55°C
Allowed Humidity:	25%~95%
Equipment Dimensions:	1370 mm(H) × 2000 mm(L) × 1080 mm(W)

V. Mechanical Design Highlights

The assembly production process of relays is quite complex, and existing production methods generally rely on manual or semi-automatic processes for mass production. Consequently, the production efficiency, yield, and quality are low. The automatic relay assembly machine, through structural design and assembly line optimization, achieves highly automated assembly. Compared to the existing manual or semi-automatic assembly methods, it offers higher production volume, increased production efficiency, and improved quality [64].

Case 42: Shock Absorber Inflator

I. Case Description

Shock Absorber Inflator is a specialized inflation device for shock absorbers, mainly used in the production and processing of inflatable shock absorbers for automobiles. The overall structure is shown in Fig. 3.84, consisting of an operating console (1), inflation device (2), clamping mechanism

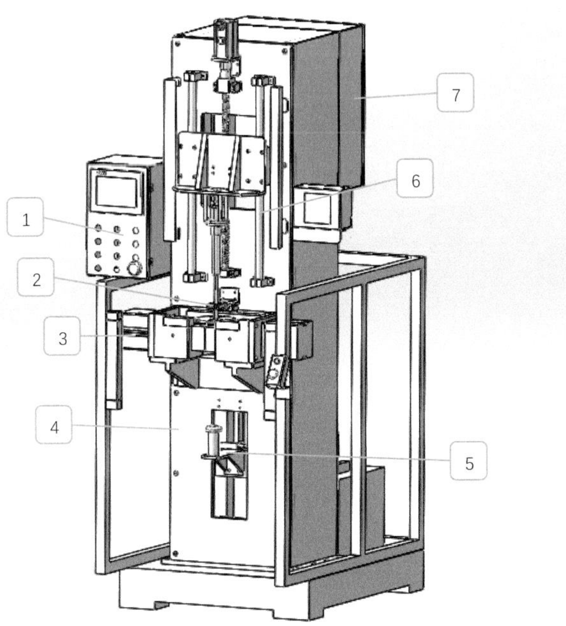

Fig. 3.84 Shock Absorber Inflator

1 - Operating Console 2 - Inflation Device 3 - Clamping Mechanism 4 - Frame 5 - Worktable 6 - Lifting Mechanism 7 - Distribution Box

(3), frame (4), worktable (5), lifting mechanism (6), and distribution box (7). The shock absorber inflator can change the traditional method of inflating shock absorbers, improving inflation production efficiency and product quality, reducing the subjectivity of manual judgment, and lowering manual labor intensity.

II. Working Principle

When the shock absorber inflator is in operation, the shock absorber to be inflated is placed on the worktable. The servo motor of the worktable is started, driving the worktable and the shock absorber to move upward to the height of the clamping mechanism. The cylinder of the clamping mechanism is then activated to clamp and secure the shock absorber, ensuring its accurate positioning. The inflation device is connected to the lifting mechanism. At this point, the connected servo motor of the lifting mechanism is activated, causing the inflation device to move downward along the lead screw. The inflation device completes the inflation and testing of the shock absorber. The shock absorber inflator integrates the inflation and testing processes into one device, using the results from the testing equipment to control the opening and closing of the inflation device, ensuring the effectiveness of the product's inflation.

III. Main Structure Introduction

The worktable, as shown in Fig. 3.85, is where the shock absorber is placed on the column (2). It is driven and controlled in its vertical movement by the servo motor (1).

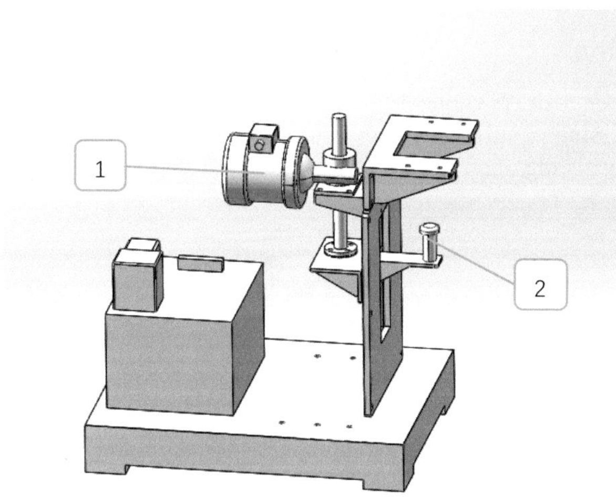

Fig. 3.85 Worktable
1 - Servo Motor 2 - Column

Clamping mechanism, as shown in Fig. 3.86, is driven and controlled by cylinders (1) and (3) to clamp and secure the shock absorber with fixture (2), ensuring the accurate positioning of the shock absorber.

The inflation device and lifting mechanism are shown in Fig. 3.87. The lifting mechanism is driven by servo motor (1), controlling the ascent and descent of the inflation device (3) via the lead screw (2).

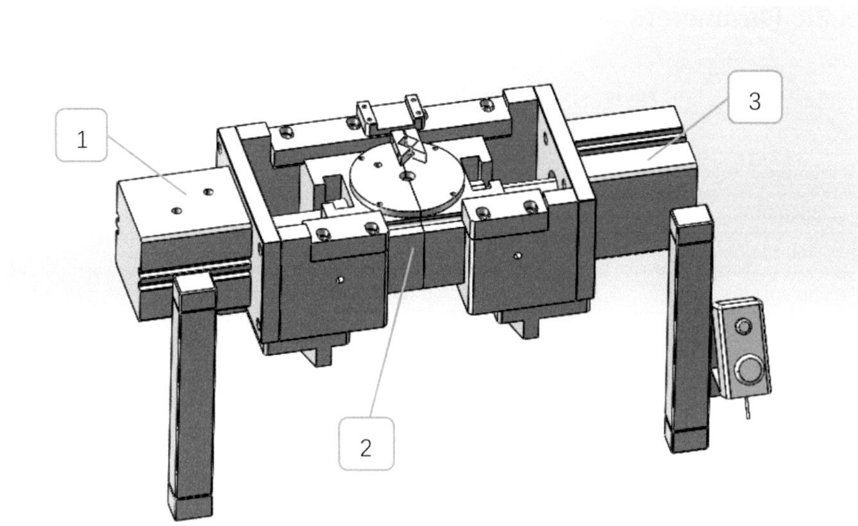

Fig. 3.86 Clamping Mechanism
1, 3 - Cylinders 2 - Fixture

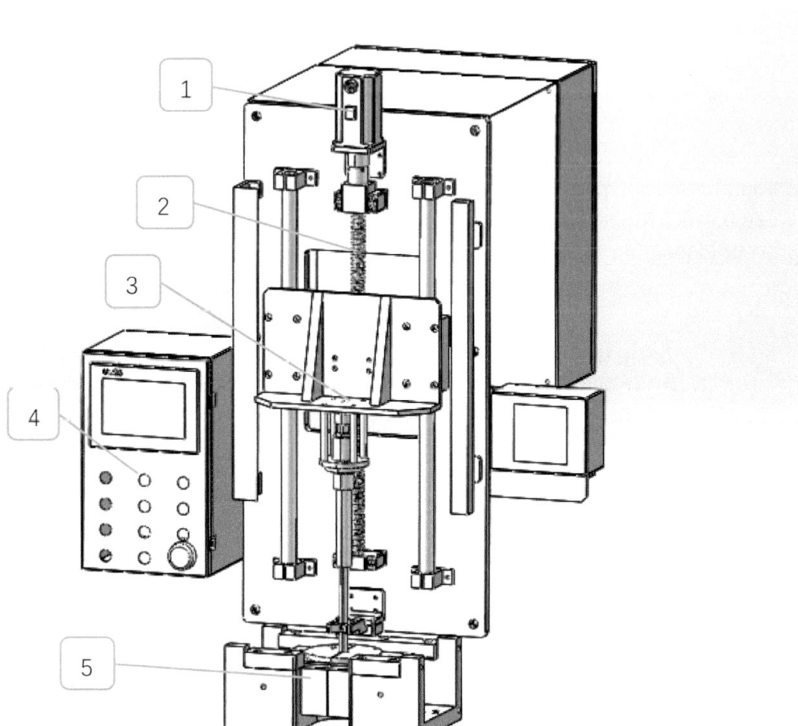

Fig. 3.87 Shows the inflation device and lifting mechanism.
1 - Servo motor 2 - Lead screw 3 - Inflation device 4 - Operating platform 5 - Clamping mechanism

IV. Specific Parameters (see Table 3.19)

Table 3.19 Shock Absorber Inflator Parameters

Yield Rate:	≥98%
Operating Pressure Range:	4-6 kg/cm^2
Ambient Temperature:	– 10~40°C
Permissible Humidity:	30%~90%
Equipment Dimensions:	2200 mm(H) × 950 mm(L) × 1300 mm(W)

V. Mechanical Design Highlights

Compared with existing technology, this shock absorber inflator adopts an automatic control mode, featuring simple operation and safe and reliable processing performance. It integrates inflation and testing into a single device, using the detection results from the testing equipment to control the opening and closing of the inflation device, ensuring the effectiveness of the inflation process. Additionally, it reduces the subjectivity of manual judgment and lowers the labor intensity, contributing to improved product quality and large-scale production [65].

VI. Precautions

(1) The operating temperature of the inflator should be between – 10°C and + 40°C, with an altitude not exceeding 1500 m.
(2) Use appropriate air intake filtration equipment and regularly clean or replace the inflator.
(3) During operation, pay attention to the wear and leakage of the inflation head's vulnerable parts. If any leakage is found, replace the vulnerable parts promptly.
(4) The compressed air pressure regulated by the main air source relief valve is generally set between 0.4-0.7 Mpa. If there is abnormal pressure, stop production and investigate the cause to prevent inaccurate measurements or production failures.

Case 43: Automatic Screw Assembly Machine for Connector Products

I. Case Description

The Automatic Screw Assembly Machine for Connector Products is designed for automatically tightening screws on connector products. The overall structure is shown in Fig. 3.88, with an enlarged view in Fig. 3.89. It mainly consists of the screw feeder (1), correction mechanism (2), frame (3), transfer mechanism (4), discharge mechanism (5), bending mechanism (6), and electric screwdriver (7). This machine offers strong operational flexibility, easy usage, and a high level of automation and intelligence.

II. Working Principle

The automatic screw assembly machine for connector products operates by connecting to the power supply. Product components are fed from the transfer mechanism, and screws are supplied by the screw feeder. The correction mechanism fixes the product in the preset position. The cylinder connected to the electric screwdriver is activated, causing the electric screwdriver to move downward and tighten the screw for the product. The transfer mechanism continues to push the product to the

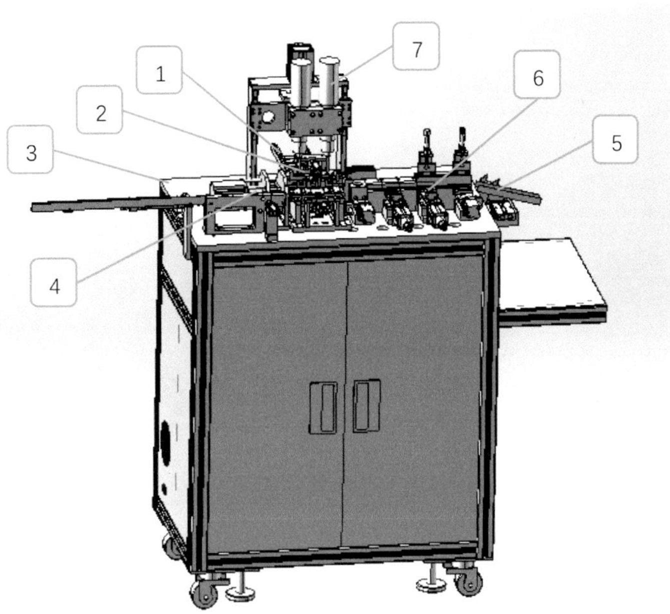

Fig. 3.88 Automatic Screw Assembly Machine for Connector Products
1 - Screw Feeder 2 - Correction Mechanism 3 - Frame 4 - Transfer Mechanism 5 - Discharge Mechanism
6 - Bending Mechanism 7 - Electric Screwdriver

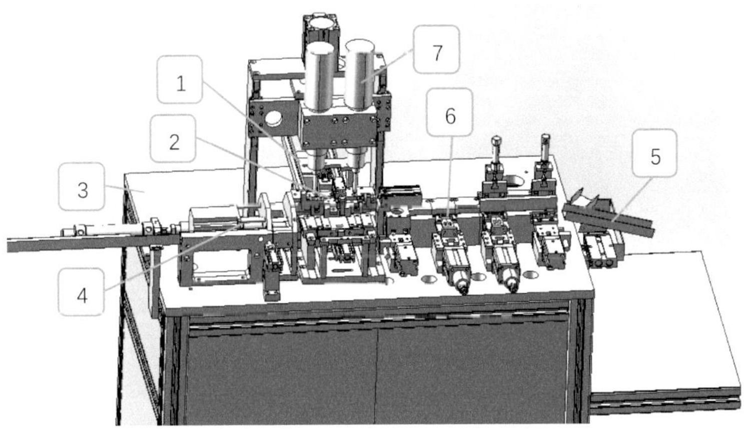

Fig. 3.89 Enlarged View of Automatic Screw Assembly Machine for Connector Products
1 - Screw Feeder 2 - Correction Mechanism 3 - Frame 4 - Transfer Mechanism 5 - Discharge Mechanism
6 - Bending Mechanism 7 - Electric Screwdriver

bending mechanism, where the two forks on the left and right sides of the product are bent. After bending is completed, the product is discharged from the discharge mechanism [66].

III. Main Structure Introduction

The electric screwdriver, as shown in Fig. 3.90, is connected to cylinder (1) and is driven and controlled by cylinder (1) to move up and down along the guide rail (3).

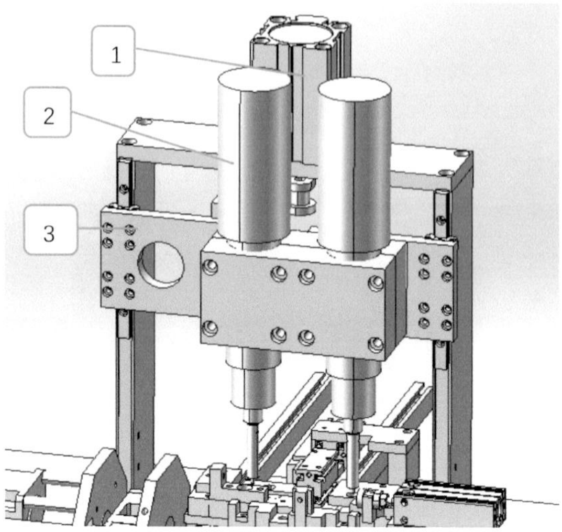

Fig. 3.90 Electric Screwdriver
1 - Cylinder 2 - Electric Screwdriver 3 - Guide Rail

Correction Mechanism as shown in Fig. 3.91 is composed of cylinder (2) that pushes block (3) and cylinder (4) that moves base (5) to fix the product in the preset position.

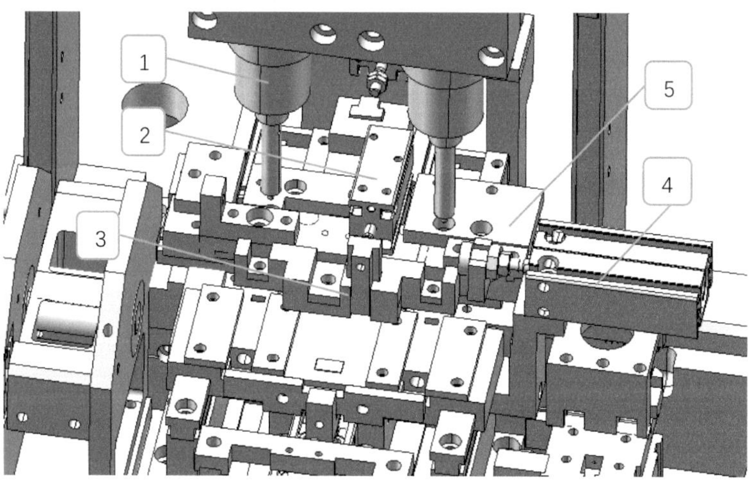

Fig. 3.91 Correction Mechanism
1 - Electric Screwdriver 2, 4 - Cylinders 3 - Stopper 5 - Base

The bending mechanism, as shown in Fig. 3.92, is driven by cylinder (1) to control the pressing block (2) to press down and fix the product. Then, the bending machine (3) bends the two prongs of the product to the left and right.

Screw Feeder as shown in Fig. 3.93 operates by generating continuous vibration through the vibrating base (3), allowing screws to be transported along the guide rail (2) to the area below the electric screwdriver (1).

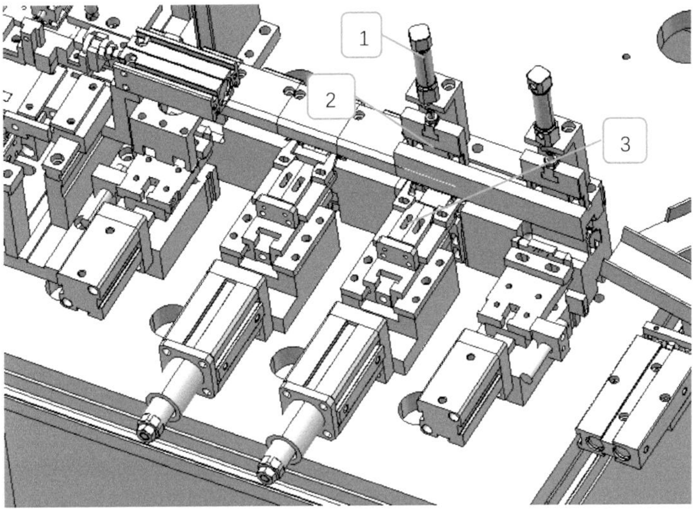

Fig. 3.92 Bending Mechanism
1 - Cylinder 2 - Pressing Block 3 - Bending Machine

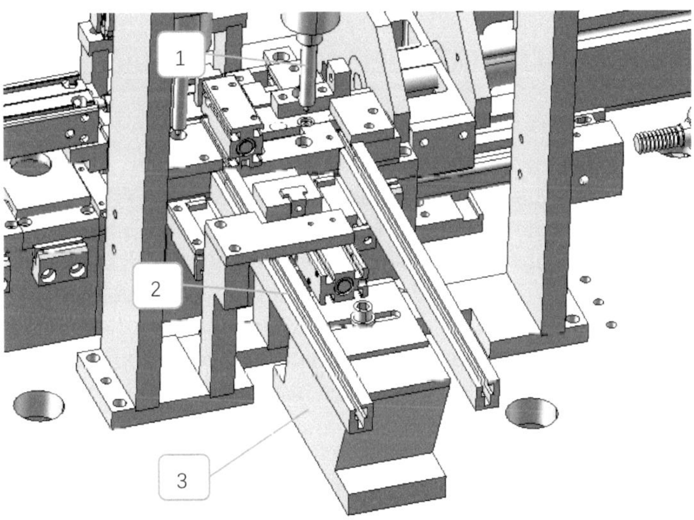

Fig. 3.93 Screw Feeder
1 - Electric screwdriver 2 - Guide rail 3 - Vibrating base

IV. Specific Parameters (see Table 3.20)

Table 3.20 Automatic Screw Assembly Machine for Connecting Products Parameters

Working Efficiency:	1800-2500 pcs/h
Operating Rate:	≥95%
Yield Rate:	≥98%
Ambient Temperature:	0~45°C
Allowable Humidity:	30%~95%
Equipment Dimensions:	1240 mm(H) × 1300 mm(L) × 450 mm(W)

V. Mechanical Design Highlights

In the assembly process of products, screwing screws is one of the most basic operations. In traditional assembly processes, manual screwdrivers are used to assemble products, which require a large number of assembly personnel [67]. This primitive manual operation method is inefficient, with low output, inconsistent operations, and unreliable product quality. The Automatic Screw Assembly Machine for Connecting Products utilizes PLC (Programmable Logic Controller) to control the transportation of product components and screws, automating the screw locking operation with electric screwdrivers. This automation saves labor time, reduces labor intensity, improves labor efficiency, and enhances the assembly quality and output of products [68].

VI. Precautions

(1) The screws used should have fewer impurities, mixing, defects like bent heads, etc., to avoid causing the machine to work improperly [69].
(2) The selected screws should be commonly used, in large quantities, and of few types to improve the efficiency of the automatic screw machine.
(3) The positions for screwing should have as uniform dimensions as possible, avoiding variations in depth, uneven surfaces, limited space, etc.
(4) When using the screw machine, try to use screws of the same specifications for the same product.

Case 44: Connector Assembly Machine

I. Case Description

Connectors, as shown in Fig. 3.94, consist of plastic parts, connecting terminals, grounding plates, and cover plates. The Connector Assembly Machine is designed for the automatic assembly of connectors, with an overall structure shown in Fig. 3.95 and a detailed view in Fig. 3.96. It mainly comprises mechanisms such as the connecting terminal insertion mechanism (1), cutting mechanism (2), grounding plate polishing mechanism (3), synchronous belt (4), feeding mechanism (5), and loading mechanism (6). The Connector Assembly Machine offers stability in quality, reduces labor costs, lowers defect rates, improves production efficiency, and standardizes production for enhanced quality [70].

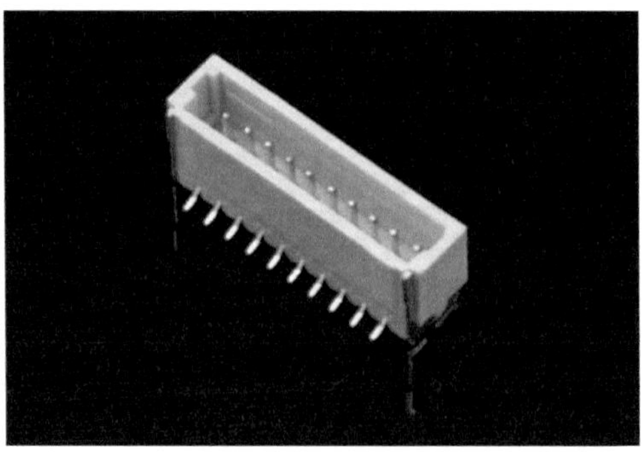

Fig. 3.94 Connectors

Fig. 3.95 Connector Assembly Machine
1 - Connecting Terminal Insertion Mechanism 2 - Cutting Mechanism 3 - Grounding Plate Polishing Mechanism
4 - Synchronous Belt 5 - Loading Mechanism 6 - Feeding Mechanism

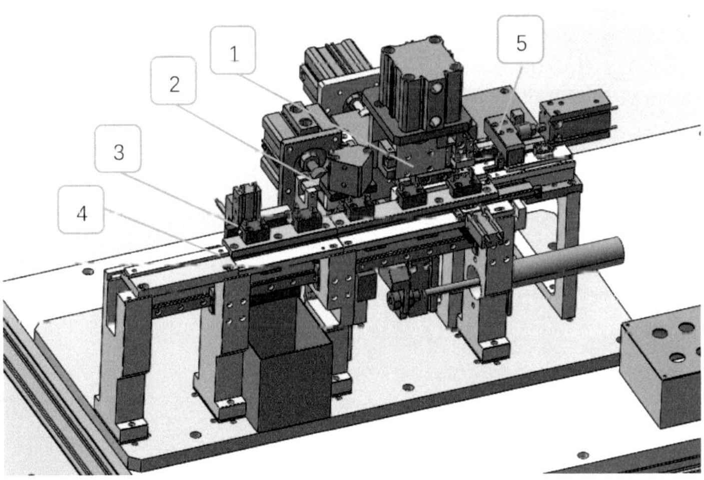

Fig. 3.96 Connector Assembly Machine - Detailed View
1 - Connecting Terminal Insertion Mechanism 2 - Cutting Mechanism 3 - Grounding Plate Polishing Mechanism
4 - Synchronous Belt 5 - Feeding Mechanism

II. Working Principle

The connector automatic assembly machine operates by introducing product components from the feeding mechanism into the material loading mechanism. The feeding mechanism translates the connector assembly below the connecting terminal insertion mechanism, where the connecting terminals are pressed into the plastic part and pushed onto the synchronous belt. The connector assembly is then picked up by the suction device on the synchronous belt, which transports it to the cutting mechanism. The cutting mechanism precisely cuts the grounding plate, and the cut connector assembly is further conveyed to the grounding plate polishing mechanism. Here, the product terminals undergo polishing operations, and the finished connector is then released.

III. Main Structure Introduction

The feeding mechanism, as shown in Fig. 3.97, is driven by servo motor (1), continuously transporting the tape and connector components to the feeding mechanism.

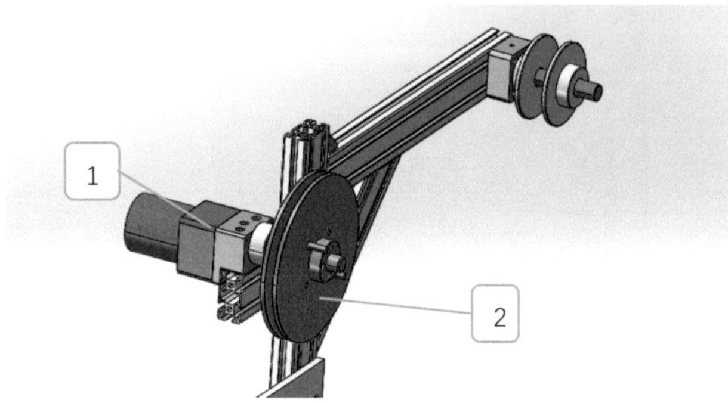

Fig. 3.97 Feeding Mechanism
1 - Servo Motor 2 - Material Tray

The feeding mechanism and terminal insertion mechanism, as shown in Fig. 3.98, are driven by cylinder (3) in the feeding mechanism, controlling the slider (4) to translate the connector component beneath the terminal insertion mechanism. The terminal insertion mechanism is driven by cylinder (1), controlling the pressing block (2) to lower and insert the terminal into the plastic part. Afterward, cylinder (5) is activated to horizontally push the connector component to the synchronous belt, where the connector component is picked up by the suction device on the synchronous belt.

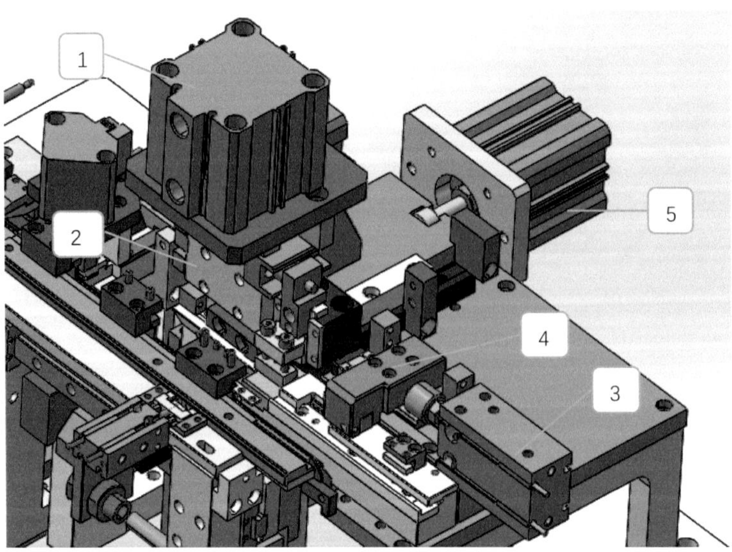

Fig. 3.98 Feeding Mechanism and Terminal Insertion Mechanism
1, 3, 5 - Cylinder 2 - Pressing block 4 - Slider

Cutting mechanism and ground piece polishing mechanism are shown in Fig. 3.99. The blade (2) of the cutting mechanism is driven by cylinder (1), which performs a fixed-size cutting on the ground piece. The grinding wheel (4) of the ground piece polishing mechanism is driven by servo motor (3), which polishes the product terminals.

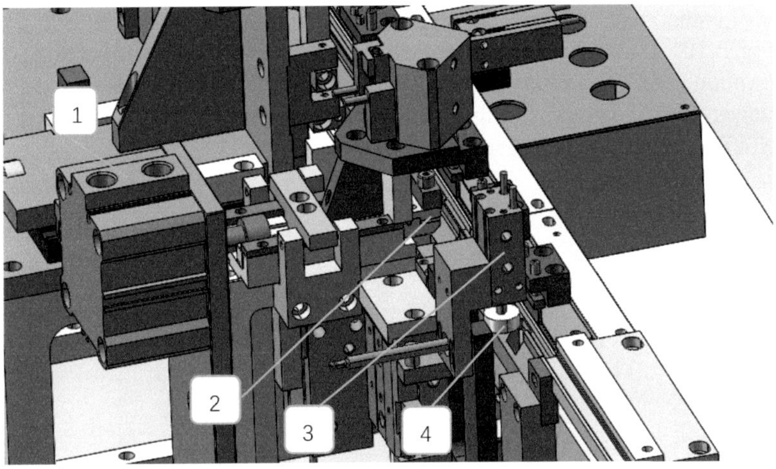

Fig. 3.99 Shows the cutting mechanism and the ground piece polishing mechanism
1 - Cylinder 2 - Blade 3 - Servo motor 4 - Grinding wheel

IV. Specific Parameters (see Table 3.21)

Table 3.21 Connector Assembly Machine Parameters

Working Efficiency:	600-1000 pcs/h
Operation Rate:	≥95%
Yield Rate:	≥99%
Ambient Temperature:	0~55°C
Allowable Humidity:	30%~95%
Equipment Dimensions:	1510 mm(H) × 1320 mm(L) × 500 mm(W)

V. Mechanical Design Highlights

Due to the small spacing between connector terminals, traditional production processes use multiple independent jigs, and multiple operators perform the assembly individually. When the demand for the same product increases, it requires continuously adding multiple manual lines and production fixtures, resulting in high costs, low efficiency, and unstable quality. The Connector Assembly Machine utilizes a PLC (Programmable Logic Controller) to control servo motors, cylinders, solenoid valves, and other components. Operators can set relevant parameters through the human-machine interface, and the machine automatically completes the entire production process of feeding, conveying, pressing, cutting, and polishing connector products based on the programmed sequence. With only one operator required to operate the machine, it achieves a production capacity of more than double that of manual lines, ensuring stable quality, significantly saving labor costs, reducing defects, improving production efficiency, and promoting standardized production with enhanced quality.

Case 45: Six-Station Assembly Machine

I. Case Description

The Six-Station Assembly Machine is suitable for the automatic assembly of products with a large number of components. The overall structure is shown in Figure 3.100, with a detailed view provided in Fig. 3.101. It mainly consists of an electric screwdriver (electric batch) (1|), a fixed base (2), a transfer robotic arm (3), a rotary table (4), and a frame (5). The Six-Station Assembly Machine offers advantages such as simple and quick operation, reduced manual labor intensity, improved production efficiency, and cost savings.

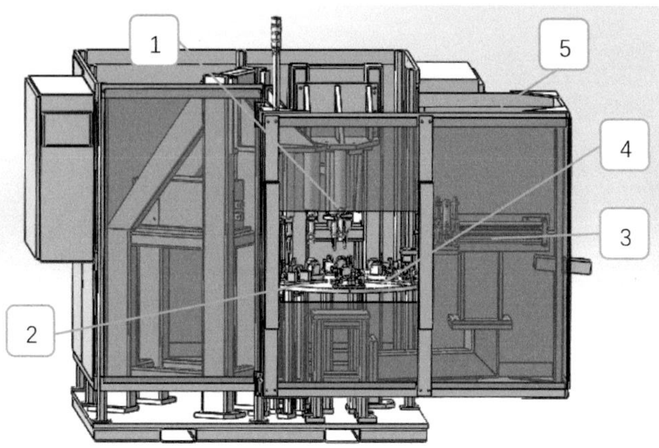

Fig. 3.100 Six-Station Assembly Machine
1 - Electric Screwdriver (Electric Batch) 2 - Fixed Base 3 - Transfer Robotic Arm 4 - Rotary Table 5 - Frame

Fig. 3.101 Partial Enlarged View of Six-Station Assembly Machine
1 - Electric Screwdriver (Electric Batch) 2 - Fixed Base 3 - Transfer Robotic Arm 4 - Rotary Table 5 - Frame

II. Working Principle

Underneath the rotary table is equipped with a rotating electric motor. When the six-station assembly machine is in operation, the transfer robotic arm transports subcomponents to the fixed base on the workstation rotary table. The rotary table automatically rotates to the next station. The positions of the electric screwdrivers are pre-set based on product requirements. Assembly is carried out by the electric screwdrivers at the third and fourth stations. The entire system is controlled by a PLC (Programmable Logic Controller), which manages precise and controlled movements and positions of the servo robotic arm and the rotary table [71].

III. Main Structure Introduction

The transfer robotic arm, as shown in Fig. 3.102, is controlled by a servo motor (3) to achieve horizontal displacement of the gripper (2). Additionally, the vertical displacement of the gripper (2) is controlled by a cylinder (1).

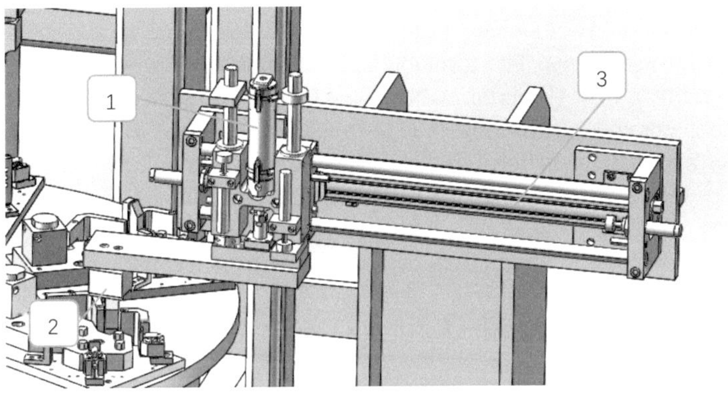

Fig. 3.102 Transfer Robotic Arm
1 - Cylinder 2 - Gripper 3 - Servo Motor

The fixed seat and turntable are shown in Fig. 3.103. Turntable (2) is driven and controlled by a rotary motor, and it has 6 positions with 6 fixed seats (1). When the six-station assembly machine is in operation, the product to be assembled is placed on the dedicated fixed seat (1) by the transfer robot arm (3).

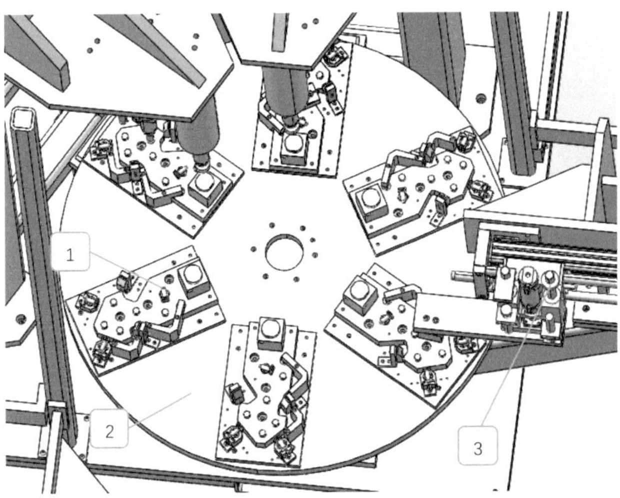

Fig. 3.103 Fixed seat and turntable
1 - Fixed seat 2 - Turntable 3 - Transfer robot arm

IV. Specific Parameters (see Table 3.22)

Table 3.22 Six-station assembly machine parameters

Efficiency:	500~800 pcs/h
Operating rate:	≥95%
Yield:	≥98%
Ambient temperature:	0~45°C
Allowable humidity:	20%~95%
Equipment dimensions:	2315 mm(H) × 3075 mm(L) × 2110 mm(W)

V. Mechanical Design Highlights

The original assembly process was manually completed, and due to the large number of components, manual assembly was laborious, time-consuming, and incurred high labor costs with low production efficiency. Additionally, visual inspection was prone to appearance defects, risks of omissions, and misjudgments. The turntable six-station assembly machine, through integrated design, is controlled by a PLC (Programmable Logic Controller) system, overseeing the actions of the transfer manipulator and turntable. Human intervention is limited to placing main components, while the transfer manipulator handles the transportation of various components to the turntable. The actions and positions are precise and controllable. The machine employs a visual touch screen for setting various operating parameters, making the operation simple and quick, reducing manual labor intensity, improving production efficiency, saving costs, and enhancing product quality.

Case 46: Automated Jump Rope Threading Machine

I. Case Description

In the production of jump ropes, the method of threading a string of small polyamide rings into jump ropes made of highly wear-resistant materials like polyamide and elastic materials such as polyurethane ensures low production costs while maintaining the elasticity and wear resistance of the jump rope. The automated jump rope threading machine is designed for automated threading of jump ropes, as shown in the overall structure in Fig. 3.104. It mainly consists of rubber rope (1), insertion needle (2), needle guide (3), first vibration conveyor (4), second vibration conveyor (5), feeding mechanism (6), and frame (7). The automated jump rope threading machine offers advantages such as stable operation, reliable performance, high production efficiency, and the capability for automated production.

II. Working Principle

In the usage process of the Automatic Jump Rope Threading Machine, the red nylon rings are first placed into the second vibration conveyor, and the blue nylon rings are placed into the first vibration conveyor. The second vibration conveyor transports the red nylon rings along the conveyor guide to the push block, while the first vibration conveyor transports the blue nylon rings along the conveyor guide to the push block. The optical fiber sensor detects the position of the red nylon rings, and the cylinder moves the red nylon rings to the right side of the insertion needle. The optical fiber sensor detects the position of the blue nylon rings, and the cylinder moves the blue nylon rings to the left side of the red nylon rings. The cylinder then moves the red nylon rings and blue nylon rings along the needle guide into the insertion needle. Through this movement, the two colors of nylon rings are continuously threaded into the rubber rope in sequence, achieving the automatic threading of two-color nylon rings into the jump rope.

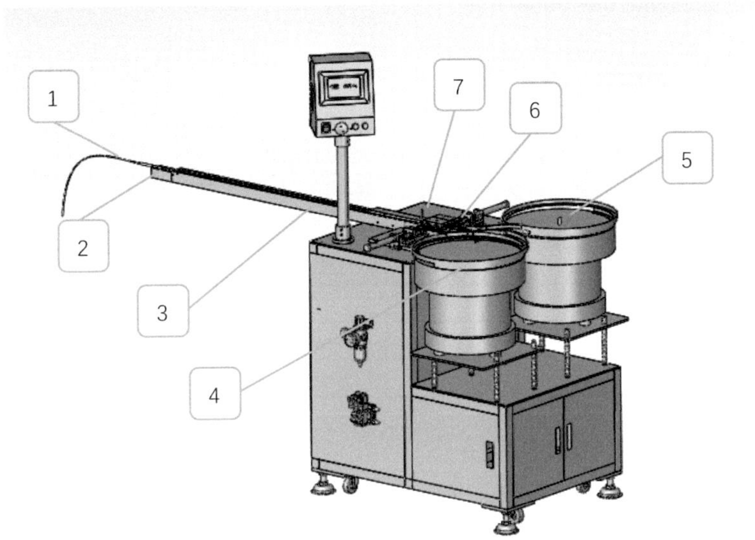

Fig. 3.104 Automatic Jump Rope Threading Machine
1 - Rubber Rope 2 - Insertion Needle 3 - Needle Guide 4, 5 - Vibration Conveyors 6 - Feeding Mechanism 7 - Frame

III. Main Structure Introduction

The feeding mechanism, as shown in Fig. 3.105, operates as follows: the second vibration conveyor (6) transports the red nylon rings along the conveyor guide to the push block (7), while the first vibration conveyor (3) transports the blue nylon rings along the conveyor guide to the push block (2). The optical fiber sensor detects the position of the red nylon rings, and cylinder (8) moves the red nylon rings to the right side of the insertion needle (4). Similarly, the optical fiber sensor detects

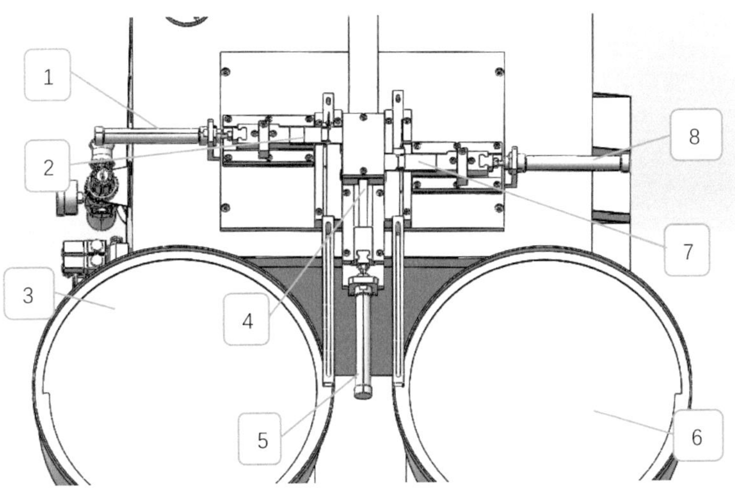

Fig. 3.105 Feeding Mechanism
1, 5, 8 - Cylinder 2, 7 - Push Block 3, 6 - Vibration Conveyor 4 - Insertion Needle

the position of the blue nylon rings, and cylinder (1) moves the blue nylon rings to the left side of the red nylon rings. Cylinder (5) then moves the red nylon rings and blue nylon rings along the needle guide into the insertion needle (4). Through this movement, the two colors of nylon rings are continuously threaded into the rubber rope.

IV. Specific Parameters (see Table 3.23)

Table 3.23 Parameters of the Automatic Jump Rope Threading Machine

Working Efficiency:	20~30 pcs/h
Operation Rate:	≥90%
Yield Rate:	≥97%
Environmental Temperature:	0~45°C
Permissible Humidity:	10%~95%
Equipment Dimensions:	1350 mm(H) × 1700 m(L) × 780 mm(W)

V. Mechanical Design Highlights

The Automatic Jump Rope Threading Machine utilizes a vibratory conveyor to automatically transport small polyamide rings, powered by cylinders for stable and reliable operation. Simultaneously, the machine can automatically thread two colors of polyamide rings in sequence into the jump rope, enhancing the appearance of the jump rope and addressing the efficiency issues present in existing technologies, thus achieving automated production.

Chapter 4

Parts Processing Equipment

Case 47: USB Shell Punching Machine

I. Case Description

The USB Shell Punching Machine is designed for hole processing in USB shells, as shown in Fig. 4.1. A detailed view is provided in Fig. 4.2. It mainly consists of the frame (1), pre-pressing mechanism (2), hole processing mechanism (3), indexing rotary mechanism (4), specialized mold (5), material fetching mechanism (6), and display screen (7). The electrical control system automates various actions and movements of the USB Shell Punching Machine.

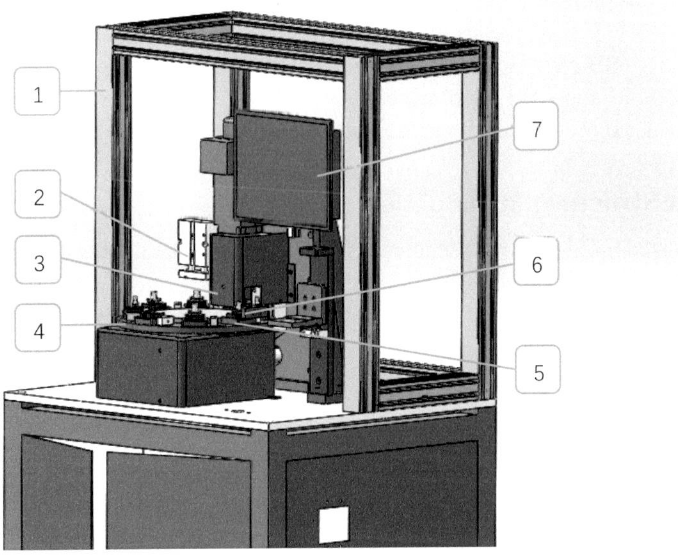

Fig. 4.1 USB Shell Punching Machine

1 - Frame 2 - Pre-pressing Mechanism 3 - Hole Processing Mechanism 4 - Indexing Rotary Mechanism 5 - Specialized Mold 6 - Material Fetching Mechanism 7 - Display Screen

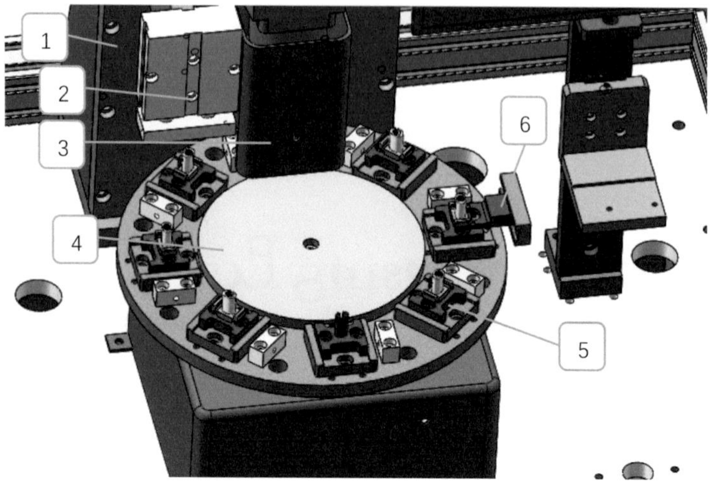

Fig. 4.2 Enlarged View of USB Shell Punching Machine
1-Frame 2-Pre-pressing Mechanism 3-Hole Processing Mechanism 4-Indexing Rotary Mechanism 5-Specialized Mold 6-Material Fetching Mechanism

II. Working Principle

When the USB Shell Punching Machine is in operation, the USB shell is mounted on the specialized mold's core positioning component. The indexing rotary mechanism's motor is activated, initiating the rotation of the cam indexer [72]. As the turntable rotates, it conveys the specialized mold and USB shell to the pre-pressing mechanism. The pre-pressing mechanism tightly secures the USB shell on the specialized mold. During the second rotation of the turntable, the specialized mold and USB shell are conveyed to the hole processing mechanism. The hole processing mechanism processes the USB shell on the specialized mold. During the third rotation of the turntable, the specialized mold and USB shell are conveyed to the material fetching mechanism. The material fetching mechanism removes the processed USB shell, completing the hole punching operation.

III. Main Structure Introduction

The specialized mold (Fig. 4.3) secures the USB shell and facilitates its removal after punching.

Fig. 4.3 Specialized Mold
1-Base 2-Core Positioning Component 3-Material Fetching Component

The pre-pressing mechanism (Fig. 4.4) uses a cylinder (1) to press down the fixed plate (2), securing the USB shell on the specialized mold.

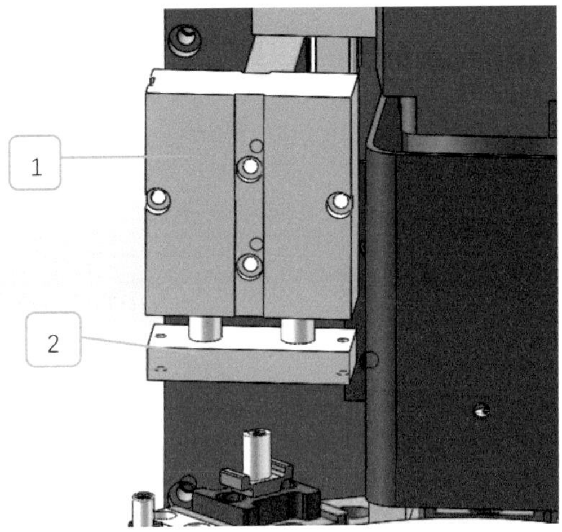

Fig. 4.4 Pre-pressing Mechanism
1 - Cylinder 2 - Fixed Plate

The hole processing mechanism (Fig. 4.5) uses a cylinder to control the up and down movement. When the USB shell is positioned below the hole processing mechanism by the indexing rotary mechanism, the cylinder lowers the hole punching machine (1) to punch the USB shell [73].

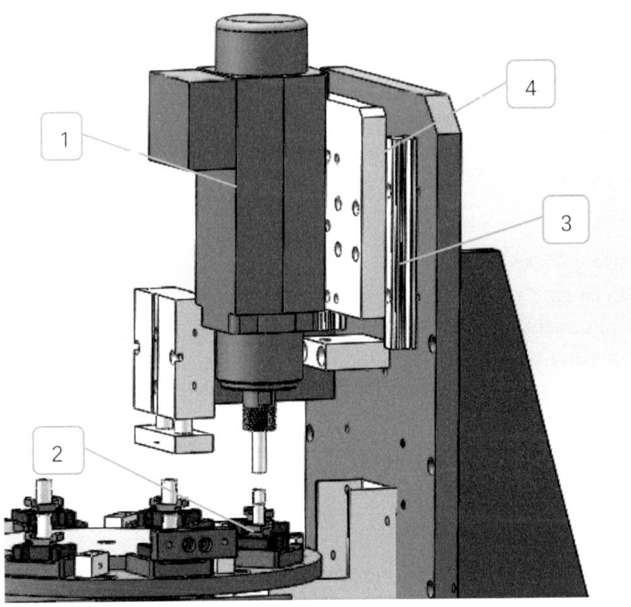

Fig. 4.5 Hole Processing Mechanism
1 - Punching Machine 2 - Specialized Mold 3 - Linear Guide 4 - Connecting Plate

IV. Specific Parameters (see Table 4.1)

Table 4.1 USB Shell Punching Machine Parameters

Processing Accuracy:	±0.01 mm
Processing Efficiency:	30~50 pcs/min
Operation Rate:	≥93%
Yield Rate:	≥99%
Operating Pressure Range:	3-5 kg/cm^2
Allowable Humidity:	20%~95%
Machine Dimensions:	1180 mm(H) × 600 mm(L) × 495 mm(W)

V. Mechanical Design Highlights and Precautions

(1) Before using the punching machine, check for debris at the punch and lower die edge.
(2) Apply lubricating oil to the upper part of the punch (moving part of the punch).
(3) Ensure the lubrication points of the punching machine are not lacking oil during continuous operation.
(4) If abnormal sounds (not normal punching machine operation sounds) occur during use, address them promptly.
(5) Do not use iron tools to strike any part of the punching machine during installation.
(6) After use, clean the punching machine, especially the punch and lower die.
(7) Apply lubricating oil to the punch and lower die, and press the punch down to its lowest position to better protect the punching machine.
(8) Operating speed: Up to 130 cycles per minute.
(9) High-quality electronic control components ensure stable performance and easy operation.
(10) The equipment features multiple safety protection devices, ensuring operator safety, prolonging its service life, and reducing product losses.

Case 48: Automatic Grinding Machine

I. Case Description

The Automatic Grinding Machine is designed for precision machining and surface polishing in the mold industry. It serves as a replacement for similar pneumatic products. The overall structure is shown in Fig. 4.6 and consists of the following mechanisms: sliding rail mechanism (1), grinding mechanism (2), processing positioning mechanism (3), control box (4), coolant collection box (5), and frame (6). The machine features simple operation, high grinding accuracy, and low noise.

II. Working Principle

Before grinding, place the workpiece on the fixed frame, secure it with the processing positioning mechanism, power on, and adjust the position of the abrasive tool. The abrasive tool starts rotating at a constant speed and begins the grinding process. Coolant is injected simultaneously. During grinding, the abrasive tool rotates at a constant speed and can translate along two rails, achieving comprehensive grinding. In the final stage, the abrasive tool slows down, moves away from the workpiece, and the coolant is collected, completing the grinding process.

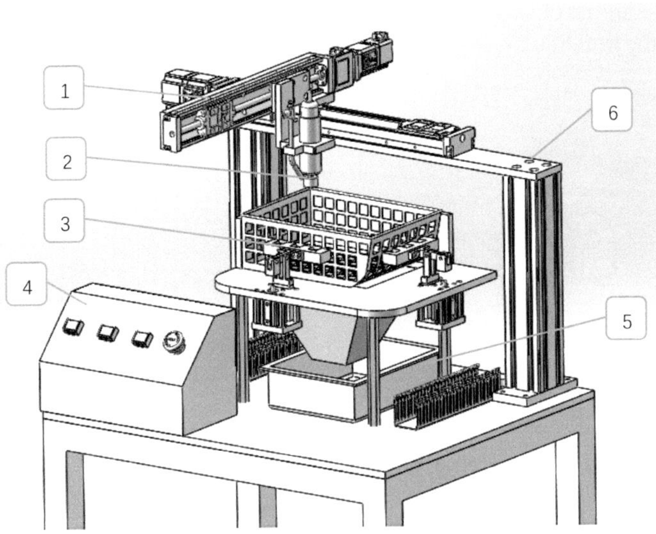

Fig. 4.6 Semi-Automatic Grinding Machine
1-Sliding Rail Mechanism 2-Grinding Mechanism 3-Processing Positioning Mechanism 4-Control Box 5-Coolant Collection Box 6-Frame

III. Main Structure Introduction

The grinding mechanism (Fig. 4.7) allows for the replacement of different types of abrasives based on requirements. Coolant is injected through the coolant injection pipe (2) for cooling during grinding [74].

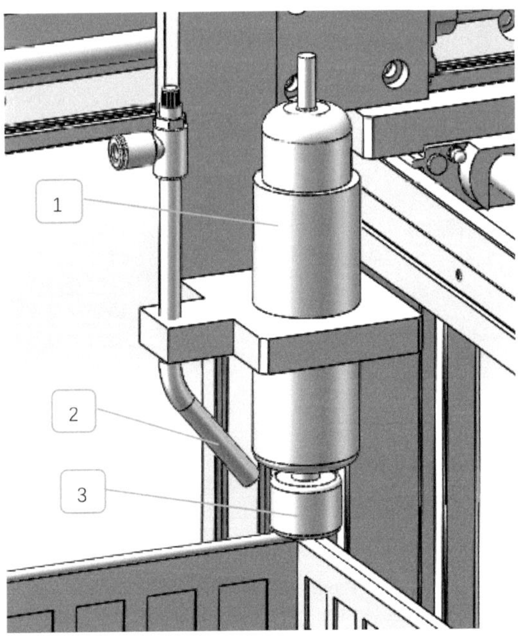

Fig. 4.7 Grinding Mechanism
1-Servo Motor 2-Coolant Injection Pipe 3-Abrasive Tool

The sliding rail mechanism (Fig. 4.8) is controlled by cylinders to achieve horizontal displacement of the grinding mechanism.

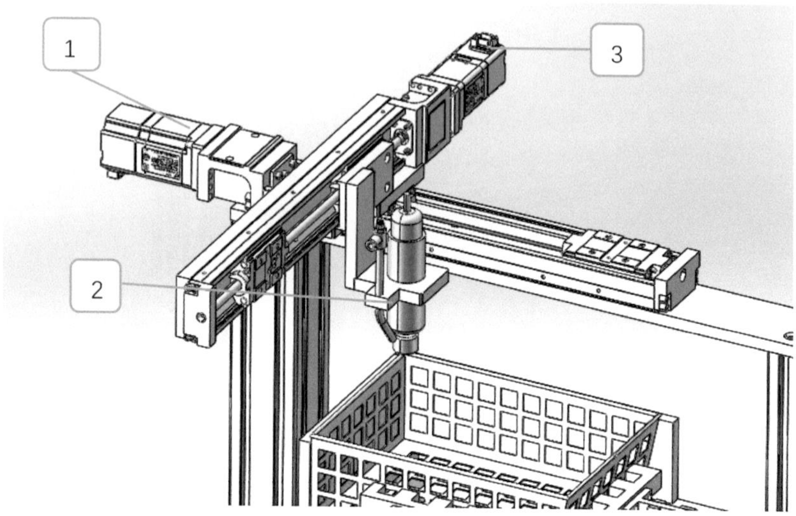

Fig. 4.8 Sliding Rail Mechanism
1, 3 - Cylinders 2 - Grinding Mechanism

The processing positioning mechanism (Fig. 4.9) holds the workpiece in the fixed frame (2). The height can be controlled by cylinders (3, 4), ensuring full contact between the product and the grinding mechanism.

Fig. 4.9 Processing Positioning Mechanism
1 - Grinding Mechanism 2 - Fixed Frame 3, 4 - Cylinders

IV. Specific Parameters (see Table 4.2)

Table 4.2 Automatic Grinding Machine Parameters

Power:	≥248W
No-Load Speed:	1000~50000 R/min (adjustable)
Maximum Torque:	8.7 N.cm
Purpose:	Burr removal and grinding
Machine Dimensions:	1130 mm(H) × 690 mm(L) × 615 mm(W)

V. Mechanical Design Highlights

The grinding tool of this machine is connected to a spring, which converts the upward tension of the spring into downward pressure on the grinding tool, maintaining elastic contact between the working part of the grinding machine and the product's edge. The force can be adjusted by adjusting the spring height [75]. The rolling bearing rolls against the contour block, driving the entire sliding mechanism and grinding tool to move horizontally back and forth. This ensures constant clamping force between the working part of the grinding machine and the part, preventing damage due to uneven forces during operation. In general, one rotation of the rotating mechanism is sufficient to remove burrs from the product. If encountering higher burrs, the number of rotations can be set to allow for additional grinding. The number of rotations can be adjusted based on the actual grinding effect [76].

VI. Precautions

(1) The power socket used must have a residual current device, and check for any damage to the power cord.
(2) The direction of grinding during operation must not face surrounding workers or any flammable or explosive materials to avoid unnecessary harm.
(3) Before use, the grinding machine must be started and tested to ensure smooth and normal operation.
(4) Do not use abrasive tools leftover from other specifications of power tools.
(5) Check whether the heat dissipation of each motor is good and whether the airflow is unobstructed. Ensure that the power is disconnected during assembly and disassembly to avoid unnecessary damage.
(6) Check if the worktable and work cover are in the same horizontal plane to avoid affecting the grinding effect.

Case 49: Magnetic Ring Tube Forming Machine

I. Case Description

The magnetic ring, as shown in Fig. 4.10, is a circular magnet used as a common interference-resistant component in electronic circuits. It effectively suppresses high-frequency noise. The Magnetic Ring Tube Forming Machine is primarily used for the automated production of magnetic ring tubes. The overall structure is shown in Fig. 4.11, consisting of mechanisms such as the horizontal screw rod (1), discharge port (2), workbench (3), feeding mechanism (4), frame (5), motor (6), and magnetic ring tube forming mechanism (7). The machine boasts stable operation, safety, reliability, and low noise.

Fig. 4.10 Magnetic Ring

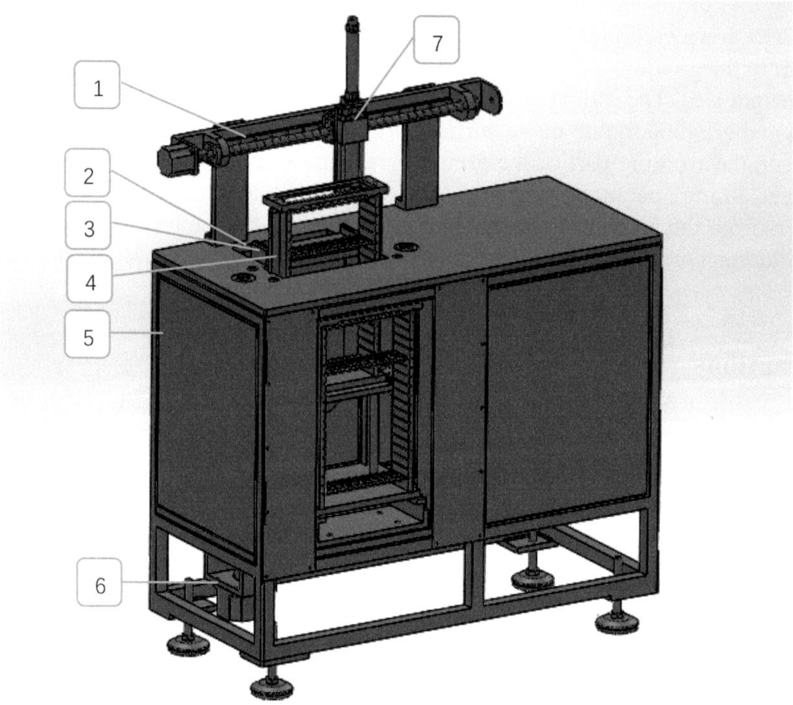

Fig. 4.11 Magnetic Ring Tube Forming Machine
1 - Horizontal Screw Rod 2 - Discharge Port 3 - Workbench 4 - Feeding Mechanism 5 - Frame 6 - Motor 7 - Magnetic Ring Tube Forming Mechanism

II. Working Principle

During operation, the Magnetic Ring Tube Forming Machine uses a screw rod perpendicular to the workbench to transport magnetic rings to the workbench. Simultaneously, the horizontal screw rod rotates, and the magnetic ring tube forming mechanism sleeves the magnetic ring, undergoes heat shrinkage forming, and the cooled magnetic ring tube is output through the discharge port, completing the magnetic ring tube forming process. This machine is suitable for piercing and sleeving magnetic rings for strip diodes, with a vertical forming shape.

III. Main Structure Introduction

Feeding Mechanism (Fig. 4.12): When the machine is operational, a shelf (6) with magnetic rings is placed on the lifting platform (4). The motor (5) lifts the lifting platform (3) using the rotating vertical screw rod (4). When the shelf (6) is level with the workbench (1), the magnetic rings on the shelf (6) are transferred to the workbench (1). After all products on the shelf (6) are transferred to the workbench (1), the shelf (6) is removed, and the next shelf (6) with magnetic rings is introduced into the lifting platform (3).

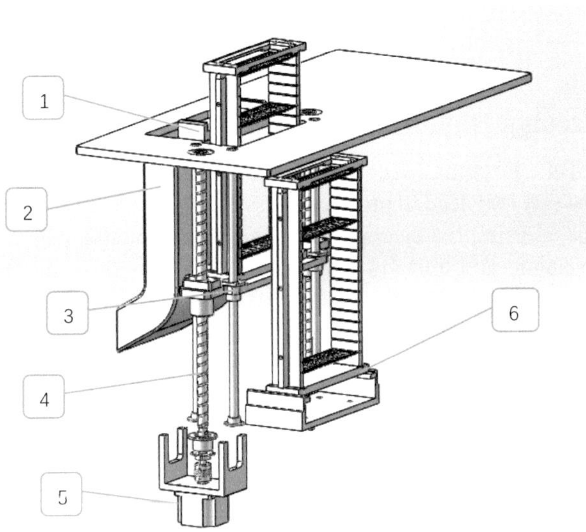

Fig. 4.12 Feeding Mechanism
1 - Workbench 2 - Discharge Port 3 - Lifting Platform 4 - Vertical Screw Rod 5 - Motor 6 - Shelf

Magnetic Ring Tube Forming Mechanism (Fig. 4.13): The magnetic ring to be processed is introduced into the workbench through the feeding mechanism (4). The magnetic ring tube forming mechanism sleeves the tube onto the magnetic ring and performs heat shrinkage forming.

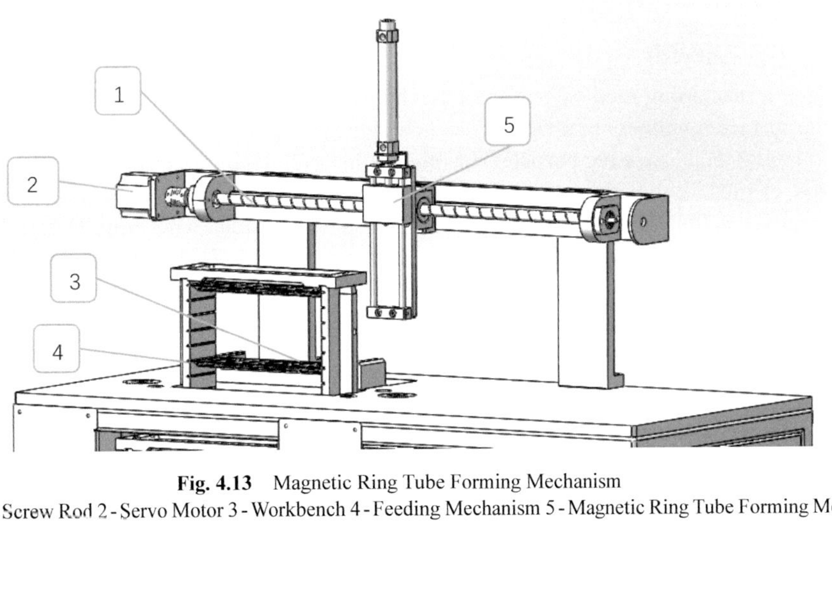

Fig. 4.13 Magnetic Ring Tube Forming Mechanism
1 - Horizontal Screw Rod 2 - Servo Motor 3 - Workbench 4 - Feeding Mechanism 5 - Magnetic Ring Tube Forming Mechanism

IV. Specific Parameters (see Table 4.3)

Table 4.3 Magnetic Ring Tube Forming Machine Parameters

Equipment Weight:	200 kg
Working Efficiency:	800-1000 pcs/h
Yield Rate:	≥98%
Utilization Rate:	≥80%
Machine Dimensions:	1350 mm(H) × 1000 mm(L) × 440 mm(W)

V. Mechanical Design Highlights

The Magnetic Ring Tube Forming Machine ensures mold accuracy through a guiding mechanism, preventing deviations that may lead to production issues. After cooling following injection molding, the workpiece is ejected using the ejection mechanism. The melted plastic enters the mold cavity through the sprue bushing, and uniform material flow from both sides enhances product quality. The lifting plate is moved upward by the pneumatic cylinder, lifting the workpiece using the lifting rod. The lifting rods are symmetrically and evenly arranged below the workpiece, ensuring even force during lifting, improving guiding and positioning accuracy, and avoiding mechanical damage caused by misalignment.

VI. Precautions

(1) The Magnetic Ring Tube Forming Machine must be placed level. If it tilts, errors in the forming machine may occur, resulting in deformed and uneven magnetic ring tubes.
(2) Fix the magnetic ring tube mold securely and adjust it. If the mold is not fixed properly, magnetic ring deformation may occur, or even damage to the mold.
(3) The workbench should be free of debris, and check whether all guide rails are lubricated and in good condition.
(4) Operators should avoid inserting hands into the forming machine to prevent accidents.

Case 50: Multi-Station Automatic Polishing Machine

I. Case Description

Polishing is a machining method that uses mechanical, chemical, or electrochemical processes to reduce the surface roughness of a workpiece, achieving a bright and smooth surface. It involves the use of polishing tools, abrasive particles, or other polishing media for surface refinement. After the production of metal components, polishing is often required to improve the surface smoothness, ensuring high-quality subsequent processes like electroplating. The Multi-Station Automatic Polishing Machine can simultaneously polish multiple products [77]. The overall structure is shown in Fig. 4.14, with an enlarged view of specific components in Fig. 4.15. It consists mainly of the electrical control box (1), polishing mechanism (2), frame (3), feeding mechanism (4), conveying mechanism (5), and discharge mechanism (6). The machine is known for its simplicity of operation and high efficiency.

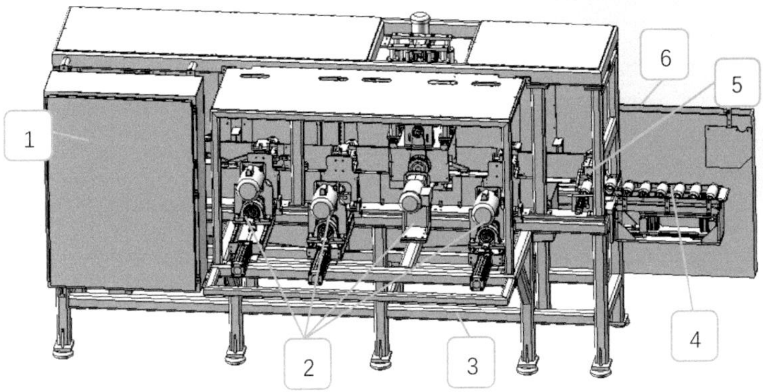

Fig. 4.14 Multi-Station Automatic Polishing Machine
1 - Electrical Control Box 2 - Polishing Mechanism 3 - Frame 4 - Feeding Mechanism 5 - Conveying Mechanism 6 - Discharge Mechanism

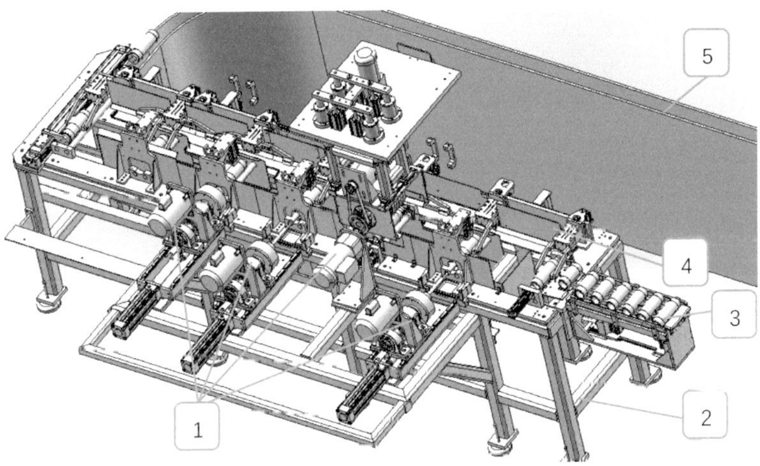

Fig. 4.15 Enlarged View of Multi-Station Automatic Polishing Machine
1 - Polishing Mechanism 2 - Frame 3 - Feeding Mechanism 4 - Conveying Mechanism 5 - Discharge Mechanism

II. Working Principle

The Multi-Station Automatic Polishing Machine, controlled by a Programmable Logic Controller (PLC), coordinates the operation of various motors for polishing. Products to be polished are introduced through the feeding mechanism. The conveying mechanism, equipped with six fixtures, lifts the product from the feeding mechanism and transports it to the polishing equipment. The internal surface polishing mechanism and external surface polishing mechanism perform polishing operations on the product's internal and external surfaces, respectively [78]. After completing the predetermined surface polishing, the product is transferred to the next station. This process is repeated at four polishing stations. The finished polished products are output through the discharge port.

III. Main Structure Introduction

Feeding Mechanism (Fig. 4.16): Driven by cylinders (2), the synchronous belt (1) moves horizontally. The synchronous belt (1) has slots to prevent surface scratching due to rolling. Cylinders (3) and (4) control the vertical displacement of the synchronous belt (1), transporting the product to the conveying mechanism.

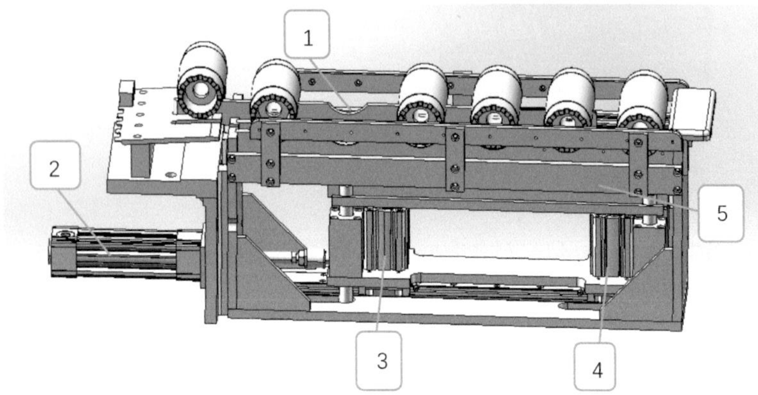

Fig. 4.16 Feeding Mechanism
1 - Synchronous Belt 2, 3, 4 - Cylinders 5 - Guardrail

Conveying Mechanism Fixture (Fig. 4.17): The lifting of the fixture (1) is controlled by cylinder (3). The fixture (1) has a spacer (2) to prevent scratching of the product.

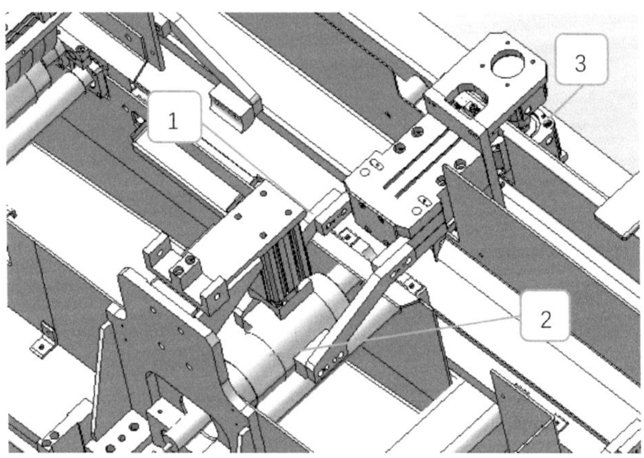

Fig. 4.17 Fixture
1 - Fixture 2 - Spacer 3 - Cylinder

Internal Surface Polishing Mechanism (Fig. 4.18): The polishing wheel (1) is driven by an electric motor (2) and is used for polishing the internal surface of the product. During operation, fixture (4) positions the product in the internal surface polishing mechanism. Cylinder (3) pushes the polishing wheel (1) to the product's internal surface, and the electric motor (2) starts rotating, polishing the internal surface as the fixture (4) rotates.

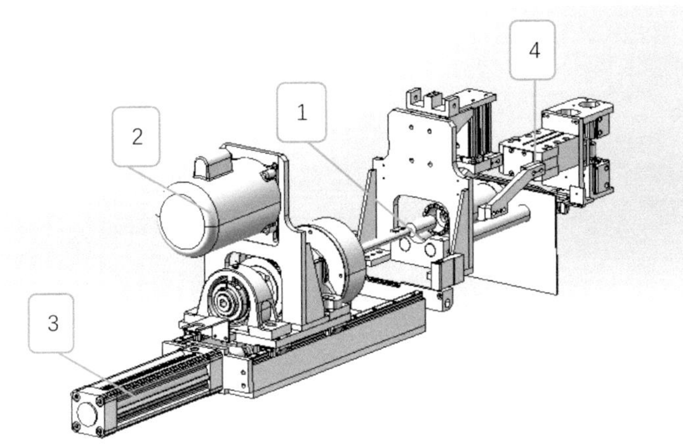

Fig. 4.18 Internal Surface Polishing Mechanism
1 - Internal Surface Polishing Wheel 2 - Electric Motor 3 - Cylinder 4 - Fixture

External Surface Polishing Mechanism (Fig. 4.19): The external surface polishing wheel (8) is driven by electric motor (2). Fixture (7) positions the product on roller (6). Electric motors (2) and (5) control the rotation of the polishing wheel (8) and roller (6), respectively. Cylinder (1) controls the gradual descent of the polishing wheel (8) until it contacts the product, polishing the external surface.

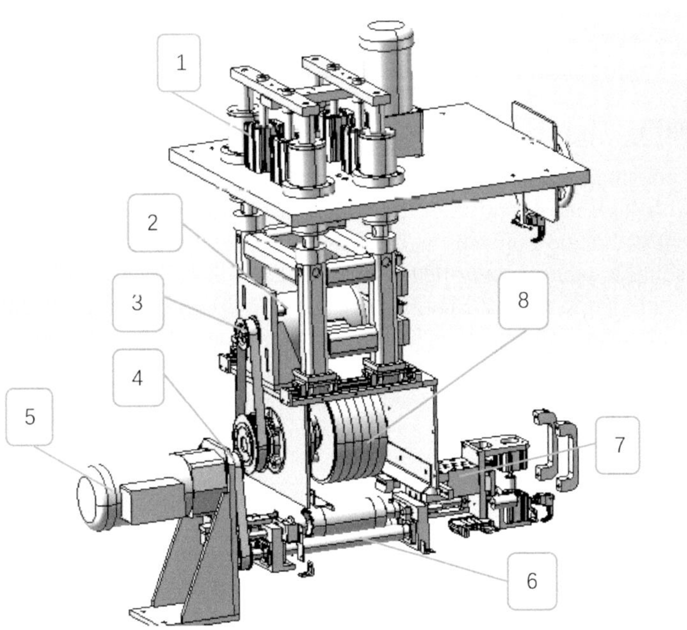

Fig. 4.19 External Surface Polishing Mechanism
1 - Cylinder 2, 5 - Electric Motor 3, 4 - Pulleys 6 - Roller 7 - Fixture 8 - External Surface Polishing Wheel

Discharge Mechanism (Fig. 4.20): Driven by cylinder (2), the finished polished product is conveyed to the discharge track (1).

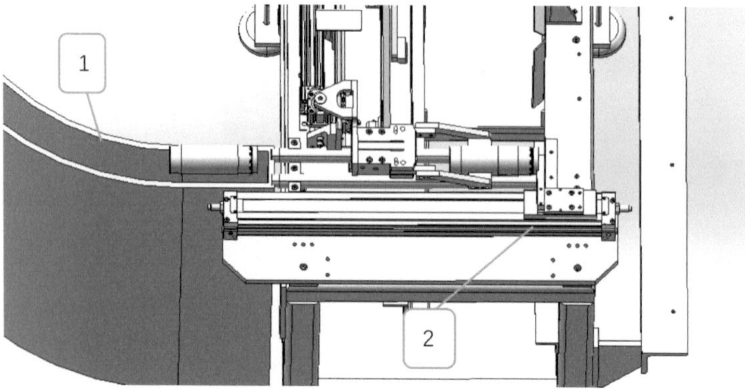

Fig. 4.20 Discharge Mechanism
1 - Discharge Track 2 - Cylinder

IV. Specific Parameters (see Table 4.4)

Table 4.4 Multi-Station Automatic Polishing Machine Parameters

Utilization Rate:	≥60%
Yield Rate:	≥98%
Wheel Diameter:	25 mm, 50 mm, 200 mm
Wheel Speed:	3200 r/min
Processing Efficiency:	70 pcs/min
Machine Dimensions:	2030 mm(H) × 3960 mm(L) × 2540 mm(W)

V. Mechanical Design Highlights

Many manual polishing processes are still prevalent in the industry. Workers operating in environments with high noise and dust for prolonged periods can have adverse effects on their health. Additionally, manual polishing is inefficient, and the quality may vary. The Multi-Station Automatic Polishing Machine, controlled by a PLC, coordinates various motors for simultaneous polishing at multiple workstations. Placed in a closed environment, workers only need to replace the polishing metal parts at preset intervals, eliminating the need for prolonged exposure to dust and noise. Some polishing stations are inclined at an angle to the workbench top, meeting the varied polishing needs of different metal components, expanding the machine's application range.

VI. Precautions

(1) Before operating the automatic polishing machine, check the power supply. The specific safety startup process involves connecting the external power supply, turning on the device's main power switch, and finally, turning on the device's power start switch.
(2) Ensure to check the motor exhaust port before work to ensure smooth airflow.
(3) Perform a safety check on the automatic polishing machine before work. Allow the machine to run empty and check whether all fasteners are secure to ensure safe production.
(4) Do not insert any part of the body into the polishing area while the automatic polishing machine is in operation to prevent injuries.
(5) If the automatic polishing machine works continuously for an extended period, and the polishing roller generates high heat, it may cause burns. Let the machine cool down before further operation.

(6) Before and after work, clean the surface of the polishing roller and wax it to maintain a smooth and dust-free surface.
(7) During use, control the pressure of the feeding roller appropriately to avoid excessive resistance, which may lead to overheating and damage.
(8) Ensure proper lubrication of rotating parts in the transmission chain to maintain accurate and reliable transmission.

Case 51: Pure Cam Structure Air Blowing Dust Removal Machine

I. Case Description

The Pure Cam Structure Air Blowing Dust Removal Machine is designed to remove dust adhering to the surface of electronic components. The overall structure is shown in Fig. 4.21, consisting mainly of the blowing mechanism (1), material-carrying track (2), drive mechanism (3), cam mechanism (4), clutch (5), air outlet pipe (6), support plate (7), dust removal mechanism (8), and conveying mechanism (9). The machine utilizes a cam mechanism to transport and carry out dust removal on products, ensuring mature stability, high efficiency, and practicality.

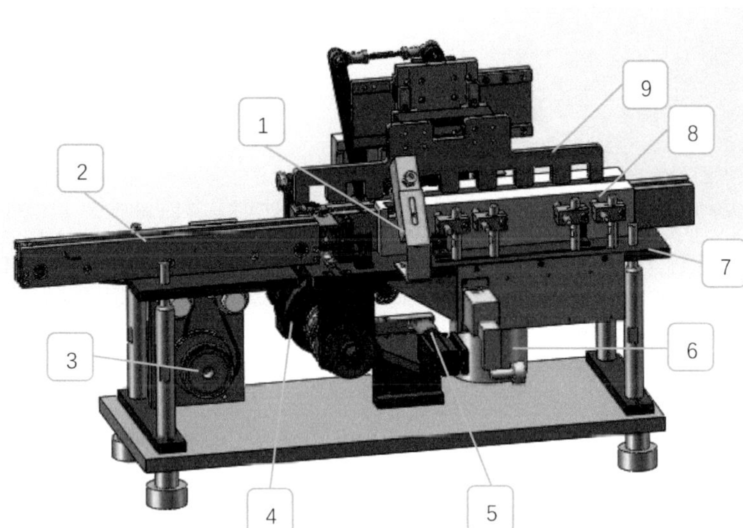

Fig. 4.21 Pure Cam Structure Air Blowing Dust Removal Machine
1-Blowing Mechanism 2-Material-Carrying Track 3-Drive Mechanism 4-Cam Mechanism 5-Clutch 6-Air Outlet Pipe 7-Support Plate 8-Dust Removal Mechanism 9-Conveying Mechanism

II. Working Principle

The Pure Cam Structure Air Blowing Dust Removal Machine is an electronic component air blowing dust removal device. During operation, the material-carrying track can be used in conjunction with other equipment. Electronic components are transported to the dust removal mechanism through the material-carrying track for dust removal. The cam mechanism automatically separates and transfers the materials. The main spindle drive structure has a clutch design. When the sensor on the material-carrying track does not detect the arrival of components at the corresponding position, the clutch is disengaged. The machine waits for the components to be transported to the specified position by the conveying mechanism before the dust removal mechanism starts working [79].

III. Main Structure Introduction

Clutch (Fig. 4.22): The main spindle drive mechanism (4) is driven by the drive mechanism (1) through the pulley (2). The clutch, controlled by cylinder (5), disengages when the sensor on the material-carrying track does not detect components at the corresponding position. The cylinder (5) pushes the toothed disc (3) on the main spindle drive mechanism (4), disconnecting the connection between the toothed discs, thereby stopping the dust removal mechanism.

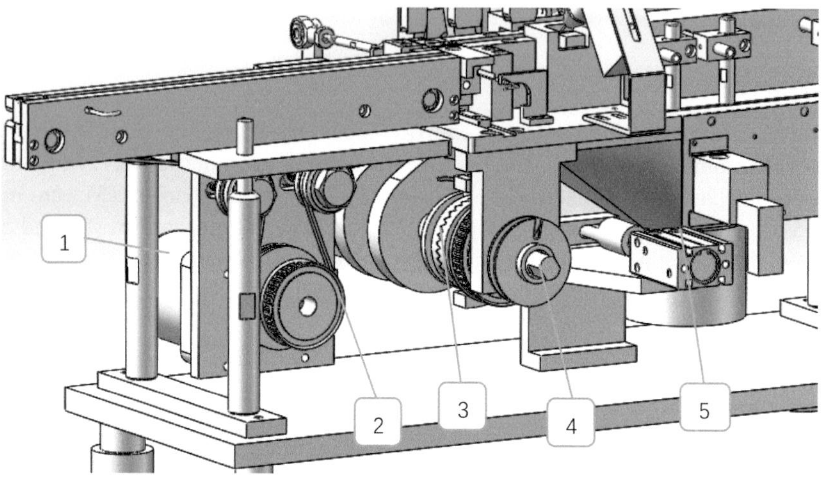

Fig. 4.22 Clutch
1 - Drive Mechanism 2 - Pulley 3 - Toothed Disc 4 - Main Spindle Drive Mechanism 5 - Cylinder

Cam Mechanism (Fig. 4.23): Driven by the drive mechanism (4) through the pulley, the cam (3) rotates synchronously. The cam (3), connected to the conveying mechanism (1) by a connecting rod (2), controls the displacement of the conveying mechanism (1), continuously transporting electronic components in need of dust removal.

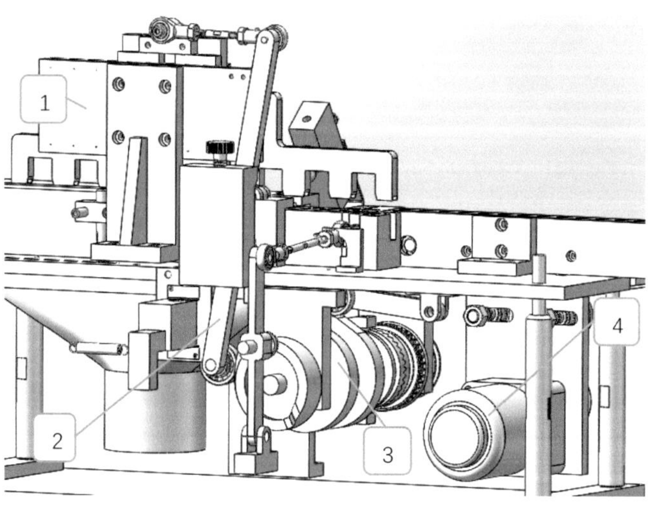

Fig. 4.23 Cam Mechanism
1 - Conveying Mechanism 2 - Connecting Rod 3 - Cam 4 - Drive Mechanism

Blowing Mechanism (Fig. 4.24): When the conveying mechanism (3) transports electronic components to the front of the air blowing port, the blowing mechanism (1) operates, blowing away the dust on the surface of the electronic components. The dust is blown out through the air outlet pipe (2).

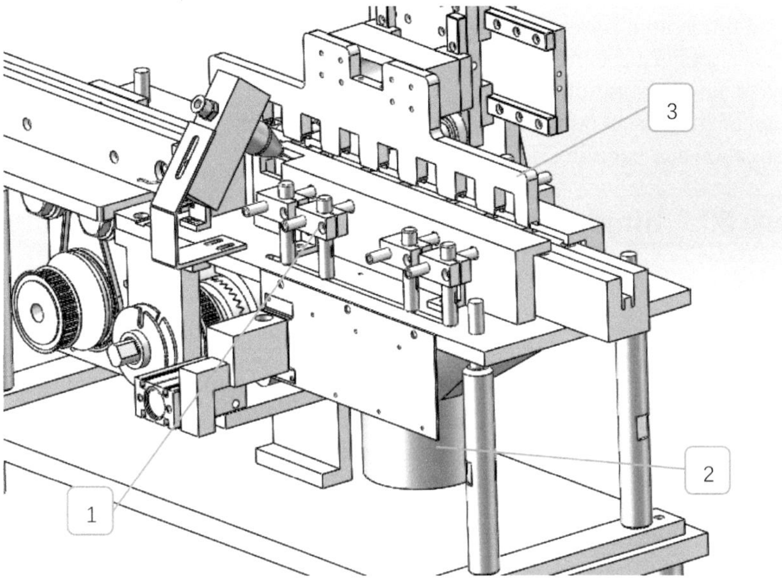

Fig. 4.24 Blowing Mechanism
1 - Blowing Mechanism 2 - Air Outlet Pipe 3 - Conveying Mechanism

IV. Specific Parameters (see Table 4.5)

Table 4.5 Pure Cam Structure Air Blowing Dust Removal Machine Parameters

Air Handling Capacity:	6690 m^3/h
Dust Removal Area:	50 m^2/h
Pressure Loss:	1470-1770 Pa
Utilization Rate:	≥93%
Environmental Temperature:	10~65°C
Permissible Humidity:	20%~95%
Machine Dimensions:	400 mm(H) × 730 mm(L) × 365 mm(W)

V. Mechanical Design Highlights

The Pure Cam Structure Air Blowing Dust Removal Machine uses a fully cam-driven conveying mechanism for moving products, intermittently conveying electronic components in a rectangular motion trajectory for dust removal. The main spindle drive structure incorporates a clutch design. When the sensor on the synchronous belt does not detect electronic components at the corresponding position, the clutch disengages, stopping the dust removal mechanism until the electronic components reach the designated position.

VI. Precautions

(1) Inspect all transmission components, such as blowers, motors, couplings, etc., to ensure that fasteners are secure.
(2) Check the oil level in the bearing seat lubrication points of the fan to ensure normal oil levels.
(3) Inspect various instruments in the electrical control cabinet, such as voltage meters, current meters, and differential pressure gauges, to ensure normal operation.
(4) Ensure that moving parts such as the fan and motor operate smoothly, without abnormal directions or severe vibrations.
(5) In the event of an accident during operation, press the emergency stop button to stop the machine for investigation and handling.

Case 52: Single-Power Stripping and Twisting Machine

I. Case Description

The Single-Power Stripping and Twisting Machine is designed for processing cables such as telephone lines, parallel lines, power lines, and computer multi-core lines. It can handle the tasks of both stripping and twisting, combining the functionalities of two separate machines. The overall structure is shown in Fig. 4.25, consisting mainly of the twisting mechanism (1), stripping mechanism (2), gantry frame (3), driving mechanism (4), and installation frame (5). The twisting mechanism (1) comprises components such as twisting assembly, lateral slider, longitudinal slider, drive rod, and guide groove. The stripping mechanism (2) includes components like the positioning mechanism, connecting parts, and cutting blade. This machine, with its single-power stripping and twisting capability, has a simple structure, reducing production costs, and features synchronous stripping and twisting.

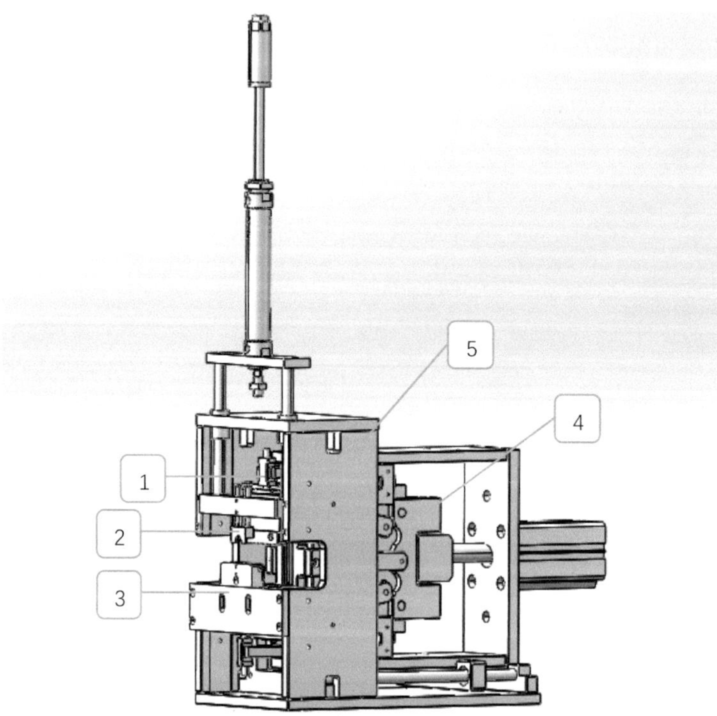

Fig. 4.25 Single-Power Stripping and Twisting Machine
1 - Twisting Mechanism 2 - Stripping Mechanism 3 - Gantry Frame 4 - Driving Mechanism 5 - Installation Frame

II. Working Principle

When the Single-Power Stripping and Twisting Machine is operational, the operator places the cable bundle's head on the positioning seat of the stripping mechanism. The positioning cylinder drives the positioning head to press down on the cable bundle's head, and the cutting blade peels off the cable bundle's head. Subsequently, the cylinder drives the connecting part to move. The connecting part moves the connecting block, which, in turn, moves one end of the first limiting piece and the second limiting piece along the first guide groove and the second guide groove, respectively. The first limiting piece and the second limiting piece drive the first vertical slider and the second vertical slider to move longitudinally relative to each other, causing the upper twisting piece and the lower twisting piece to come into contact with each other. Simultaneously, the connecting block moves the first fixed plate and the second fixed plate. The first drive rod's first limiting column moves along the third guide groove, and the second drive rod's second limiting column moves along the fourth guide groove. The first drive rod and the second drive rod respectively drive the first horizontal slider and the second horizontal slider to move laterally. This allows the upper twisting piece and the lower twisting piece to move laterally, and the directions of movement of the upper and lower twisting pieces are always opposite. Eventually, the cable bundle's insulation is stripped, and the cores inside the cable bundle are twisted together.

III. Main Structure Introduction

Driving Mechanism (Fig. 4.26): When the Single-Power Stripping and Twisting Machine is operational, cylinder (6) drives connecting part (7) to move. Connecting part (7) moves connecting block (5), which moves one end of the first limiting piece (2) and the second limiting piece (3) along the first guide groove 1 and the second guide groove (4), respectively. The first limiting piece (2) and the second limiting piece (3) drive the first vertical slider and the second vertical slider to move longitudinally, causing the upper twisting piece and the lower twisting piece to come into contact with each other [80].

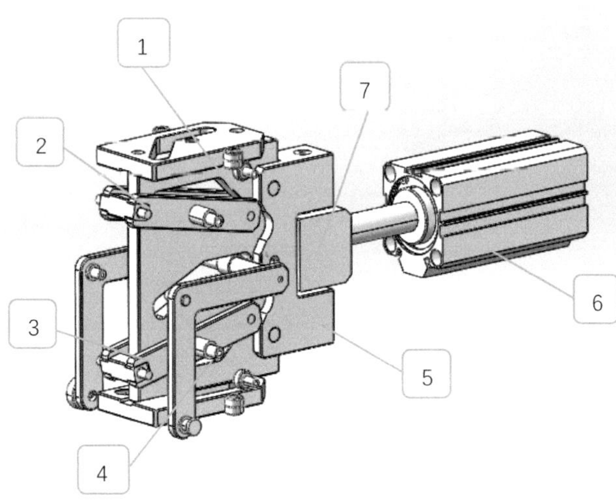

Fig. 4.26 Driving Mechanism
1-First Guide Groove 2-First Limiting Piece 3-Second Limiting Piece 4-Second Guide Groove 5-Connecting Block 6-Cylinder 7-Connecting Part

Stripping Mechanism (Fig. 4.27): When the Single-Power Stripping and Twisting Machine is operational, the operator places the cable bundle's head on the positioning seat (3) of the stripping mechanism. Positioning cylinders (1) and (4) drive positioning head (2) to press down on the cable bundle's head, and cutting blade (5) peels off the cable bundle's head.

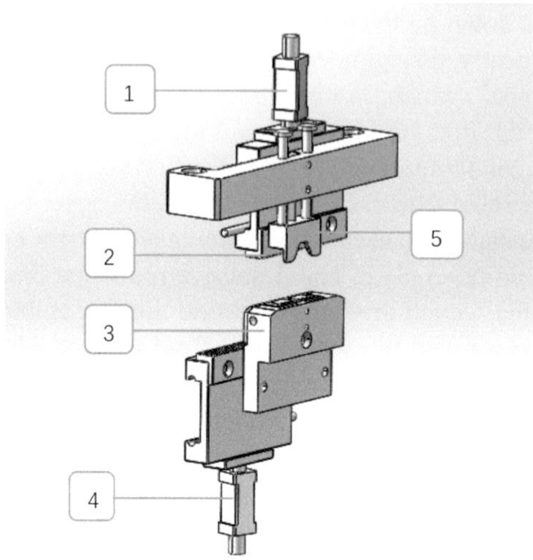

Fig. 4.27 Stripping Mechanism
1, 4 - Positioning Cylinders 2 - Positioning Head 3 - Positioning Seat 5 - Cutting Blade

Twisting Mechanism (Fig. 4.28): During operation, connecting block (8) moves the first fixed plate (2) and the second fixed plate (5). The first drive rod (1)'s first limiting column moves along the third guide groove (9), and the second drive rod (6)'s second limiting column moves along the fourth guide groove (7). The first drive rod (1) and the second drive rod (6) respectively drive the first horizontal slider (3) and the second horizontal slider (4) to move laterally. This allows the upper twisting piece and the lower twisting piece to move laterally, and the directions of movement of the upper and lower twisting pieces are always opposite. Eventually, the cable bundle's insulation is stripped, and the cores inside the cable bundle are twisted together.

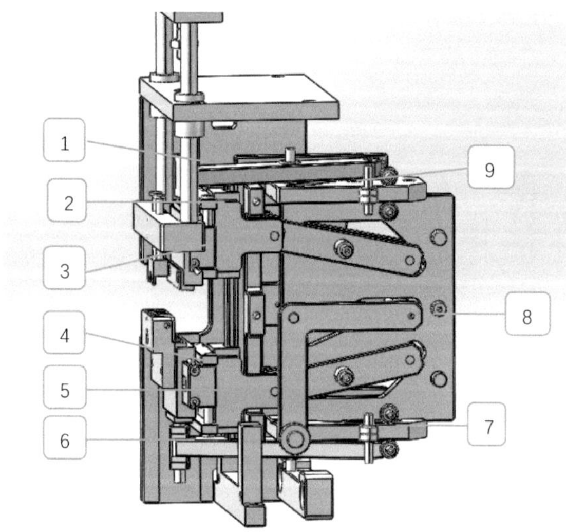

Fig. 4.28 Twisting Mechanism
1 - First Drive Rod 2 - First Fixed Plate 3 - First Horizontal Slider 4 - Second Horizontal Slider 5 - Second Fixed Plate 6 - Second Drive Rod 7 - Fourth Guide Groove 8 - Connecting Block 9 - Third Guide Groove

IV. Specific Parameters (see Table 4.6)

Table 4.6 Single-Power Stripping and Twisting Machine Parameters

Utilization Rate:	≥93%
Yield Rate:	≥98%
Applicable Cables:	Two-core, three-core cables (up to 1.5 mm^2 wires)
Stripping Length:	0.1-30 mm
Air Pressure:	0.5-0.8 MPa
Machine Dimensions:	740 mm(H) × 180 mm(L) × 470 mm(W)

V. Mechanical Design Highlights

Traditional equipment typically uses two separate machines for stripping and twisting. Although some devices have dual functionalities, combining stripping and twisting, their structures are often overly complex, requiring multiple cylinders to coordinate the processes, thereby increasing production costs. The Single-Power Stripping and Twisting Machine mounts the twisting mechanism on the driving mechanism, and the gantry frame is located at the front end of the installation frame. Both the upper and lower twisting mechanisms are installed on the gantry frame and cooperate with the driving mechanism. This achieves single-power stripping and twisting, resulting in a simple structure and reduced production costs.

Case 53: Integrated Capacitor Lead Cutting and Bending Machine

I. Case Description

In most electronic products, capacitors are soldered onto the mainboard. Capacitors typically consist of a head and two parallel pins. When soldering capacitors, it's often necessary to bend the pins in specific ways based on factors like the board's space. The Integrated Capacitor Lead Cutting and

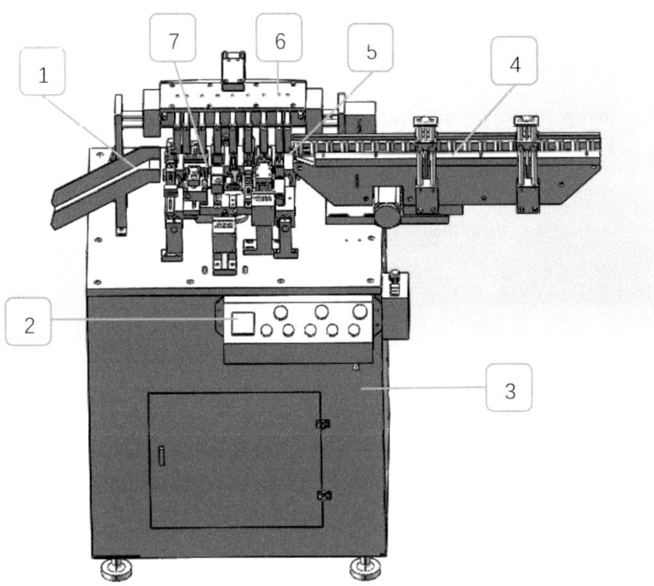

Fig. 4.29 Integrated Capacitor Lead Cutting and Bending Machine
1-Discharge Port 2-Control Panel 3-Frame 4-Feeding Mechanism 5-Cutting Mechanism 6-Handling Mechanism 7-Bending Mechanism

Bending Machine is designed for precisely bending capacitor pins for subsequent soldering. The overall structure is shown in Fig. 4.29, consisting mainly of the discharge port (1), control panel (2), frame (3), feeding mechanism (4), cutting mechanism (5), handling mechanism (6), and bending mechanism (7). The bending mechanism (7) includes a left mold, a mold movable left and right, and a cylinder. The cutting mechanism (5) consists of cutting tools and a cylinder, used for cutting the two pins of the capacitor. This machine excels in high precision pin-cutting, speed, stable operation, durability, and low noise.

II. Working Principle

The handling mechanism's fixture can move horizontally and vertically. During operation, the fixture grips the capacitor from the feeding mechanism and transfers it to the cutting mechanism. After the cutting mechanism completes the pin-cutting action, the fixture grips the capacitor from the cutting mechanism and transfers it to the bending mechanism. The bending mechanism bends the two pins according to the shaping molds, completing the operation of cutting and bending the capacitor pins.

III. Main Structure Introduction

Feeding Mechanism (Fig. 4.30): Used to transport capacitors, adjusted for pin direction, in batches to the cutting and bending machine. The feeding mechanism is driven by an electric motor (2), which rotates the pulley (3) to transport capacitors. The press plate (1) and guardrail (4) secure the capacitors, ensuring that the capacitor pins' direction is always downward.

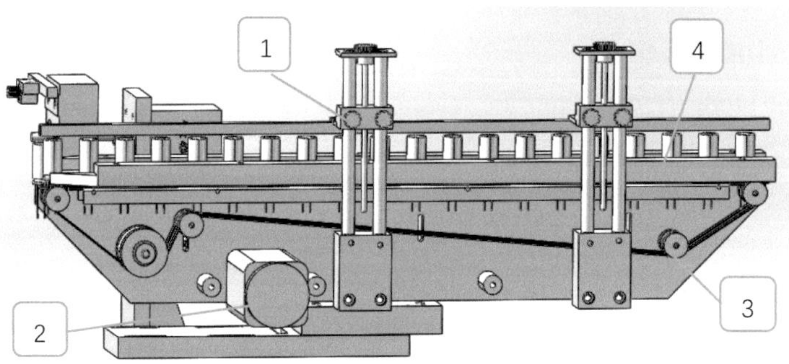

Fig. 4.30 Feeding Mechanism
1 - Press Plate 2 - Electric Motor 3 - Pulley 4 - Guardrail

Handling Mechanism (Fig. 4.31): Comprises cylinders (1, 4), fixture (2), guide rail (3), and is responsible for moving the capacitor to the cutting and bending stations. Cylinder (1) controls the vertical movement of the fixture, and cylinder (4) controls the horizontal movement along the guide rail.

Cutting Mechanism (Fig. 4.32): During operation, the handling mechanism transports the capacitor to the front of the cutting mechanism. Cylinder (3) of the cutting mechanism activates, pushing the cutting tool (2) open in preparation for cutting the capacitor pins. Simultaneously, the handling mechanism's cylinder activates, guiding the capacitor pins into the cutting mechanism. When cylinder (3) retracts, the cutting tool (2) trims the excess portion of the pins.

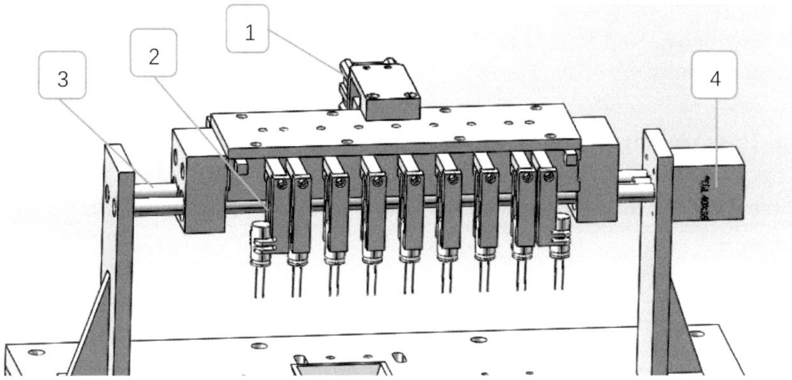

Fig. 4.31 Handling Mechanism
1, 4 - Cylinders 2 - Fixture 3 - Guide Rail

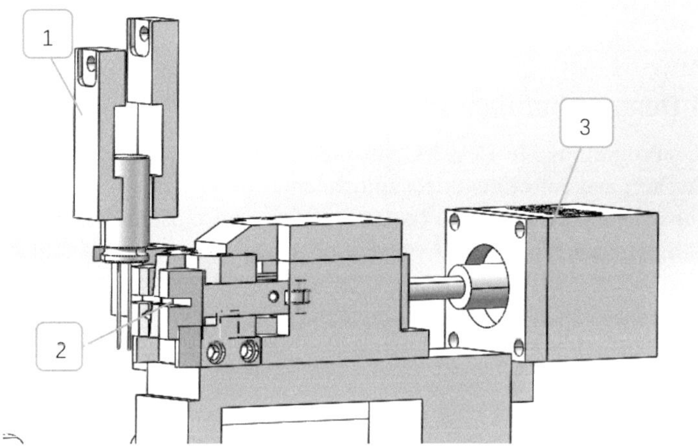

Fig. 4.32: Cutting Mechanism
1 - Fixture 2 - Cutting Tool 3 - Cylinder

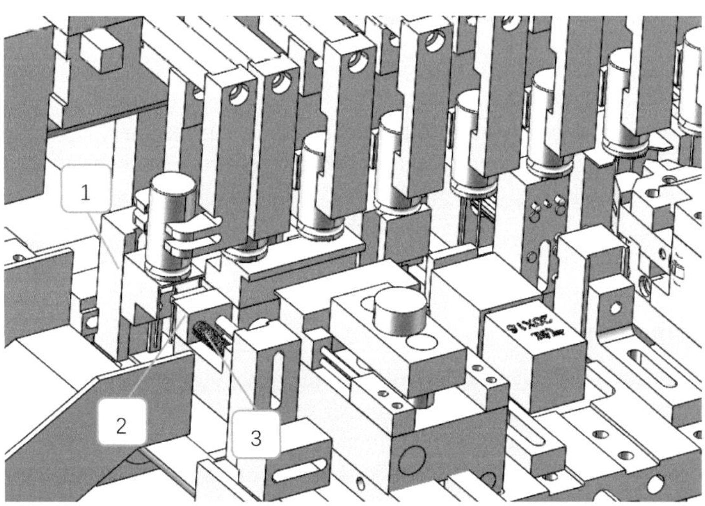

Fig. 4.33 Bending Mechanism
1 - Left Mold 2 - Movable Mold 3 - Spring

Bending Mechanism (Fig. 4.33): Composed of a left mold (1) and a mold movable left and right (2), driven by a cylinder. A spring (3) is incorporated into the movable mold to ensure stable and efficient bending operations.

IV. Specific Parameters (see Table 4.7)

Table 4.7 Integrated Capacitor Lead Cutting and Bending Machine Parameters

Working Air Pressure:	4 kg/cm^2
Production Speed:	80-110 pcs/min
Lead Spacing:	2.0 mm 2.5 mm 5 mm 7.50 mm 10 mm 15 mm 20 mm 25 mm
Cutting/Bending Modes:	Long/Short leads, 90-degree bending, inward bending, outward bending, right-angle, etc.
Cutting Length:	2.5-28 mm
Machine Dimensions:	1080 mm(H) × 1240 mm(L) × 670 mm(W)

V. Mechanical Design Highlights

(1) Utilizes PLC (Programmable Logic Controller) control for strong stability, easy-to-learn operation interface, and convenient data adjustment.
(2) Rack-and-pinion mechanism for the fixture opening and closing system ensures consistent centerlines during fixture replacement, preventing product damage and facilitating specification changes.
(3) The machine can automatically identify component polarity and adjust the direction accordingly.
(4) The cutting mechanism is integral, allowing easy adjustment and significant time savings in machine setup.
(5) The machine has an automatic counting function for packaging according to the specified quantity per bag.
(6) The machine incorporates an anti-static function.

Case 54: Air Drum Dryer

I. Case Description

The air drum dryer is a core component of the air drum drying machine. It is used to effectively dry products by installing impellers on a rotating shaft. The overall structure is shown in Fig. 4.34, consisting mainly of a cylinder (1), slide rail (2), base plate (3), electric motor (4), rotating shaft (5), pulley (6), and positioning pillar (7). The air drum dryer has the advantages of a simple structure, reliable operation, and low cost.

II. Working Principle

This device is typically installed in an air drum drying machine. When the drying machine is in operation, the air drum dryer is fixed, the cylinder extends the rotating shaft into the air drum channel, and the electric motor rotates. The rotation of the shaft and impellers is driven by the pulley. The positioning pillar is used to fix the position of the rotating shaft, preventing the impellers and rotating shaft from falling off and effectively reducing the noise generated by the equipment.

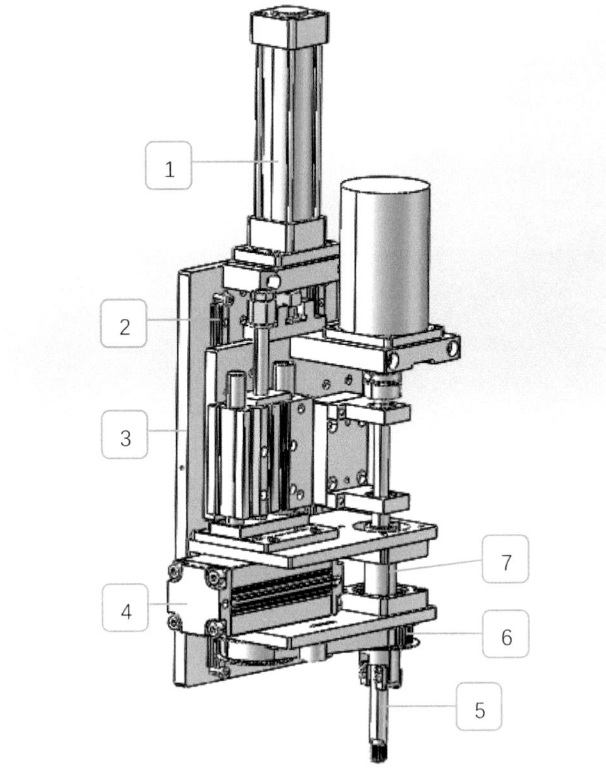

Fig. 4.34 Air Drum Dryer
1 - Cylinder 2 - Slide Rail 3 - Base Plate 4 - Electric Motor 5 - Rotating Shaft 6 - Pulley 7 - Positioning Pillar

III. Main Structure Introduction

Drive Mechanism and Positioning Pillar (Fig. 4.35): When the air drum drying machine is in operation, the electric motor (1) provides power, and the rotating shaft (2) rotates through the pulley (3).

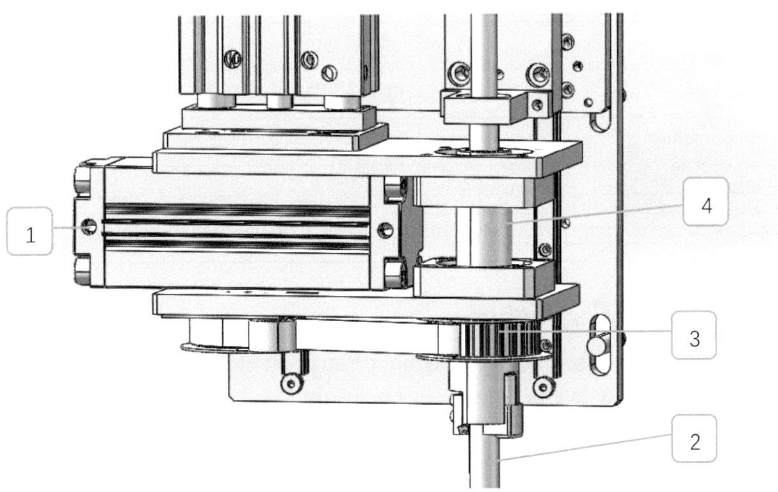

Fig. 4.35 Drive Mechanism and Positioning Pillar
1 - Electric Motor 2 - Rotating Shaft 3 - Pulley 4 - Positioning Pillar

Slide Rail Mechanism (Fig. 4.36): Driven by a cylinder (1), it controls the lifting of the air drum drying machine.

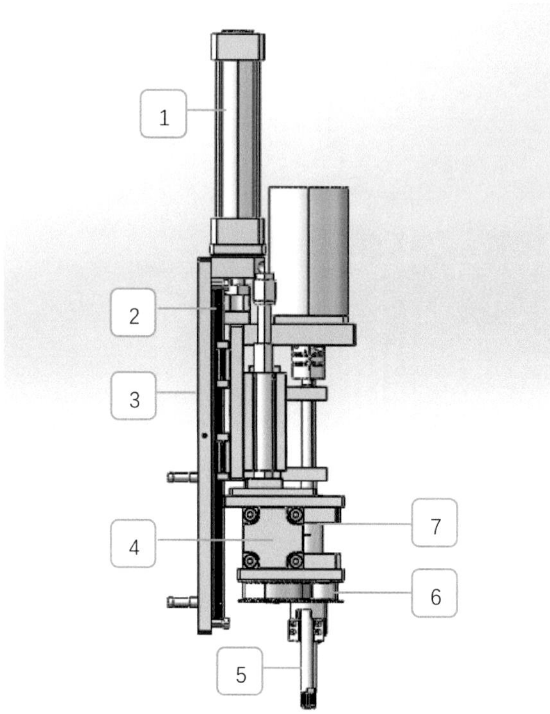

Fig. 4.36 Slide Rail Mechanism
1 - Cylinder 2 - Slide Rail 3 - Base Plate 4 - Electric Motor 5 - Rotating Shaft 6 - Pulley 7 - Positioning Pillar

IV. Specific Parameters (see Table 4.8)

Table 4.8 Air Drum Dryer Parameters

Main Shaft Speed:	24000 r/min
Rated Air Volume:	100 m^3/min
Adjustable Flow Range:	45%~100%
Environmental Temperature:	0~45°C
Permissible Humidity:	10%~95%
Machine Dimensions:	820 mm(H) × 330 mm(L) × 230 mm(W)

V. Mechanical Design Highlights

The air drum dryer rapidly rotates the drying drum through an electric motor, effectively removing moisture from the objects. A heating grid can be set up for additional heating. The air blower introduces heat from the heating grid into the drying drum, accelerating the drying process and improving efficiency.

Case 55: Switch Terminal Bending Machine

I. Case Description

Switch terminals, as shown in Fig. 4.37, are products designed for use in various miniature circuit breakers, residual current circuit breakers, air switches, etc. The switch terminal bending machine is a CNC bending machine specifically designed for bending the pins of switch terminals. The overall structure is shown in Fig. 4.38, and a top view of the terminal bending machine is shown in Fig. 4.39. It consists mainly of the frame (1), handling robot (2), clamp feeding mechanism (3), pushing mechanism (4), terminal feeding mechanism (5), display screen (6), and bending mechanism (7). The switch terminal bending machine features high precision, quick die change, high efficiency, and user-friendly operation.

Fig. 4.37 Switch Terminals

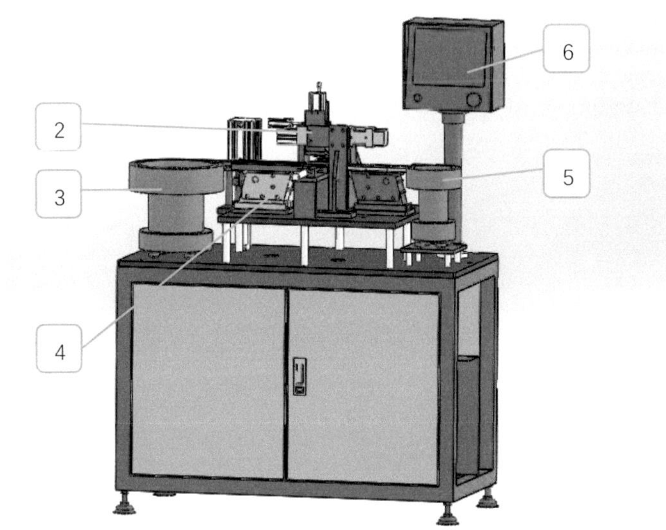

Fig. 4.38 Slider Switch Terminal Bending Machine
2-Handling Robot 3-Clamp Feeding Mechanism 4-Pushing Mechanism 5-Terminal Feeding Mechanism 6-Display Screen 7-Bending Mechanism

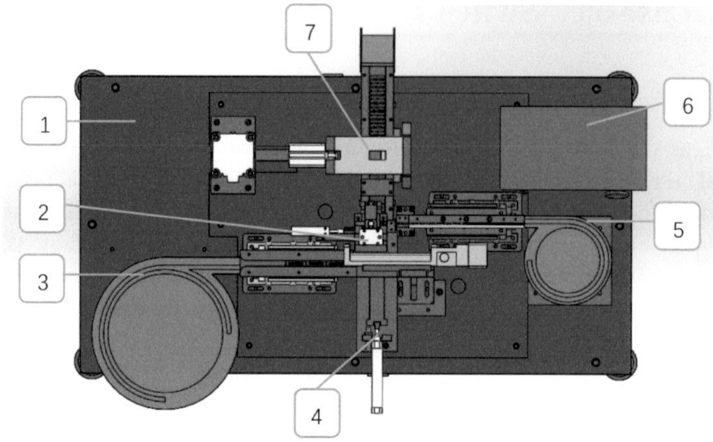

Fig. 4.39 Top View of the Terminal Bending Machine
1-Frame 2-Handling Robot 3-Clamp Feeding Mechanism 4-Pushing Mechanism 5-Terminal Feeding Mechanism 6-Display Screen 7-Bending Mechanism

II. Working Principle

The switch terminal bending machine works by feeding the clamp from the left-side clamp feeding mechanism into the bending mechanism through the pushing mechanism, which is perpendicular to the feeding track. Simultaneously, the terminal feeding mechanism on the right side inputs the pins to be bent. The handling robot picks up the pins and places them in the specified position on the clamp. The pushing mechanism synchronously pushes the clamp containing the pins to the bending mechanism. The bending mechanism, driven by a cylinder, bends the pins in batches using coordinated action with the bending roller [81].

III. Main Structure Introduction

Clamp Feeding Mechanism and Pushing Mechanism (Fig. 4.40): The clamp is neatly arranged by the vibration conveyor (1) and sent into the pushing track (3). The pushing mechanism is driven by a cylinder (2) and synchronously pushes the clamp into the bending mechanism.

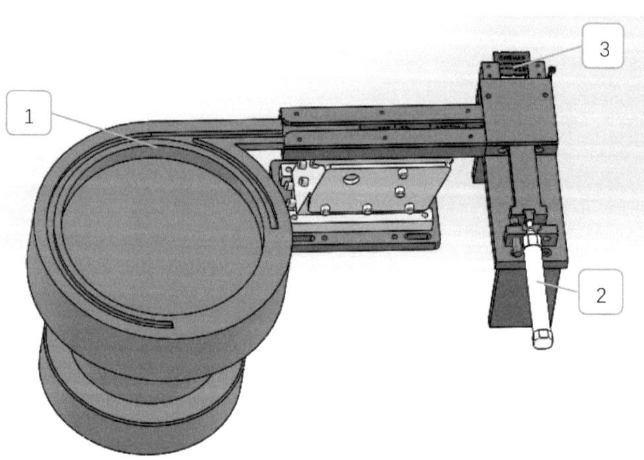

Fig. 4.40 Clamp Feeding Mechanism and Pushing Mechanism
1-Vibration Conveyor 2-Cylinder 3-Pushing Track

Pin Feeding Mechanism and Handling Robot (Fig. 4.41): The pins are neatly arranged by the vibration conveyor (2) and sent into the track (3). The handling robot, controlled by cylinders (1, 5, 6), horizontally displaces, vertically displaces, and rotates to orderly place the pins in the clamp above the pushing track [82].

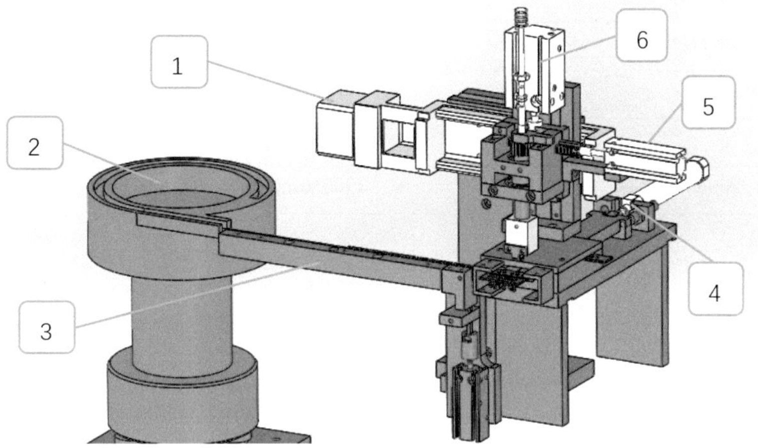

Fig. 4.41 Terminal Feeding Mechanism and Handling Robot
1, 5, 6 - Cylinders 2 - Vibration Conveyor 3 - Track 4 - Pushing Mechanism

Bending Mechanism (Fig. 4.42): Controlled by a cylinder (4), the bending roller (1) is driven by the lever support (3) to press against the lower die (2), bending the pins.

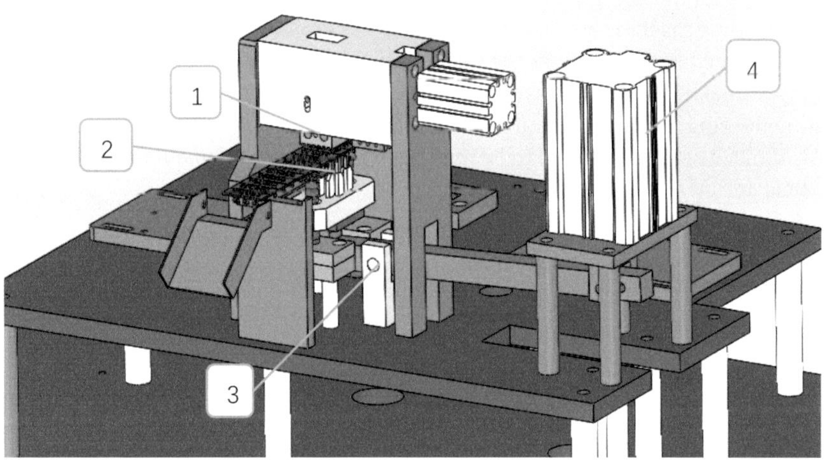

Fig. 4.42 Bending Mechanism
1 - Bending Roller 2 - Lower Die 3 - Lever Support 4 - Cylinder

IV. Specific Parameters (see Table 4.9)

Table 4.9 Switch Terminal Bending Machine Parameters

Nominal Pressure:	500 kN
Working Table Size:	550 mm(L) × 445 mm(W)
Maximum Opening Height:	5 mm
Working Efficiency:	1800-2500 pcs/h
Yield Rate:	≥98%
Machine Dimensions:	1300 mm(H) × 950 mm(L) × 500 mm(W)

V. Mechanical Design Highlights

In the process of terminal production, many bending mechanisms directly bend the pins using a pressure block. However, this direct method can lead to sudden force on the pins, causing them to break during bending. The switch terminal bending machine fixes the bending roller above the pushing track. The lever support drives the pins upward, not only increasing the bending force but also gradually increasing the force on the pins. This design ensures that the pins won't break due to sudden force.

VI. Precautions

(1) Strictly follow machine safety operating regulations and wear proper labor protection equipment.
(2) Before starting the bending machine, carefully check if the electric motor, switches, lines, and grounding are normal and secure.
(3) Check the coincidence and sturdiness of the upper and lower dies and whether the positioning devices meet the processing requirements.
(4) After the bending machine starts, run it empty for 1-2 minutes. If there are abnormal sounds or faults, stop immediately, eliminate the faults, and only resume work when everything is normal.
(5) During operation, if abnormal noises or oil leakage are detected, stop the machine immediately. Operating the equipment with issues is strictly prohibited.
(6) If the workpiece or mold is found to be misaligned, stop the machine for correction. Correcting with hands during rotation is strictly prohibited to avoid hand injuries.
(7) Check if there are any foreign objects on the material support, material stop, and slide block. If any foreign objects are found, clean them.
(8) The bending force of the workpiece should not exceed 500 KN, as calculated by the bending force formula.

Case 56: Special CNC Milling Machine

I. Case Description

A special CNC milling machine is capable of precise milling under the control of program code. The overall structure is shown in Fig. 4.43, consisting mainly of a positioning mechanism (1), working table (2), protective cover (3), bed (4), and milling device (5). The special CNC milling machine has advantages such as small size, flexible installation, and high work efficiency.

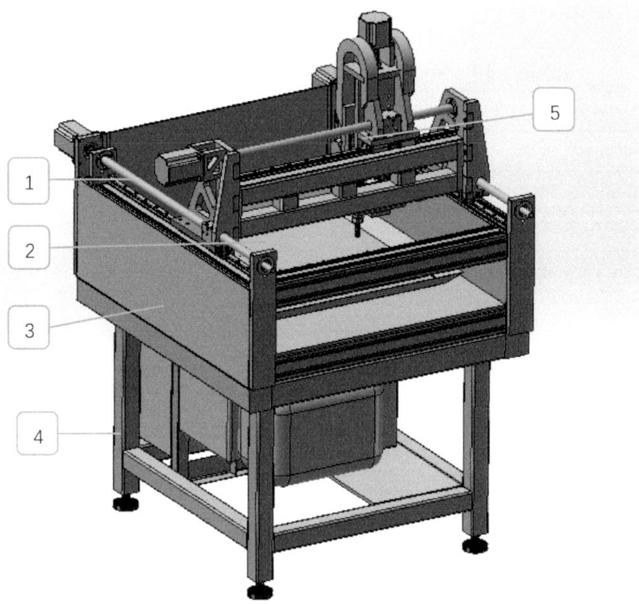

Fig. 4.43 Special CNC Milling Machine
1 - Positioning Mechanism 2 - Working Table 3 - Protective Cover 4 - Bed 5 - Milling Device

II. Working Principle

The special CNC milling machine has multiple mounting tables. When machining a part, the part is mounted on one of the tables. The table is then fixed to the working table for processing. Meanwhile, the operator can prepare another part on a different table. After completing the machining of one part, the table with the finished part can be fixed to the working table for processing. This allows for simultaneous part installation, thereby improving the efficiency of part processing.

III. Main Structure Introduction

Positioning Mechanism and Milling Device (Fig. 4.44): The positioning mechanism is driven by servo motors (4, 6, 7) to control the vertical displacement in the horizontal plane of the milling device. A cylinder (1) controls the vertical movement of the milling device, and an electric motor (2) drives the milling head to rotate for machining [83].

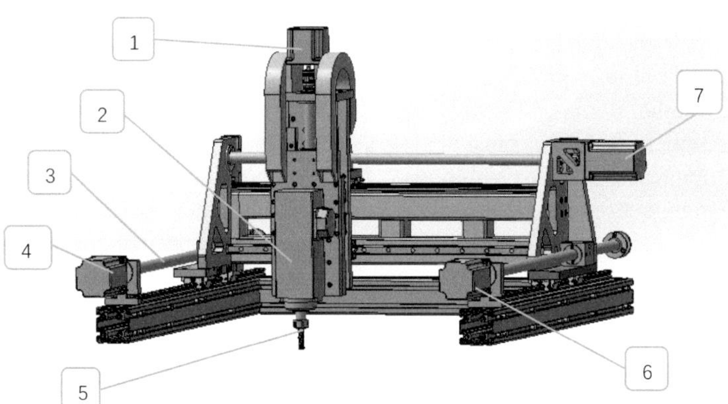

Fig. 4.44 Milling Device
1 - Cylinder 2 - Electric Motor 3 - Lead Screw 4, 6, 7 - Servo Motors 5 - Milling Head

IV. Specific Parameters (see Table 4.10)

Table 4.10 Special CNC Milling Machine Parameters

Working Table:	600 mm(L) × 600 mm(W)
Milling Head Travel Range:	600 mm(L) × 750 mm(W)
Distance from the Spindle's Bottom to the Table Surface:	20-100 mm
Spindle Speed Range/Levels:	6000 rpm
Maximum Diameter of Vertical Milling:	25 mm
Maximum Hole Diameter:	30 mm
Maximum Rapid Traverse Speed:	6000 mm/min
Positioning Accuracy:	0.02 mm
Repeat Positioning Accuracy:	0.01 mm
Minimum Setting Unit:	0.01 mm
Main Motor Rated Power:	3.7 kW
Machine Dimensions:	1585 mm(H) × 960 mm(L) × 1135 mm(W)

V. Mechanical Design Highlights

The machine has a worktable on the bed. A connecting frame is fixed to the front end face of the worktable, housing an electric motor with its output shaft oriented vertically upward. Two connecting rods, symmetrically set on both sides of the output shaft, are connected to the output shaft. The end of the connecting rods away from the output shaft can be fixed with a table. The upper end surface of the table can hold the part to be machined. The table can be rotated to the working table position and aligned with the upper end face of the working table using the electric motor [84]. This setup enables the simultaneous installation of parts during part processing, thus improving efficiency.

VI. Precautions

(1) When machining parts, always close the protective door. It's prohibited to insert heads or hands into the protective door openings. Opening the protective door during the machining process is strictly forbidden.

(2) Avoid any contact with rotating spindles, workpieces, or other moving parts by hands or any other means. Do not touch the cutting edge and chips with hands; use hooks or brushes to clean chips.

(3) The CNC milling machine is a large precision device. Apart from work tools and workpieces on the worktable, it's strictly prohibited to stack any tools, clamps, cutting tools, measuring tools, workpieces, or other items on the machine.

(4) During operation, the operator must not leave the machine. Keep a high level of concentration, observe the running state of the machine tool. If any abnormality or accident occurs, immediately stop the program, cut off the power, and refrain from any other operations [85].

(5) It is strictly forbidden to change speed during the operation. Speed change or tool change must be done with the machine completely stopped and the switch in the "OFF" position to prevent accidents.

(6) When temporarily pausing to measure the workpiece size during program operation, wait until the machine tool is completely stopped and the spindle has stopped rotating before conducting measurements to avoid accidents.

Case 57: DIY CNC Engraving Machine

I. Case Description

The DIY CNC engraving machine is capable of relief carving, flat carving, and hollow carving on materials such as aluminum alloy, copper, MDF, wood, jade, glass, plastic, and acrylic. The overall structure is shown in Fig. 4.45, comprising mainly the direction-adjustable slider (1), engraving head (2), worktable (3), guide rail (4), base (5), load-bearing frame (6), and positioning mechanism (7). The engraving head (2) consists of a locking cover head, machine cavity, automatic tool changer, drill bit, and rotating center shaft. The DIY CNC engraving machine is known for its simplicity, fast processing speed, high precision, and stability [86].

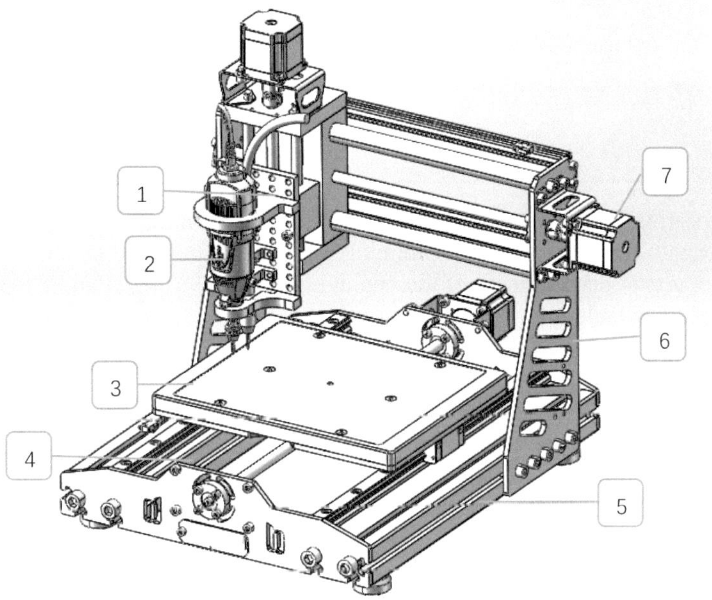

Fig. 4.45 DIY CNC Engraving Machine
1 - Direction-Adjustable Slider 2 - Engraving Head 3 - Worktable 4 - Guide Rail 5 - Base 6 - Load-Bearing Frame 7 - Positioning Mechanism

II. Working Principle

During operation, the engraving machine user opens the locking cover head, installs the required tool in the automatic tool changer, starts the motor, and activates the drill bit's rotation around the rotating center shaft. Simultaneously, the positioning mechanism controls the axial movement of the engraving head, carving the material under the action of the squeezing and centrifugal forces of the working blades [87].

III. Main Structure Introduction

The worktable and base are illustrated in Fig. 4.46. The worktable is placed on the base and guided by rails (1), driven and controlled by cylinder (3).

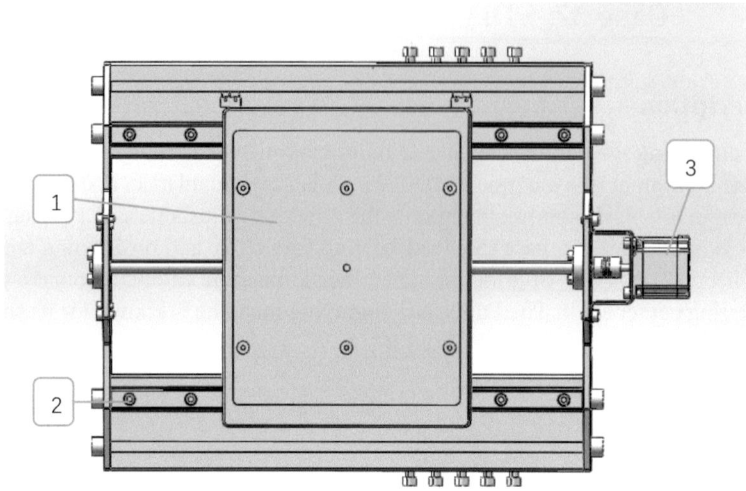

Fig. 4.46 Worktable and Base
1 - Worktable Surface 2 - Cylinder 3 - Guide Rail

The load-bearing frame and positioning mechanism are depicted in Fig. 4.47. The load-bearing frame (3) is fixed to the base, and the positioning mechanism, mounted on the load-bearing frame, controls the horizontal displacement of the engraving head (2) using cylinder (4) and the vertical displacement using cylinder (1).

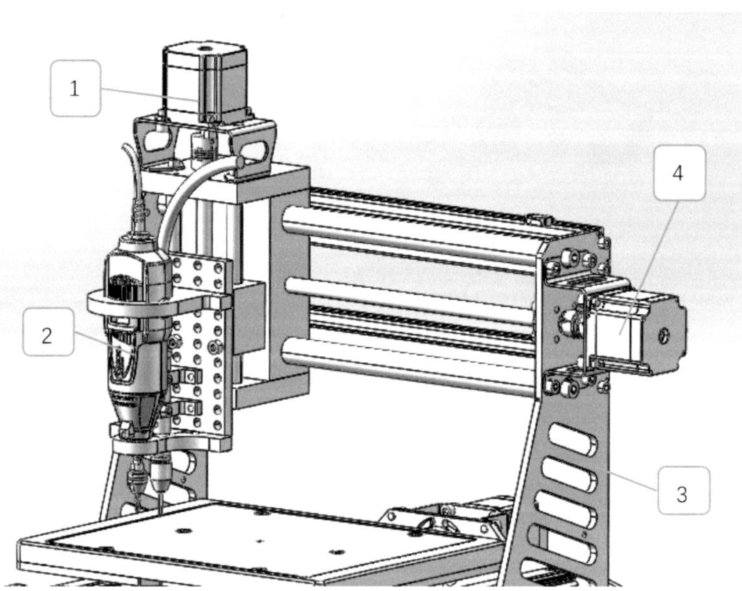

Fig. 4.47 Load-Bearing Frame and Positioning Mechanism
1, 4 - Cylinder 2 - Engraving Head 3 - Load-Bearing Frame

IV. Specific Parameters (see Table 4.11)

Table 4.11 DIY CNC Engraving Machine Parameters

Jump Frequency:	300 Hz
Z-Axis Max Speed:	Depending on the material hardness, typically between 1000-2500 mm/min
Single-Axis Acceleration:	600 mm/min
Curve Acceleration:	800 mm/min
Processing Range:	90 mm(H) × 250 mm(L) × 210 mm(W)
Machine Dimensions:	470 mm(H) × 575 mm(L) × 500 mm(W)

V. Mechanical Design Highlights and Precautions

(1) The Y-axis transmission system of the CNC engraving machine is fixed to the base, ensuring smoother carving. However, attention should be paid to input parameters, especially for the Y-axis direction.

(2) Ensure the cooling water is clean and the water pump is working properly for continuous operation periods below 10 hours a day. Do not allow the spindle motor to run dry. Change the cooling water regularly to prevent overheating. In winter, if the working environment temperature is too low, consider using antifreeze in the water tank.

(3) After each use, clean the platform and dust from the transmission system thoroughly. Lubricate and oil the transmission system (X, Y, Z axes) regularly (weekly). Note: Use machine oil for the X, Y, Z axis guide rods, and high-speed grease for the lead screws. In winter, if the working environment temperature is too low, clean and wash the PRTT ball screw and guide rails (square or round) with gasoline before applying machine oil to prevent excessive resistance in the machine's transmission parts.

(4) When maintaining electrical components, ensure power is cut off. Only proceed when the monitor is blank, and the main circuit power indicator light is off.

(5) Check the fasteners every three months. Tighten screws on both sides of the gantry, screws securing the lead screw nuts, and screws fastening the motors.

Case 58: Bearing Transfer Equipment

I. Case Description

The bearing transfer equipment is designed for the processing of bearings, as shown in Fig. 4.48. The top view is illustrated in Fig. 4.49. It consists mainly of the frame (1), correction mechanism (2), clamping support frame (3), product gripping mechanism (4), workstation turntable (5), and electric drill (6). The bearing transfer equipment is known for its simplicity of operation and high efficiency.

II. Working Principle

During operation, the product gripping mechanism grabs the product, the workstation turntable rotates automatically, transporting the product to the processing position. The clamping support frame, driven by a cylinder, presses down to stabilize the turntable. The electric drill is then activated to process the product. After completion, the clamping support frame rotates back, the workstation turntable continues to rotate, and the processed product is output.

Fig. 4.48 Bearing Transfer Equipment
1-Frame 2-Correction Mechanism 3-Clamping Support Frame 4-Product Gripping Mechanism 5-Workstation Turntable 6-Electric Drill

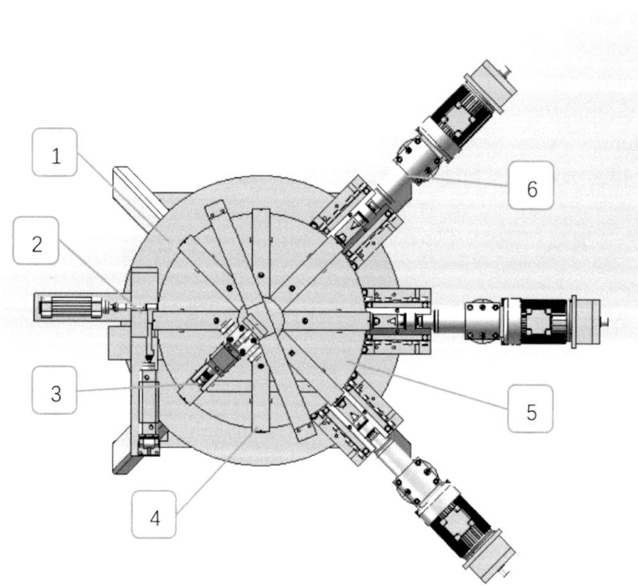

Fig. 4.49 Top View of Bearing Transfer Equipment
1-Frame 2-Correction Mechanism 3-Clamping Support Frame 4-Product Gripping Mechanism 5-Workstation Turntable 6-Electric Drill

III. Main Structure Introduction

The product gripping mechanism, shown in Fig. 4.50, features a triangular groove design to prevent bearing slippage. The clamping support frame (1) provides pressure to grip the bearing. Inside the product gripping mechanism (4), there is a spring (3) that automatically rebounds when the clamping support frame (1) rotates, releasing the bearing.

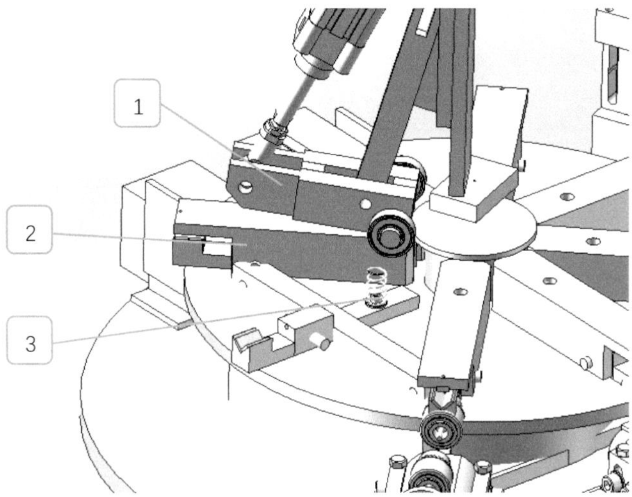

Fig. 4.50 Product Gripping Mechanism
1 - Clamping Support Frame 2 - Product Gripping Mechanism 3 - Spring

The correction mechanism, as seen in Fig. 4.51, utilizes cylinders (2) and (3) to rotate and fix the bearing gripped by the product gripping mechanism (1) when transported to the correction mechanism by the workstation turntable (4).

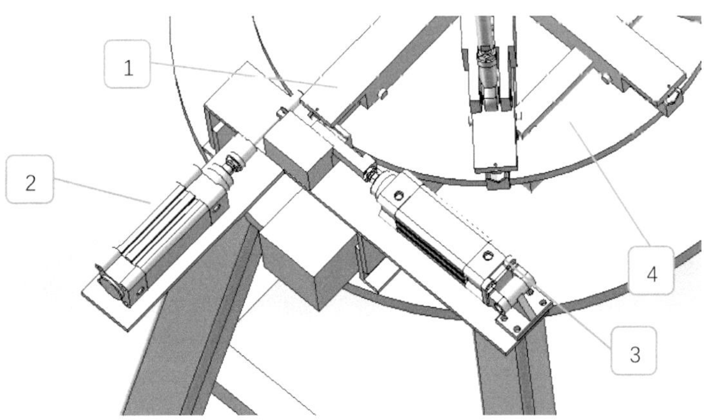

Fig. 4.51 Correction Mechanism
1 - Product Gripping Mechanism 2, 3 - Cylinders 4 - Workstation Turntable

The clamping support frame, depicted in Fig. 4.52, is controlled by cylinder (1) to apply pressure on the product gripping mechanism (2). When the cylinder (1) extends, the product gripping mechanism (2) releases the bearing, and when the cylinder (1) retracts, the product gripping mechanism (2) clamps the bearing.

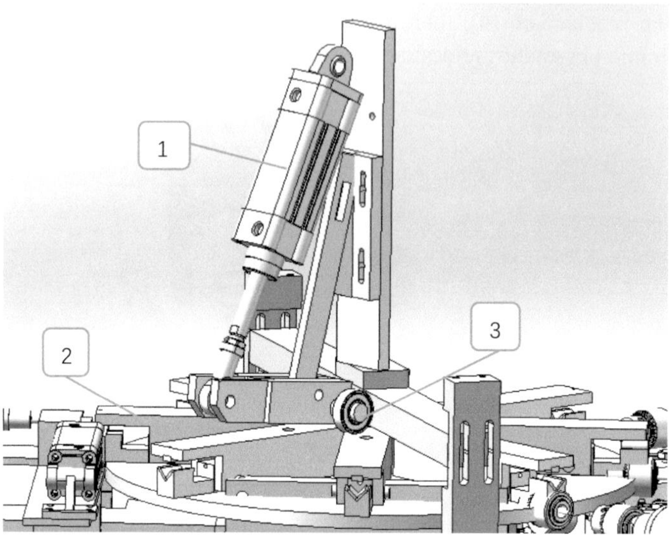

Fig. 4.52 Clamping Support Frame
1 - Cylinder 2 - Product Gripping Mechanism 3 - Roller

IV. Specific Parameters (see Table 4.12)

Table 4.12 Bearing Transfer Equipment Parameters

Working Efficiency:	600-800 pcs/h
Utilization Rate:	≥93%
Qualified Rate:	≥96%
Environmental Temperature:	0~55°C
Permitted Humidity:	25%~95%
Equipment Dimensions:	1540 mm(H) × 1180 mm(L) × 1350 mm(W)

V. Mechanical Design Highlights and Precautions

(1) The bearing transfer equipment achieves synchronized bearing positioning and processing tasks by combining the clamping support frame with the workstation turntable, significantly improving operational efficiency.
(2) The product gripping mechanism has an internal spring that gradually rebounds as the clamping support frame loosens, preventing sudden opening and bearing drop.
(3) Avoid placing tools, clothing, or other miscellaneous items on the machine to prevent accidents.
(4) Strictly prohibit reaching hands into the machine while it is running to avoid danger.
(5) When servicing the machine internally, always turn off the power first.

Case 59: Mini Adjustable Circular Saw

I. Case Description

The mini adjustable circular saw is suitable for cutting materials such as stone and wood. The overall structure is shown in Fig. 4.53, with a detailed enlarged view in Fig. 4.54. It consists mainly of a workbench (1), correction mechanism (2), control panel (3), armature (4), housing (5), circular saw (6), guard plate (7), and rotating bracket (8). The mini adjustable circular saw reduces the machine's size and weight, making it easy to carry and operate.

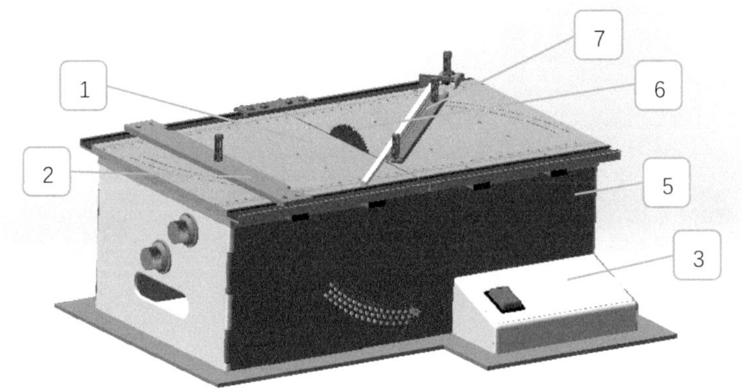

Fig. 4.53 Mini Adjustable Circular Saw
1 - Workbench 2 - Correction Mechanism 3 - Control Panel 5 - Housing 6 - Circular Saw 7 - Guard Plate

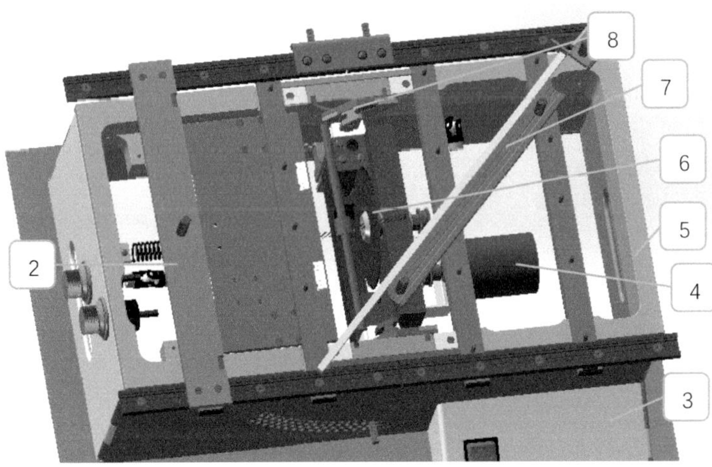

Fig. 4.54 Enlarged View of Mini Adjustable Circular Saw
2 - Correction Mechanism 3 - Control Panel 4 - Armature 5 - Housing 6 - Circular Saw 7 - Guard Plate 8 - Rotating Bracket

II. Working Principle

When the mini adjustable circular saw starts, the armature begins to rotate, driving the pulley to rotate. Through pulley transmission, the output shaft also starts rotating, achieving the function of output. The cutting angle can be adjusted by the rotating bracket (8), allowing the saw to produce different angled cuts. Different circular saws can be fixed on the output shaft using a pressure plate and hexagonal socket head screws to cut wood or stone.

III. Main Structure Introduction

The adjustable bracket mechanism, shown in Fig. 4.55, utilizes a connection head (3) connected to the rotating shaft (4) via a spring. When adjusting the cutting angle of the circular saw, the tension of the spring is controlled by rotating the knob (2) [88].

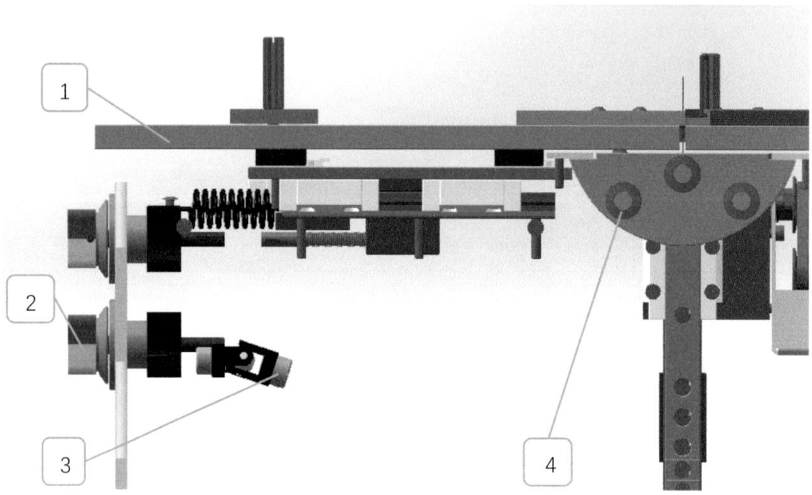

Fig. 4.55 Adjustable Bracket Mechanism
1-Workbench 2-Knob 3-Connection Head 4-Rotating Shaft

The rotating bracket and circular saw are illustrated in Fig. 4.56. The adjustable circular saw (1) is mounted on the rotating bracket (2), and it is driven by the armature (3) through the pulley (4). By adjusting the angle of the rotating bracket, different cutting angles can be achieved.

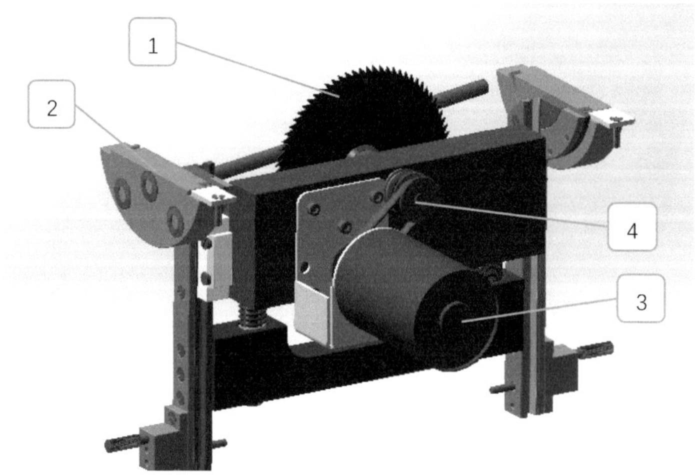

Fig. 4.56 Rotating Bracket and Circular Saw
1-Circular Saw 2-Rotating Bracket 3-Armature 4-Pulley

IV. Specific Parameters (see Table 4.13)

Table 4.13 Mini Adjustable Circular Saw Parameters

No-load Speed:	3500 rpm
Rated Power:	1 kVA
Maximum Blade Diameter:	115 mm
Environmental Temperature:	0~65°C
Permitted Humidity:	10%~95%
Equipment Dimensions:	150 mm(H) × 440 mm(L) × 330 mm(W)

V. Mechanical Design Highlights and Precautions

(1) The adjustable circular saw can manually adjust the cutting angle by rotating the knob. The spring linkage between the connection head and the rotating shaft effectively prevents accidents such as breakage or fracture due to excessive stress.

(2) The adjustable circular saw features both vertical and inclined blocks on the horizontal plane to meet various cutting needs.

(3) The equipment should have operating instructions specifying the conditions and precautions for machine operators. The circuit breaker should not be casually closed to prevent accidents.

(4) The equipment itself should have a switch control (no toggle switch to prevent accidental collision startup), and the distance from the switchbox to the equipment should not exceed 3m for quick power cutoff in case of malfunctions.

(5) The saw blade must be flat and firm, with appropriate tooth spacing. The saw blade should not have continuous missing teeth or cracks.

(6) Safety guards must be complete. The thickness and position of the feeding knife should be suitable to prevent pinching during cutting. The saw blade guard should be fixed above the saw blade and not rotated arbitrarily during use.

(7) After the saw blade rotates, cutting should only commence after the speed is normal. The thickness of the cut material should not touch the edge of the fixed saw blade guard.

(8) When the cut material is approaching the end of the saw, use the lower hand to pull the material. Do not push directly with the upper hand to avoid pushing the saw blade into empty space.

(9) For long pieces of wood, two people should cooperate in the operation. When the material exceeds 20cm beyond the saw blade, the lower hand can accept the material. When returning the material, wait until the wood completely leaves the saw blade before feeding it back. The operation should not be too early or too fast to prevent the material from hitting the saw blade.

(10) When cutting off wood and short pieces, use a push stick. Do not feed directly with your hand, and the feeding speed should not be too fast. Use a hand hook for the lower hand when receiving material. Short materials less than 50cm are prohibited from being cut on the saw.

(11) When changing the saw blade and performing maintenance, the circuit should be cut off by pulling the circuit breaker. Work should only proceed after the rotation comes to a complete stop.

(12) The cut material should be neatly stacked, and sawdust on the table and within the working range should be cleared with a broom. Do not wipe the table directly with your hands.

Case 60: 3D Printer

I. Case Description

This 3D printer supports a maximum molding size of 300 mm(H) × 250 mm(L) × 280 mm(W) and is suitable for manufacturing models in mold making, industrial design, and direct production of certain products. The overall structure is depicted in Fig. 4.58, consisting of components such as sliding guide rails (1), material tray (2), print head (3), frame (4), and worktable (5). The 3D printer utilizes three-dimensional printing technology to produce small quantities of items more quickly, flexibly, and cost-effectively compared to traditional methods.

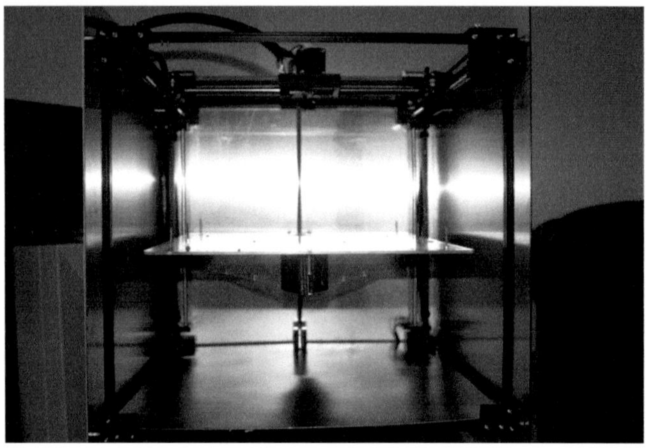

Fig. 4.57 3D Printer Product Model

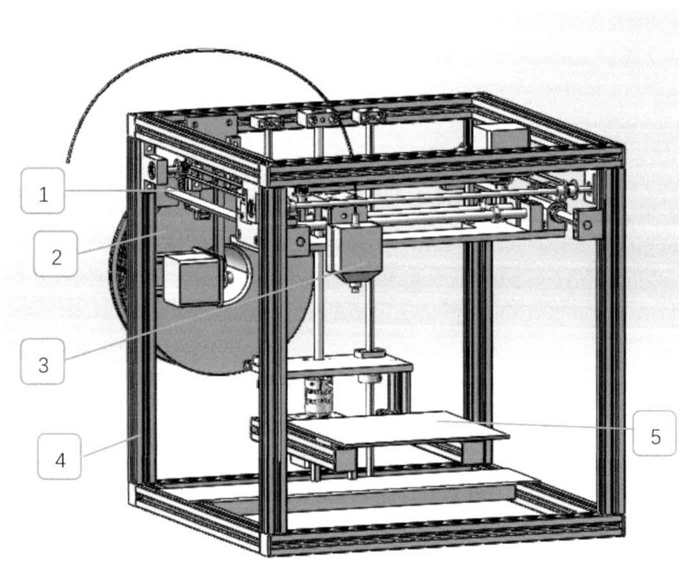

Fig. 4.58 3D Printer
1 - Sliding Guide Rails 2 - Material Tray 3 - Print Head 4 - Frame 5 - Worktable

II. Working Principle

When the 3D printer is in operation, the printing material on the material tray is guided by the guiding wheel to the print head for feeding. The print head pulls the material during printing, causing the material to move upward, reducing the pressure on the material tray. When the print head stops, the material tray stops rotating, minimizing the chance of material lines loosening. The printer employs the Fused Deposition Modeling (FDM) method, melting and depositing layers of thermoplastic material to create the final product [89].

III. Main Structure Introduction

The print head, shown in Fig. 4.59, features a heater (1) to heat the thermoplastic material and feed it through the nozzle (2).

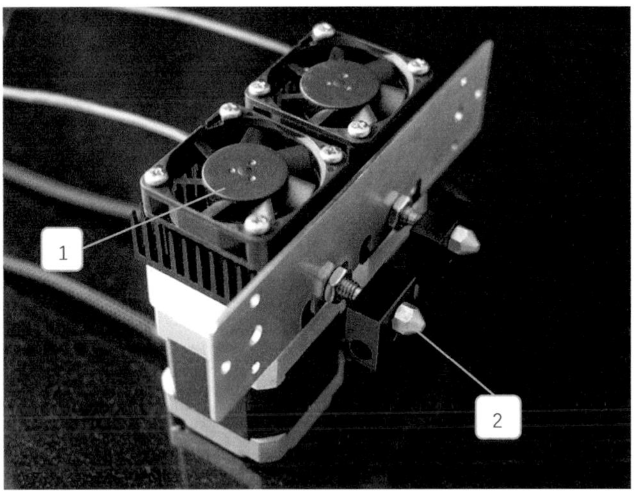

Fig. 4.59 Print Head
1 - Heater 2 - Nozzle

The sliding guide rails mechanism, illustrated in Fig. 4.60, is controlled by servo motors (1, 4) to move the print head (3) horizontally in the Y axis direction and in the X-axis direction.

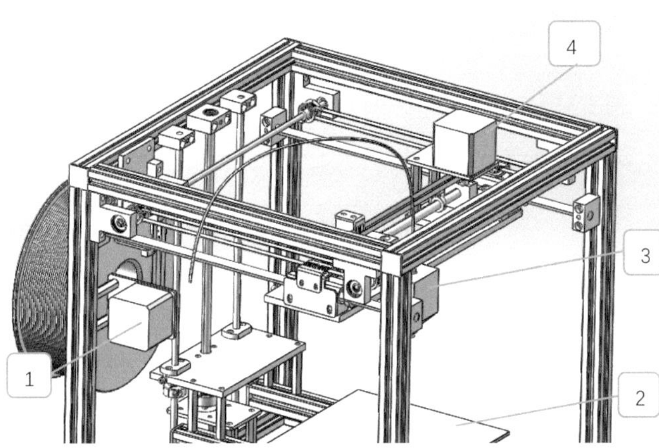

Fig. 4.60 Sliding Guide Rails Mechanism
1, 4 - Servo Motors 2 - Worktable 3 - Print Head

The worktable, as seen in Fig. 4.61, is controlled by a servo motor (2). During 3D printing, the worktable gradually descends, and the print head prints layer by layer.

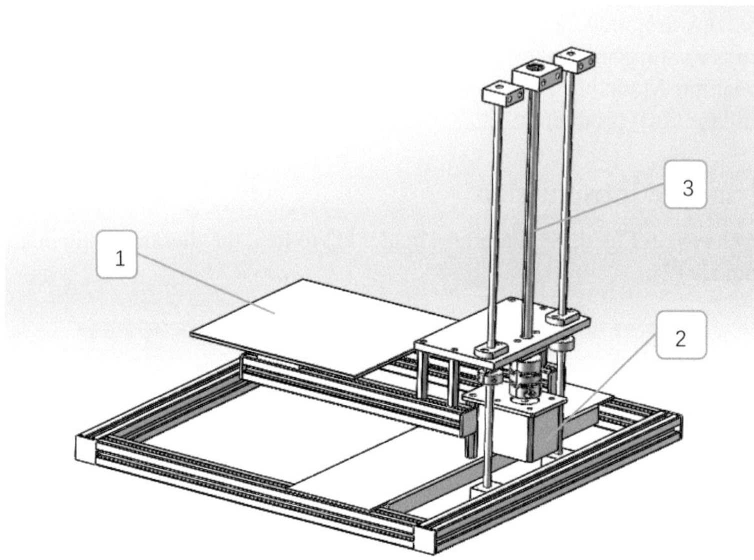

Fig. 4.61 Worktable
1 - Worktable Surface 2 - Servo Motor 3 - Z-Axis Slide Rail

IV. Specific Parameters (see Table 4.14)

Table 4.14 3D Printer Parameters

Maximum Molding Size:	300 mm(H) × 250 mm(L) × 280 mm(W)
Print Materials:	PLA, ABS, Laywood, Taulman Nylon
Filament Diameter:	1.75 mm
Nozzle Diameter:	0.35 mm
All Screws/Nuts:	M3 size
Frame:	8 mm Thick Stainless Steel
Stepper Motors:	High-torque standard NEMA 17
Electronics:	Modified RAMPS 1.4
Equipment Dimensions:	500 mm(H) × 520 mm(L) × 500 mm(W)

V. Mechanical Design Highlights

The 3D printer employs FDM technology, allowing for the rapid production of prototypes and functional parts. It uses servo motors for precise movement control, and the print head mechanism prevents material lines from loosening, enhancing efficiency [90].

VI. Precautions

(1) When changing filament, ensure the previous filament is completely removed to prevent nozzle clogs.
(2) Properly set the Z-axis distance between the print head and the print bed.
(3) Keep the worktable as level as possible during printer operation.
(4) Regularly clean and lubricate moving parts to maintain machine cleanliness.

Case 61: Tube Expanding Machine

I. Case Description

The purpose of the tube expanding machine is to process smaller-diameter pipes into larger-diameter pipes or shape near-cylindrical shells into cylindrical forms. For example, it can shape the casing of a compressor into a cylindrical form. The overall structure is depicted in Fig. 4.62, consisting of components such as the inlet (1), die mechanism (2), power mechanism (3), frame (4), and outlet (5). The tube expanding machine is known for its stable operation, high efficiency, and precision in production.

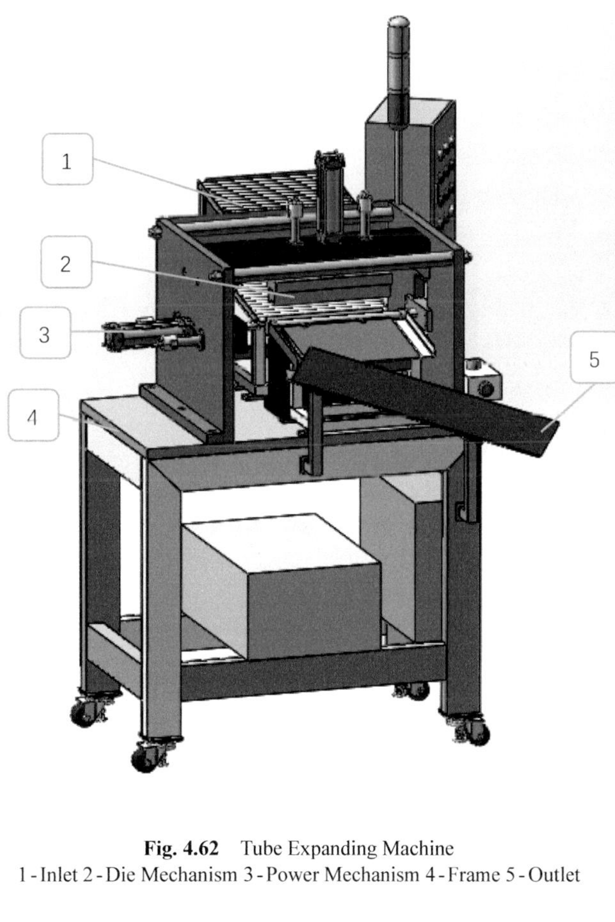

Fig. 4.62 Tube Expanding Machine
1 - Inlet 2 - Die Mechanism 3 - Power Mechanism 4 - Frame 5 - Outlet

II. Working Principle

When the tube expanding machine is in operation, the pipes to be processed are arranged and sequentially fed into the die mechanism through the inlet. When the pipe is placed on the lower die, the upper die's cylinder is driven downward, controlling the upper die to press down and fit tightly with the lower die. At this point, the hydraulic cylinder is activated, controlling the piston rod to squeeze the inner wall of the pipe, expanding it to the specified requirements [91].

III. Main Structure Introduction

The die mechanism, shown in Fig. 4.63, is driven by the cylinder (1) for the upper die (2) and the cylinder (4) for the lower die (3). When the pipe is placed on the lower die (3), the cylinder (1) for the upper die (2) is activated to control the upper die (2) to press down and snugly fit with the lower die (3).

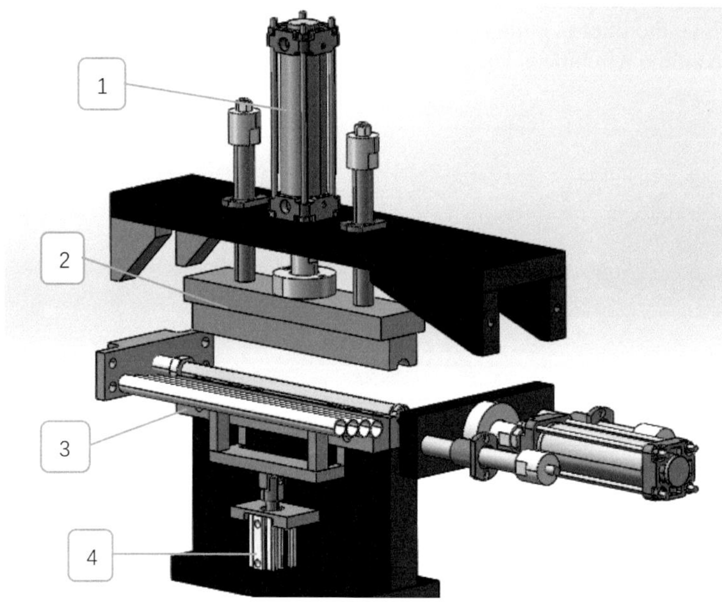

Fig. 4.63 Die Mechanism
1, 4 - Cylinders 2 - Upper Die 3 - Lower Die

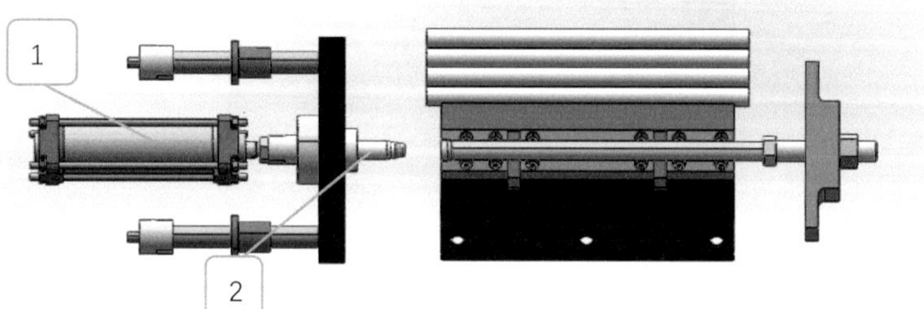

Fig. 4.64 Power Mechanism
1 - Hydraulic Cylinder 2 - Extending Piston Rod

The power mechanism, illustrated in Fig. 4.64, fixes the pipe in place using the die mechanism. The hydraulic cylinder (1) is then activated to control the extending piston rod (2) to squeeze the inner wall of the pipe, expanding it to the specified inner diameter [92].

IV. Specific Parameters (see Table 4.15)

Table 4.15 Tube Expanding Machine Parameters

Working Efficiency:	300-500 pcs/h
Processing Precision:	±0.05 mm
Operating Rate:	≥70%
Yield Rate:	≥98%
Operating Temperature:	0~55°C
Allowed Humidity:	30%~95%
Equipment Dimensions:	1965 mm(H) × 1165 mm(L) × 1105 mm(W)

V. Mechanical Design Highlights

The tube expanding machine employs a simple mechanical structure to achieve mechanical positioning. Compared to existing optical positioning technology, it can more accurately determine the endpoint of the piston rod's movement, ensuring reliable control of the tube expanding dimensions, especially for precise expansion sizes with an error fluctuation range of 0.05mm.

VI. Precautions

(1) Ensure an adequate safety operating space in the workplace.
(2) Before operation, perform an unloaded test run, confirm normal operation, and then proceed to the next process.
(3) Select the core bars and core heads based on the processed pipe diameter and thickness and install them in sequence.
(4) Clamp the components, ensuring concentric alignment of the steel pipe and the piston rod.
(5) Adjust the process parameters, start the machine in sequence, and begin production.
(6) During operation, monitor temperature, speed, dimensions, and surface quality of the steel pipe.
(7) After completing the operation, leave the machine in a ready-to-produce state, turn off the power, and perform routine maintenance.

Case 62: Crankcase Drilling Machine

I. Case Description

The crankcase drilling machine is used for drilling holes on multiple faces of the crankcase during the machining process. The overall structure is depicted in Fig. 4.65, with a detailed view in Fig. 4.66. It consists of components such as the frame (1), worktable (2), chip collection trough (3), fixing mechanism (4), drilling mechanism (5), and enclosed baffle (6). This machine is known for its quick repositioning, user-friendly operation, reduced labor intensity, and improved machining precision due to centralized processing of crankcases.

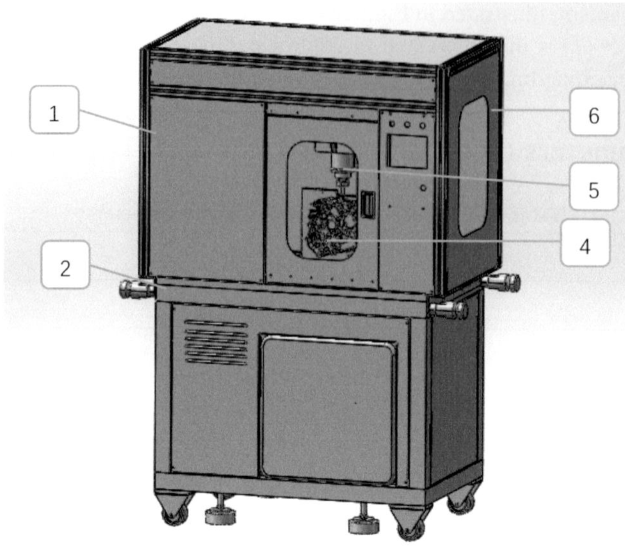

Fig. 4.65 Crankcase Drilling Machine
1 - Frame 2 - Worktable 4 - Fixing Mechanism 5 - Drilling Mechanism 6 - Enclosed Baffle

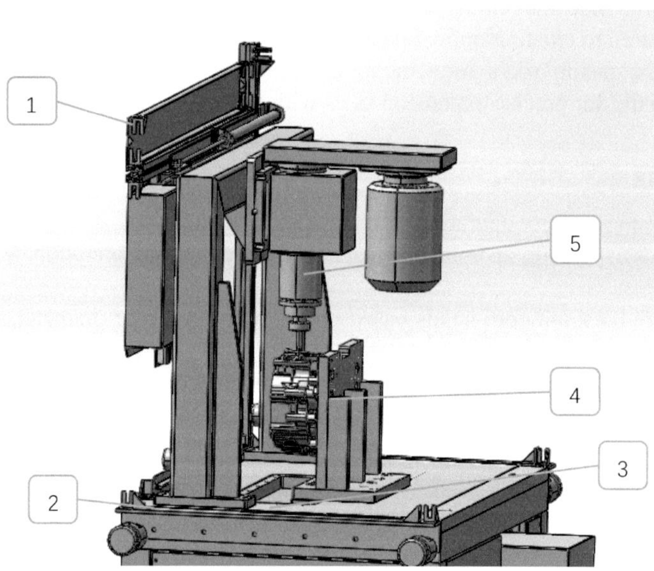

Fig. 4.66 Detailed View of Crankcase Drilling Machine
1 - Frame 2 - Worktable 3 - Chip Collection Trough 4 - Fixing Mechanism 5 - Drilling Mechanism

II. Working Principle

During operation, the crankcase is securely clamped onto the fixing mechanism. The drilling component is then activated to move downward, drilling holes in the crankcase. The drilling component can use a single drill head for sequential drilling or be equipped with a multi-axis unit for simultaneous multi-hole processing. Chips generated during drilling fall into the chip collection trough below and are expelled through the chip export channel.

III. Main Structure Introduction

As shown in Fig. 4.67, the drilling mechanism controls the drilling power head (2) to perform drilling operations on the securely fixed crankcase.

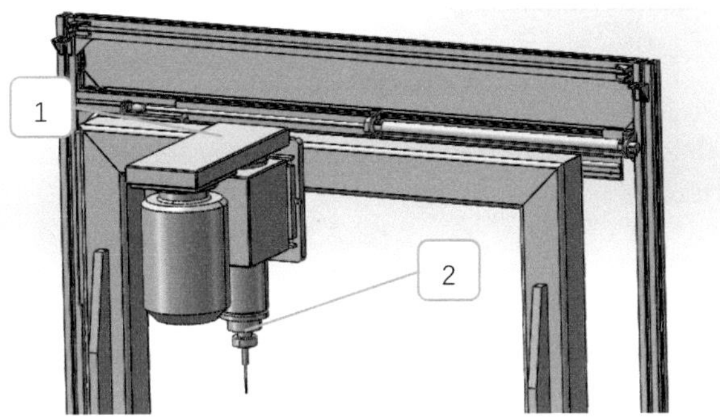

Fig. 4.67 Drilling Mechanism
1 - Drilling Mechanism 2 - Drilling Power Head

Fixing Mechanism: Fig. 4.68 illustrates the fixing mechanism with eight locating pins (2) on the positioning plate (1) to securely fix the crankcase, preventing accidental detachment during drilling.

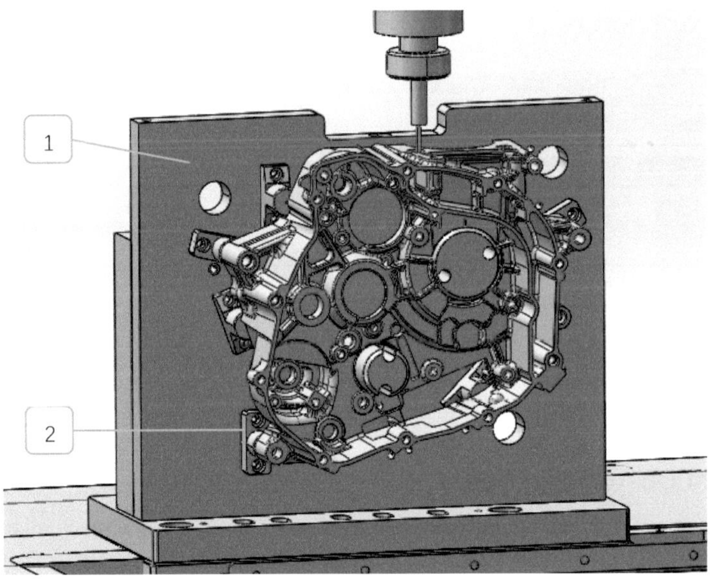

Fig. 4.68 Fixing Mechanism
1 - Positioning Plate 2 - Locating Pins

The chip collection trough, as shown in Fig. 4.69, has a sloped bottom surface. Chips generated during drilling are directed through the chip export channel (3) and discharged into the chip collection container (2).

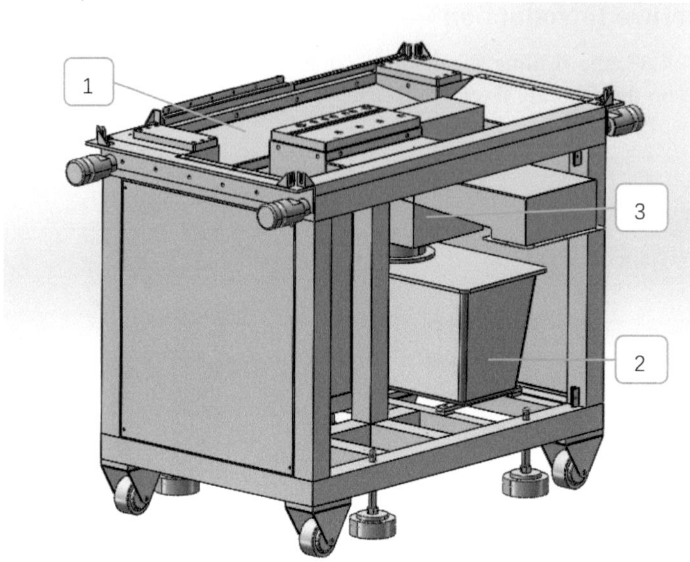

Fig. 4.69 Chip Collection Trough
1 - Chip Collection Trough 2 - Chip Collection Container 3 - Chip Export Channel

IV. Specific Parameters (see Table 4.16)

Table 4.16 Crankcase Drilling Machine Parameters

Positioning Accuracy:	0.02 mm
Repeat Positioning Accuracy:	0.01 mm
Minimum Setting Unit:	0.01 mm
Operating Rate:	≥93%
Yield Rate:	≥98%
Operating Temperature:	−10~55°C
Allowed Humidity:	25%~95%
Equipment Dimensions:	1970 mm(H) × 1500 mm(L) × 995 mm(W)

V. Mechanical Design Highlights

The chip collection trough features an electromagnetic iron debris removal mechanism at the bottom. An electromagnetic rod connected to a power supply with an automatic timer switch generates a strong magnetic field when powered on. This effectively attracts metallic debris generated during drilling. When power is cut, the electromagnetic rod loses its magnetism, allowing the attracted metallic debris to fall into the chip export channel along the inclined bottom surface of the chip collection trough. The debris is then discharged from the rear chip discharge port. In practical applications, there is a chip collection container below the discharge port.

VI. Precautions

(1) Avoid any contact with rotating drilling power heads or other moving parts using hands or any other means. Iron debris must be cleaned using iron hooks or brushes.

(2) Do not stack any tools, clamps, blades, measuring tools, workpieces, or other items on the machine except for the fixtures and workpieces on the worktable.
(3) Close the enclosed baffle of the crankcase drilling machine during the machining process. Observe the machine's operating status. In case of abnormalities or accidents, immediately terminate program operation, cut off power, and refrain from further operations.

Case 63: Automotive Parts Packaging Machine

I. Case Description

The automotive parts packaging machine is designed to automatically sleeve plastic covers onto automotive components, improving efficiency. The overall structure is shown in Fig. 4.70, and the top view is presented in Fig. 4.71. It consists of packaging mechanism (1), plastic cover dispensing mechanism (2), frame (3), vibratory conveyor (4), and automotive parts conveyor mechanism (5). This machine is characterized by high automation, high efficiency, and reduced labor requirements.

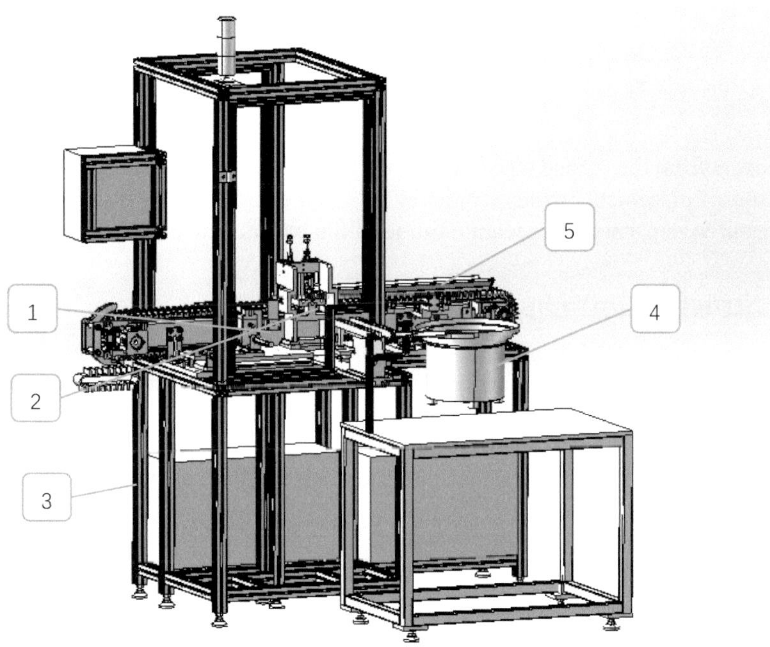

Fig. 4.70 Automotive Parts Packaging Machine
1-Packaging Mechanism 2-Plastic Cover Dispensing Mechanism 3-Frame 4-Vibratory Conveyor 5-Automotive Parts Conveyor Mechanism

II. Working Principle

During operation, automotive parts are placed on the first placement frame of the automotive parts conveyor mechanism and sent to the packaging mechanism. The plastic cover conveyor mechanism, consisting of the plastic cover dispensing mechanism and the vibratory conveyor, delivers plastic covers to the plastic cover dispensing mechanism. The plastic cover dispensing mechanism releases plastic covers onto the second placement frame. Simultaneously, the automotive parts conveyor mechanism transports both automotive parts and plastic covers. The packaging mechanism places

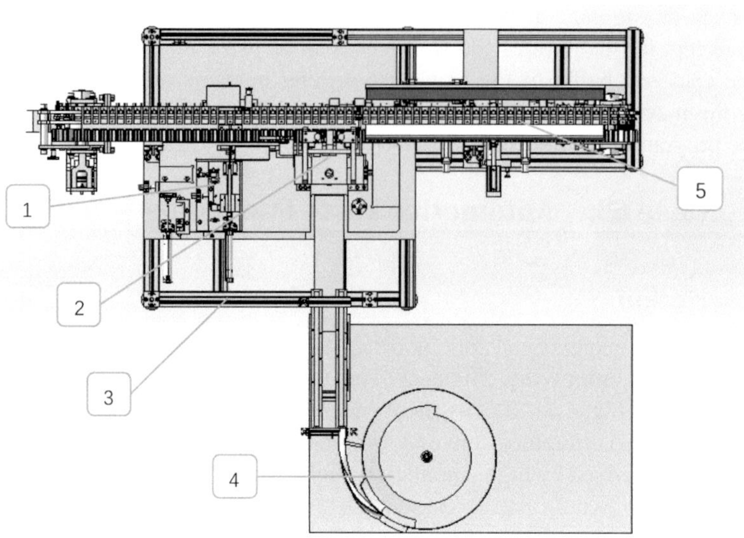

Fig. 4.71 Top View of Automotive Parts Packaging Machine
1-Packaging Mechanism 2-Plastic Cover Dispensing Mechanism 3-Frame 4-Vibratory Conveyor 5-Automotive Parts Conveyor Mechanism

the plastic covers from the second placement frame onto the automotive parts on the first placement frame. The use of placement frames facilitates precise positioning and installation of automotive parts and plastic covers, ensuring accurate fitting of the covers onto the automotive components.

III. Main Structure Introduction

Automotive Parts Conveyor Mechanism: as shown in Fig. 4.72, it is driven by a servo motor (3). The automotive parts conveyor mechanism includes the first placement frame (1) and the second placement frame (2), used to transport automotive parts and plastic covers, respectively. The mechanism also features a tensioning device (4), driven by a cylinder (6) through an extendable piston rod (5), allowing adjustment of the tensioning degree of the conveyor mechanism.

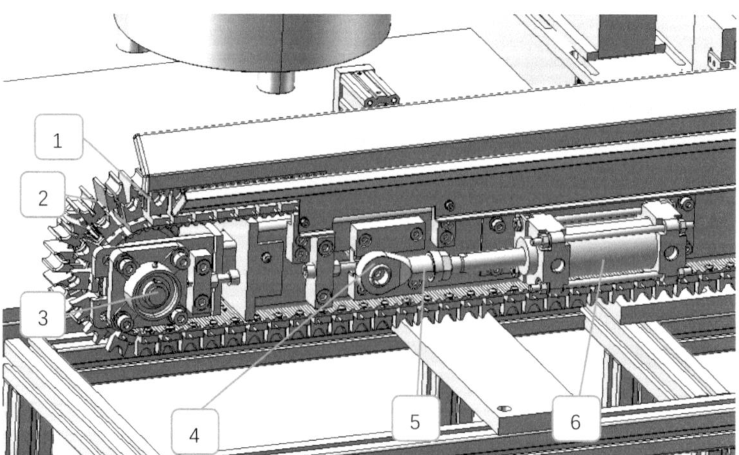

Fig. 4.72 Automotive Parts Conveyor Mechanism
1-First Placement Frame 2-Second Placement Frame 3-Servo Motor 4-Tensioning Device 5-Piston Rod 6-Cylinder

Plastic Cover Dispensing Mechanism: as shown in Fig. 4.73, plastic covers are arranged and transported by the vibratory conveyor to the plastic cover dispensing mechanism. This mechanism, located above the second placement frame, is driven by a cylinder (1) controlling the opening and closing mechanism (2). The opening mechanism (2) consists of two symmetrically arranged support blocks. When the support blocks are extended outward, plastic covers on them fall onto the second placement frame. The opposing faces of the support blocks are inclined to facilitate the descent of plastic covers onto the second placement frame.

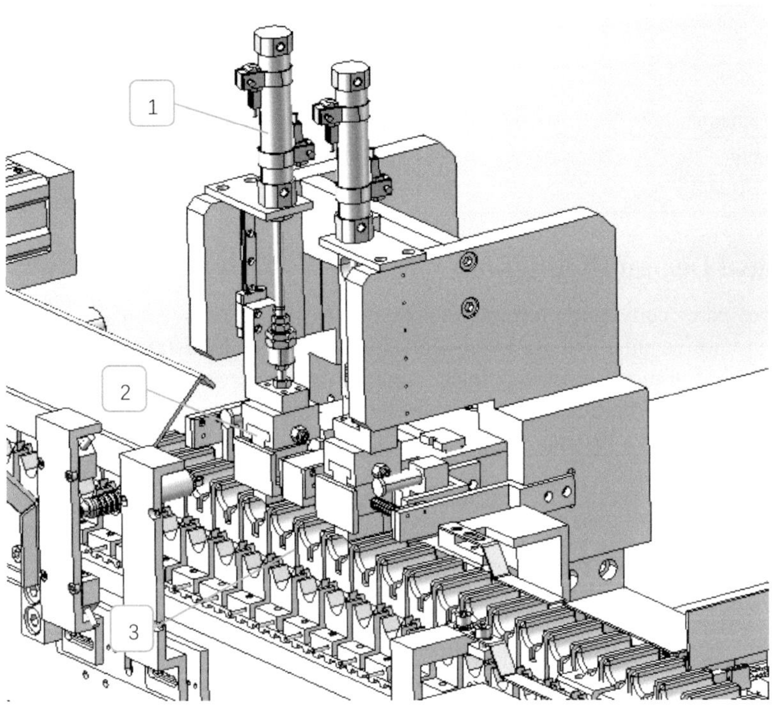

Fig. 4.73 Plastic Cover Dispensing Mechanism
1 - Cylinder 2 - Opening Mechanism 3 - Second Placement Frame

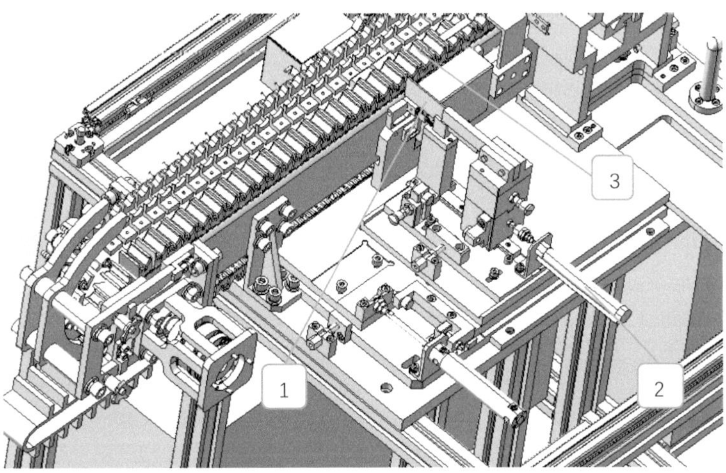

Fig. 4.74 Packaging Mechanism
1 - Push Plate 2 - Cylinder 3 - Automotive Parts Conveyor Mechanism

Packaging Mechanism: as shown in Fig. 4.74, it is driven by a cylinder (2) to push the push plate (1), packaging the plastic covers from the second placement frame onto the automotive parts on the first placement frame.

IV. Specific Parameters (see Table 4.17)

Table 4.17 Automotive Parts Packaging Machine Parameters

Working Efficiency:	1800-3000 pcs/h
Operating Rate:	≥90%
Yield Rate:	≥98%
Operating Temperature:	0~45°C
Allowed Humidity:	30%~95%
Equipment Dimensions:	2010 mm(H) × 1780 mm(L) × 1450 mm(W)

V. Mechanical Design Highlights

The automotive parts conveyor mechanism features the first and second placement frames for transporting automotive parts and plastic covers. Additionally, an adjustment mechanism to prevent conveyor belt misalignment is incorporated, comprising fixed and movable baffles. These are respectively placed on both sides of the automotive parts conveyor mechanism and are adjusted by a cylinder to control the distance between the fixed and movable baffles. This ensures that the conveyor mechanism operates between the fixed and movable baffles, avoiding misalignment and ensuring the normal operation of the automotive parts conveyor mechanism.

VI. Precautions

(1) Before operating the machine, operators must thoroughly read the relevant operation instructions.
(2) Maintenance personnel must read the machine's maintenance instructions before performing any maintenance or upkeep work.
(3) In case of any accidents during machine operation, the power must be turned off first before taking appropriate measures.
(4) While the machine is running, operators must not place their hands on the rolling parts of the front and rear conveyors to prevent injuries.
(5) In the event of an alarm or other malfunction, press the emergency stop switch first. After clearing the fault, release the emergency stop switch and return to automatic operation.

Case 64: Automotive Wheel Rim Processing Machine

I. Case Description

Automotive steel rims, also known as wheel hubs, play a crucial role in enhancing specific aspects of a vehicle's performance. Currently, automotive wheel rims are categorized into steel rims and alloy rims based on material. Among them, steel rims, with advantages such as simple manufacturing processes, relatively low costs, and strong resistance to metal fatigue, have a significant presence in the automotive component production and processing field. In the process of steel rim machining, issues arise due to uneven stress distribution caused by a single stress point on the rim, leading to decreased product quality. Additionally, long preparation times for auxiliary processing steps need improvement in production efficiency. The automotive wheel rim processing machine is designed for the production and processing of automotive wheel rims, addressing issues of production efficiency and cable confusion. The overall structure is shown in Fig. 4.76, consisting mainly of positioning detection mechanism (1), rim clamping plate (2), drill bit mounting sleeve (3), frame (4), lifting adjustment mechanism (5), and laser cutting mechanism (6). The automotive wheel rim processing machine boasts advantages such as uniform stress distribution and high processing efficiency.

Fig. 4.75 Automotive steel rims

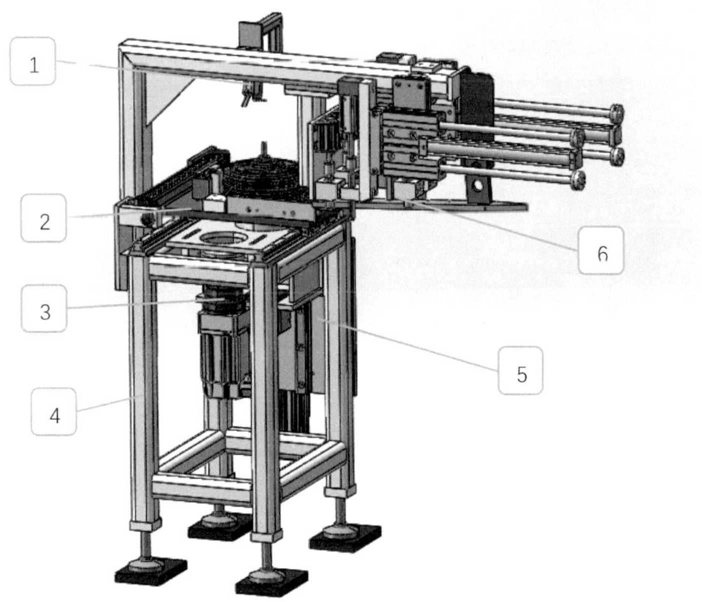

Fig. 4.76 Automotive Wheel Rim Processing Machine
1-Positioning Detection Mechanism 2-Rim Clamping Plate 3-Drill Bit Mounting Sleeve 4-Frame 5-Lifting Adjustment Mechanism 6-Laser Cutting Mechanism

II. Working Principle

During the operation of the automotive wheel rim processing machine, the steel rim blank is placed on the rim clamping plate and clamped. The laser cutting mechanism is controlled to move directly above the rim clamping plate. Simultaneously, the drill bit is retracted into the drill bit mounting sleeve, and the lifting adjustment mechanism is controlled to lower the drill bit mounting sleeve to a specified height. Laser cutting is performed on the steel rim using laser nozzles on the outer side of the laser cutting sleeve [93]. After cutting, the laser cutting sleeve returns to its original position. The lifting adjustment mechanism controls the drill bit mounting sleeve to rise to a certain height, and then the drill bit is extended. The rim clamping plate is moved back and forth on the guide rail to adjust the drilling position. After adjustment, the servo motor is activated, and the drill bit performs the drilling operation.

III. Main Structure Introduction

Rim Clamping Plate: as shown in Fig. 4.77, the rim clamping plate consists of a clamp (3) that secures the steel rim blank material. It is controlled by a servo motor (1) to move back and forth on the guide rail (2).

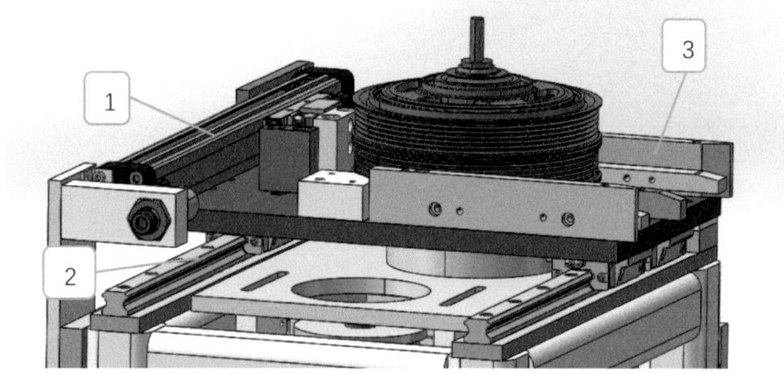

Fig. 4.77 Rim Clamping Plate
1 - Servo Motor 2 - Guide Rail 3 - Clamp

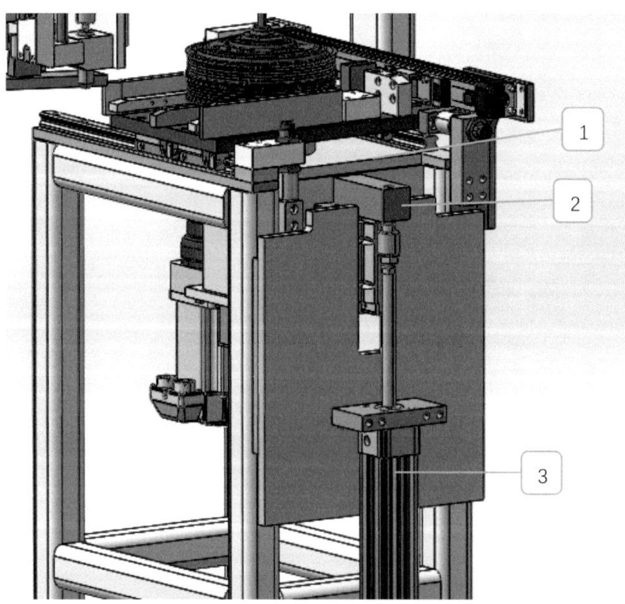

Fig. 4.78 Lifting Adjustment Mechanism
1 - Drill Bit Mounting Sleeve 2 - Fixed Block 3 - Cylinder

Lifting Adjustment Mechanism: as shown in Fig. 4.78, the fixed block (2) is fixed in position relative to the drill bit mounting sleeve (1). It is driven by a cylinder (3) to control the up-and-down movement of the fixed block, thereby controlling the up-and-down movement of the drill bit mounting sleeve.

Drill Bit Mounting Sleeve: as shown in Fig. 4.79, it includes servo motors (2 and 3). The lifting adjustment mechanism controls the up-and-down movement of the drill bit mounting sleeve (1). Servo motor 2 drives the drill bit for drilling, and servo motor 3 controls the rotation of the steel rim, adjusting the angle of the rim.

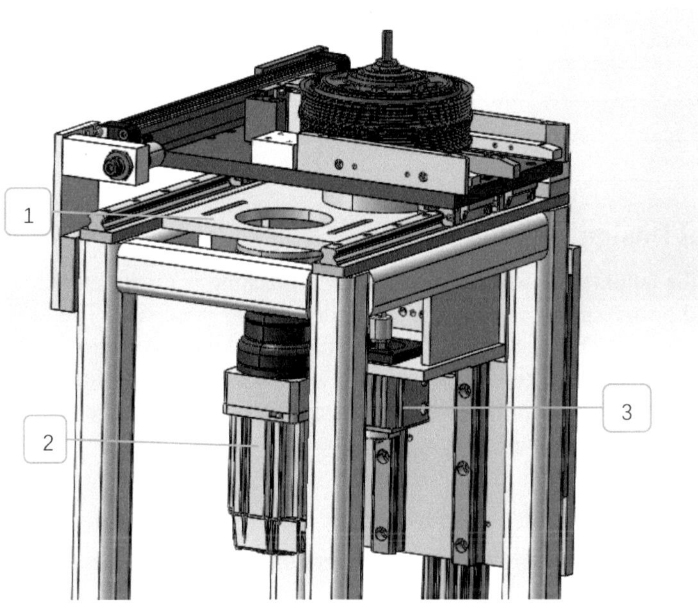

Fig. 4.79 Drill Bit Mounting Sleeve
1 - Drill Bit Mounting Sleeve 2 and 3 - Servo Motors

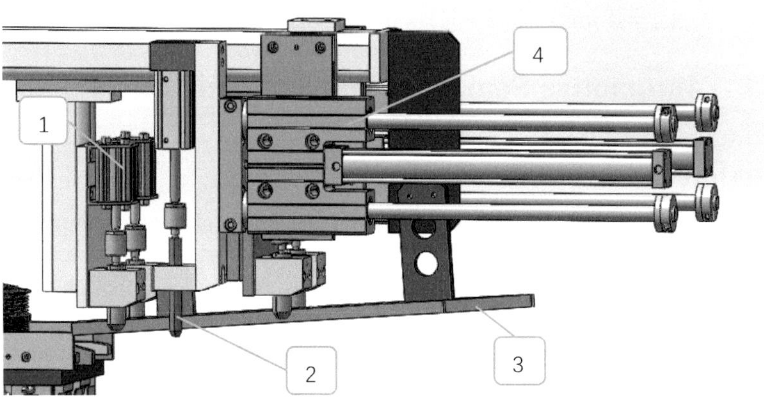

Fig. 4.80 Laser Cutting Mechanism
1 and 4 - Cylinder 2 - Laser Cutting Head 3 - Pry Bar

Laser Cutting Mechanism: as shown in Fig. 4.80, cylinder 1 controls the angle of the pry bar (3), which, in turn, controls the angle of the steel rim blank material to ensure cutting quality. The horizontal displacement of the laser cutting head (2) is controlled by cylinder 4.

IV. Specific Parameters (see Table 4.18)

Table 4.18 Automotive Wheel Rim Processing Machine Parameters

Operating Rate:	≥75%
Yield Rate:	≥98%
Operating Temperature:	0~55°C
Allowed Humidity:	25%~90%
Pressure Range:	4-6 kg/cm^2
Equipment Dimensions:	1155 mm(H) × 1095 mm(L) × 490 mm(W)

V. Mechanical Design Highlights

The left side of the automotive wheel rim processing machine is equipped with a cable box, and the cable box and power supply interface are installed on the same side. The cable box is used to organize all the required cables, avoiding cable confusion and inconvenience. Additionally, the rim clamping plate secures the steel rim blank material, and the laser cutting sleeve's outer laser nozzle allows the shaping of the steel rim blank by removing excess material. The laser cutting efficiency is high, and there are no burrs, improving cutting quality.

VI. Precautions

(1) Operators must strictly adhere to the safety operating procedures of the automotive wheel rim processing machine.
(2) Proper labor protective equipment must be worn, and approved protective eyewear must be worn near the laser beam.
(3) In case of anomalies during processing, immediately stop the automotive wheel rim processing machine and promptly troubleshoot or report to supervisors.
(4) Maintain cleanliness, orderliness, and absence of oil stains on the bed and surrounding areas. Workpieces, sheets, and waste materials should be stacked according to regulations.

Case 65: Automotive Seal Strip Punching and Cutting Machine

I. Case Description

With the rise of the automotive industry, the production and processing of automotive components have been significantly promoted. Among them, automotive seal strips play a crucial role in filling various intervals and gaps between body components. They provide functions such as shock absorption, waterproofing, dust prevention, sound insulation, and decoration. The automotive seal strip punching and cutting machine is designed for the automated processing of automotive seal strips. The overall structure is shown in Fig. 4.81, and a magnified view is shown in Fig. 4.82. It consists mainly of a frame (1), fixed cutting mechanism (2), fixed seal strip holder (3), electrical control box (4), advancing mechanism (5), and movable cutting mechanism (6). This equipment offers advantages such as high work efficiency, smooth cutting surfaces, and precise punching and cutting positioning.

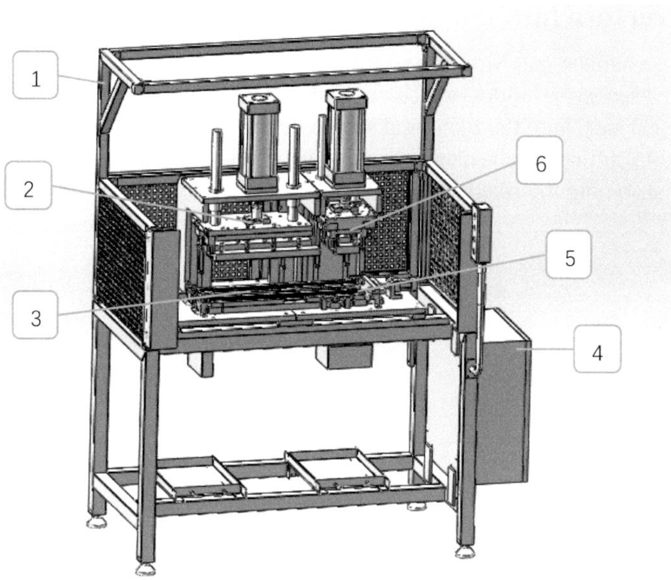

Fig. 4.81 Automotive Seal Strip Punching and Cutting Machine
1-Frame 2-Fixed Cutting Mechanism 3-Fixed Seal Strip Holder 4-Electrical Control Box 5-Advancing Mechanism 6-Movable Cutting Mechanism

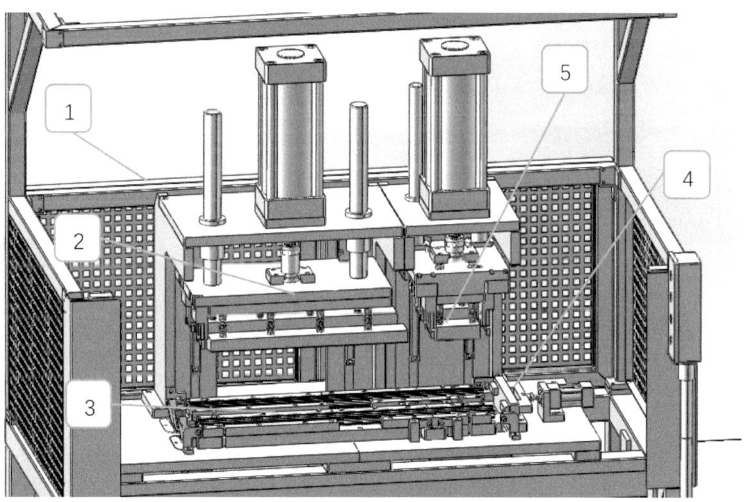

Fig. 4.82 Magnified View of Automotive Seal Strip Punching and Cutting Machine
1-Frame 2-Fixed Cutting Mechanism 3-Fixed Seal Strip Holder 4-Advancing Mechanism 5-Movable Cutting Mechanism

II. Working Principle

During the operation of the automotive seal strip punching and cutting machine, the seal strip is threaded onto the fixed cutting mechanism and the movable cutting mechanism. The left end of the seal strip is cut by the fixed cutting blade, ensuring a smooth left end. The advancing mechanism pushes the seal strip, ensuring that the left end of the seal strip is level. Subsequently, the cylinder of the fixed cutting mechanism drives the fixed pressure plate to press the seal strip onto the fixed seal strip holder. The movable cutting mechanism is controlled to move downward, cutting the right end of the seal strip, thus obtaining the desired size of the seal strip [94].

III. Main Structure Introduction

Fixed Cutting Mechanism and Movable Cutting Mechanism: as shown in Fig. 4.83, the seal strip is fixed on the fixed seal strip holder (4). The cylinder (1) drives and controls the fixed pressure plate (2) to press the seal strip onto the fixed seal strip holder (4). The fixed cutting blade (3) cuts the seal strip. The movable cutting mechanism is driven by cylinder (7), and the movable pressure plate (6) presses the seal strip onto the fixed seal strip holder (4). The movable cutting blade (5) cuts the seal strip [95].

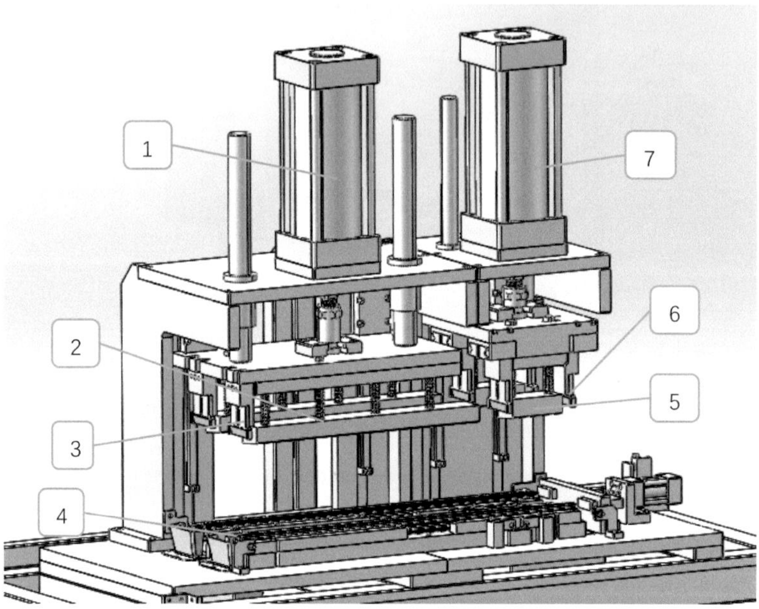

Fig. 4.83 Fixed Cutting Mechanism and Movable Cutting Mechanism
1 and 7-Cylinder 2-Fixed Pressure Plate 3-Fixed Cutting Blade 4-Fixed Seal Strip Holder 5-Movable Cutting Blade 6-Movable Pressure Plate

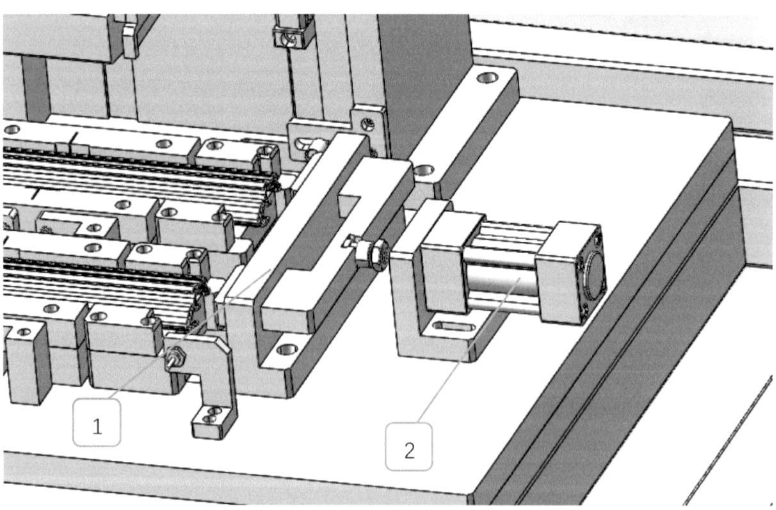

Fig. 4.84 Advancing Mechanism
1-Advancing Plate 2-Cylinder

Advancing Mechanism: as shown in Fig. 4.84, the cylinder (2) drives and controls the advancing plate (1) to push the seal strip, ensuring cutting accuracy.

IV. Specific Parameters (see Table 4.19)

Table 4.19 Automotive Seal Strip Punching and Cutting Machine Parameters

Operating Rate:	≥93%
Yield Rate:	≥98%
Cutting Accuracy:	±0.02 mm
Minimum Setting Unit:	0.01 mm
Operating Temperature:	0~45°C
Allowed Humidity:	25%~75%
Equipment Dimensions:	2015 mm(H) × 1700 mm(L) × 1000 mm(W)

V. Mechanical Design Highlights

The automotive seal strip punching and cutting machine controls the distance between two cutting mechanisms on the frame to determine the length of the seal strip to be cut. The fixed seal strip holder supports and locks the seal strip placed between the two cutting mechanisms, ensuring cutting accuracy. Additionally, the fixed seal strip holder has a buffer groove inside, improving the smoothness of the cutting surface and avoiding rigid contact with the cutting blade, thereby extending the equipment's lifespan and reducing maintenance costs.

VI. Precautions

(1) Unauthorized operation of the cutting machine without training and permission is strictly prohibited.
(2) During cutting actions, hands should be kept away from the cutting blade or cutting board. It is strictly forbidden to assist the die with hands to avoid danger.
(3) When replacing a new cutting blade with a different height, it should be re-set according to the specified method.
(4) Avoid overloading to prevent machine damage and reduce the lifespan of the equipment.

Case 66: Automotive Windshield Automatic Sealing Machine

I. Case Description

The windshield of an automobile serves the dual purpose of protecting the occupants and ensuring airtightness with excellent sound insulation. The quality of windshield installation directly affects the overall vehicle's sealing and safety. It is a crucial process in automotive assembly. The automotive windshield automatic sealing machine is designed for the automated gluing of windshields, ensuring uniform gluing across the entire windshield [96]. The overall structure is shown in Fig. 4.85, consisting of a gluing robot (1), a fixed frame (2), support bars (3), and a gluing mechanism (4). This machine boasts high production efficiency, even gluing, and stable connection between the windshield and the frame.

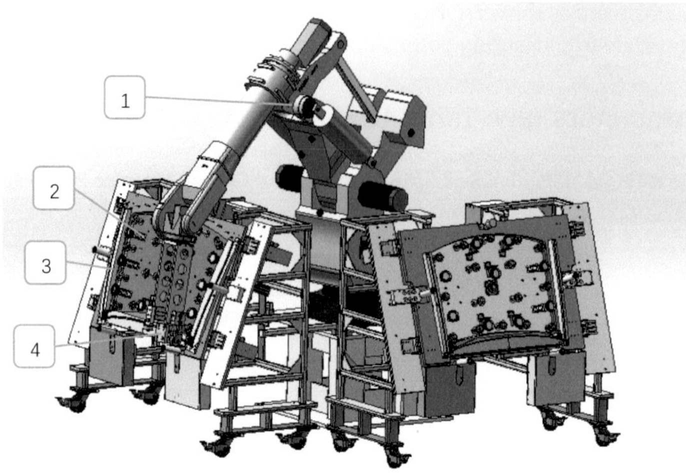

Fig. 4.85 Automotive Windshield Automatic Sealing Machine
1 - Gluing Robot 2 - Fixed Frame 3 - Support Bars 4 - Gluing Mechanism

II. Working Principle

During the operation of the automotive windshield automatic sealing machine, the gluing robot's gluing trajectory is planned using a teaching system. The gluing mechanism is then fixed on the robotic arm. The windshield is secured on the fixed frame, ensuring complete alignment with the support bars. The gluing robot moves, and the gluing mechanism, guided by the pre-set trajectory, uniformly applies glue to the edges of the windshield.

III. Main Structure Introduction

Gluing Robot: as shown in Fig. 4.86, the robot's base (3) is fixed on the ground. The robotic arm, along with the gluing mechanism, is stationary and primarily driven by a servo motor (1), with the gluing trajectory planned using a teaching system.

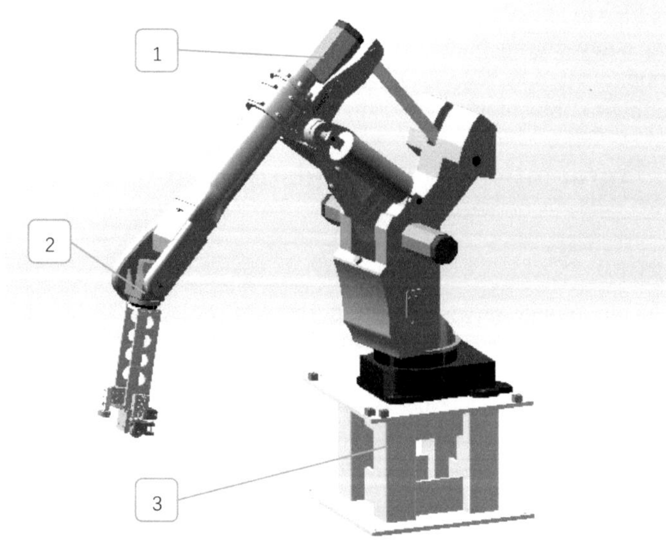

Fig. 4.86 Gluing Robot
1 - Servo Motor 2 - Robotic Arm 3 - Robot Base

Fixed Frame and Gluing Mechanism: as shown in Fig. 4.87, the windshield is fixed on the fixed frame using suction cups (1). The edges align with the support bars (3) and tightly adhere. Between the windshield and the support frame, multiple pressure blocks (2) are placed to prevent the windshield from cracking during gluing. The gluing mechanism (4) is fixed along the support bars (3) to apply glue to the windshield [97].

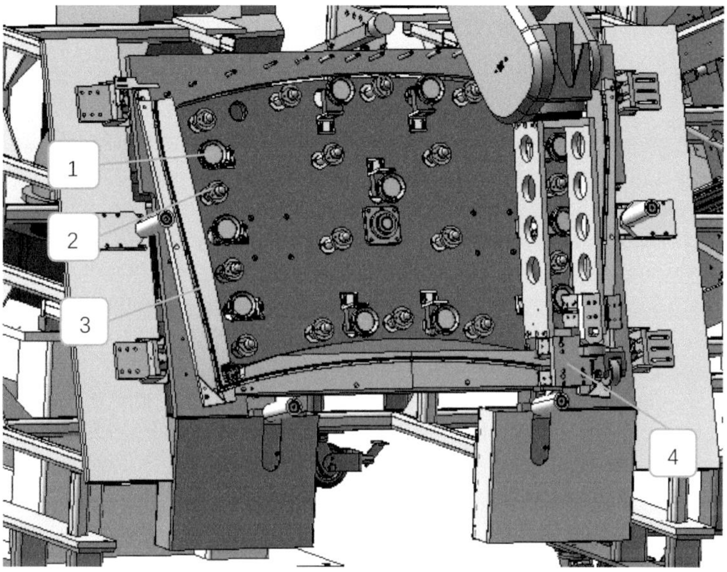

Fig. 4.87 Fixed Frame and Gluing Mechanism
1 - Suction Cups 2 - Pressure Blocks 3 - Support Bars 4 - Gluing Mechanism

IV. Specific Parameters (see Table 4.20)

Table 4.20 Automotive Windshield Automatic Sealing Machine Parameters

Yield Rate:	≥98%
Operating Temperature:	0~55°C
Allowed Humidity:	30%~95%
Equipment Dimensions:	3000 mm(H) × 3110 mm(L) × 3280 mm(W)
Glue Material:	Polymer modified with alkoxysilane
Curing Method:	Moisture curing
Density:	Approximately 1.4g/cm^3
Dry Time:	Approximately 10 minutes (20°, 50% relative humidity)
Working Time:	Approximately 15 minutes (20°, 50% relative humidity)
24-Hour Curing Thickness:	Approximately 3 mm (20°, 50% relative humidity)
Shrinkage Rate:	<3%
Hardness SHORE A:	Approximately 65
Initial Strength:	Approximately 1700 P
Tensile Strength (100%):	Approximately 2.2 MPa
Tear Strength:	Approximately 12 MPa

V. Mechanical Design Highlights

Automotive windshields are generally designed as large curved surfaces with certain curvatures in all directions. These curved glass structures pose high technical requirements both in terms of the manufacturing process and the fitting process, involving issues such as vehicle type, strength, thermal insulation, and assembly. The automotive windshield automatic sealing machine automates the gluing process using a gluing robot, ensuring high efficiency, uniform gluing, and preventing damage to the glass during gluing. This avoids unnecessary financial losses [97].

VI. Precautions

(1) The quantitative pump is controlled by a frequency-controlled electric motor, and its flow can be automatically adjusted based on the gluing gun's operating speed.
(2) The quantitative pump inlet is equipped with a quick shut-off valve, allowing for instant glue interruption to prevent damage when the equipment is not gluing.
(3) The quantitative pump system is connected to the main control cabinet via communication cables. The main control system controls the system's operation and the opening of the gluing gun. The quantitative pump should maintain stable glue pressure to the gluing gun, avoiding interruptions and excessive glue.
(4) The glue supply pipeline should have heating and insulation functions to ensure that the glue at the outlet of the gluing gun maintains a certain temperature. The delivery pipeline should have sufficient pressure resistance. The pipeline should have good sealing properties to prevent faults in the entire system after long periods of inactivity.

Case 67: Automotive Connecting Rod Precision Machining Machine Tool

I. Case Description

The automotive connecting rod, as shown in Fig. 4.88, is a crucial component in the automotive engine. It connects the piston and the crankshaft, converting the reciprocating motion of the piston into the rotary motion of the crankshaft, thus transferring the force acting on the piston to the crankshaft to output power. The connecting rod is subject to forms of damage such as fatigue fracture and excessive deformation. The working conditions of the connecting rod require it to have high

Fig. 4.88 Automotive Connecting Rod

strength, fatigue resistance, sufficient stiffness, and toughness [98]. The automotive connecting rod precision machining machine tool is designed for the precision machining of automotive connecting rods. The overall structure is shown in Fig. 4.89, and a top view is shown in Fig. 4.90. It consists mainly of milling device (1), servo feed motor (2), operating panel (3), frame (4), boring machine (5), precision boring fixture (6), fixed mechanism (7), and electrical control box (8). The machine tool offers advantages such as high machining efficiency, high automation level, and good product quality.

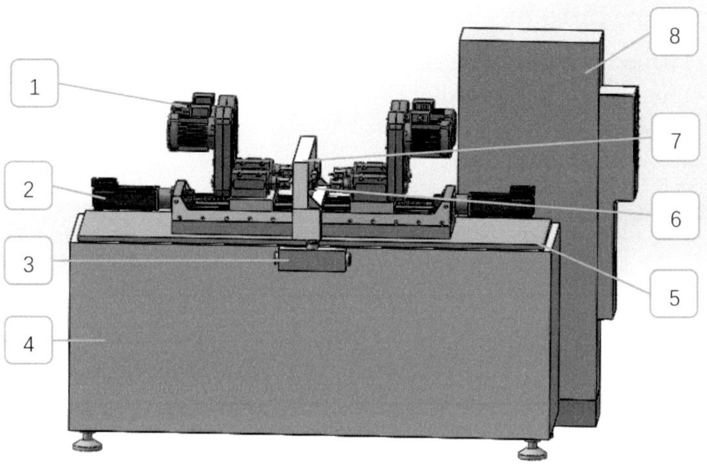

Fig. 4.89 Automotive Connecting Rod Precision Machining Machine Tool
1 - Milling Device 2 - Servo Feed Motor 3 - Operating Panel 4 - Frame 5 - Boring Machine 6 - Precision Boring Fixture 7 - Fixed Mechanism 8 - Electrical Control Box

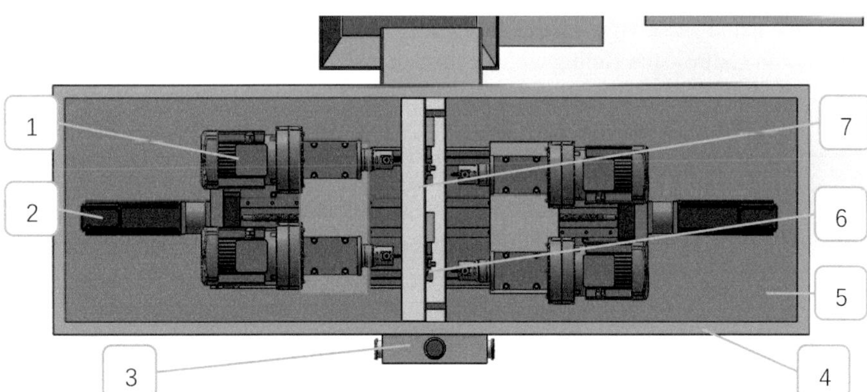

Fig. 4.90 Top View of Automotive Connecting Rod Precision Machining Machine Tool
1 - Milling Device 2 - Feed Mechanism 3 - Operating Panel 4 - Frame 5 - Boring Machine 6 - Precision Boring Fixture 7 - Fixed Mechanism

II. Working Principle

After installing and debugging the automotive connecting rod precision machining machine tool, the connecting rod to be machined is mounted and secured on the precision boring fixture. Pressing the start button on the operating panel, the CNC system sends pre-stored machining commands to the servo motor. The servo motor moves towards the fixed mechanism, driving the boring tool mounted on the milling device closer to the connecting rod on the precision boring fixture. At the same time, the servo feed motor of the milling device controls the rotation of the power head, driving the boring tool to rotate and begin boring the holes in the connecting rod. After the servo feed motor completes the hole according to the machining command, it retracts based on the stop machining command, moving the boring tool away from the connecting rod. The operator then removes the machined connecting rod, prepares the next one, and mounts it on the precision boring fixture for machining [99].

III. Main Structure Introduction

Precision Boring Fixture: as shown in Fig. 4.91, the precision boring fixture consists of left and right fixtures and a rotating fixture in the middle. The cylinder in the right fixture clamps the connecting rod on both sides, and the rotating fixture is used to secure the middle part of the connecting rod, preventing accidents during machining.

Fig. 4.91 Precision Boring Fixture
1 - Left Fixture 2 - Rotating Fixture 3 - Right Fixture

Milling Device and Feed Mechanism: as shown in Fig. 4.92, the feed mechanism is driven by servo motor (5) and controls the sliding table (4) to move towards the fixed mechanism, thus driving the boring tool (3) closer to the connecting rod on the precision boring fixture. The milling device is controlled by the servo feed motor (1) to rotate the power head (2), thereby driving the boring tool (3) to rotate and bore holes in the connecting rod.

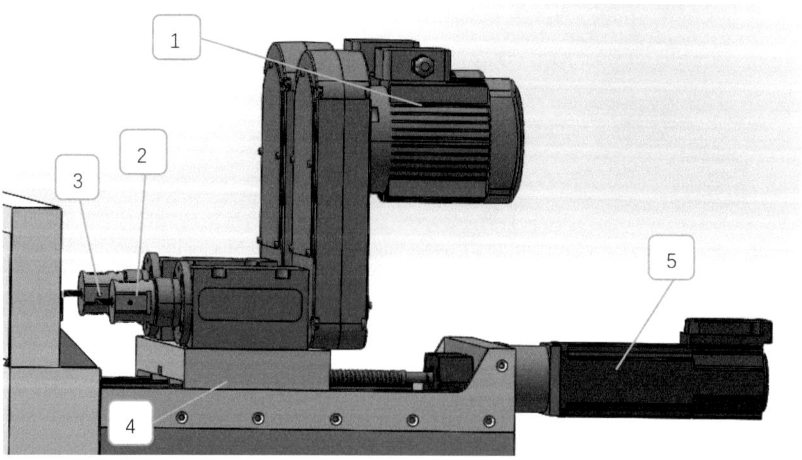

Fig. 4.92 Milling Device and Feed Mechanism
1 and 5 - Servo Motor 2 - Power Head 3 - Boring Tool 4 - Sliding Table

IV. Specific Parameters (see Table 4.21)

Table 4.21 Automotive Connecting Rod Precision Machining Machine Tool Parameters

Working Efficiency:	25-35 pcs/h
Positioning Accuracy:	0.02 mm
Repeatability Accuracy:	0.01 mm
Minimum Setting Unit:	0.01 mm
Operating Rate:	≥80%
Yield Rate:	≥95%
Operating Temperature:	0~55°C
Allowed Humidity:	25%~90%
Equipment Dimensions:	1480 mm(H) × 2155 mm(L) × 650 mm(W)

V. Mechanical Design Highlights

A key feature in the design is the presence of adjustable insert strips embedded in the gap between the boring machine and the sliding table. The sliding table has a T-shaped notch on the bottom, and this notch matches the size of the adjustable insert strips. This design not only facilitates the linear movement of the sliding table but also allows it to withstand significant forces vertically and horizontally. This ensures that the sliding table does not wobble during hole boring, guaranteeing the quality of the product. Additionally, the milling device of the automotive connecting rod precision machining machine tool has a detachable and extendable rod on the side. After precision boring the large and small holes in the connecting rod, the hydraulic cylinder's extendable rod first extends a certain distance. This action moves the connecting rod's large and small holes away from the boring tool on the sliding table. The boring tool then retracts, eliminating any tool marks and ensuring product quality.

VI. Precautions

(1) Avoid any contact with rotating spindles, workpieces, or other moving parts by hand or any other means.
(2) Strictly prohibit the stacking of any tools, clamps, blades, measuring tools, workpieces, or other debris on the machine tool except for placing fixtures and workpieces on the worktable.
(3) During the machining process, the operator must not leave the machine tool without authorization and should keep a high level of concentration to observe the machine tool's operation. If any abnormality or accident occurs, immediately stop the program, cut off the power, and refrain from further actions.
(4) Absolute prohibition of changing speeds during machine tool operation. During speed changes or tool changes, the machine tool must be completely stopped, and the switch should be in the "OFF" position to prevent machine tool accidents.
(5) When pausing for measurements during program execution, wait for the machine tool to come to a complete stop and the main spindle to cease rotation before taking measurements to avoid personal accidents.

Case 68: Parts Cleaning Machine

I. Case Description

The parts cleaning machine is designed to clean stains and residues on the surface of products. The overall structure is shown in Fig. 4.93, consisting mainly of a worktable (1), frame (2), cleaning tank (3), drive shaft (4), cleaning mechanism (5), feed port (6), and propulsion mechanism (7). The cleaning machine offers advantages such as effective cleaning, high speed, and stable operation.

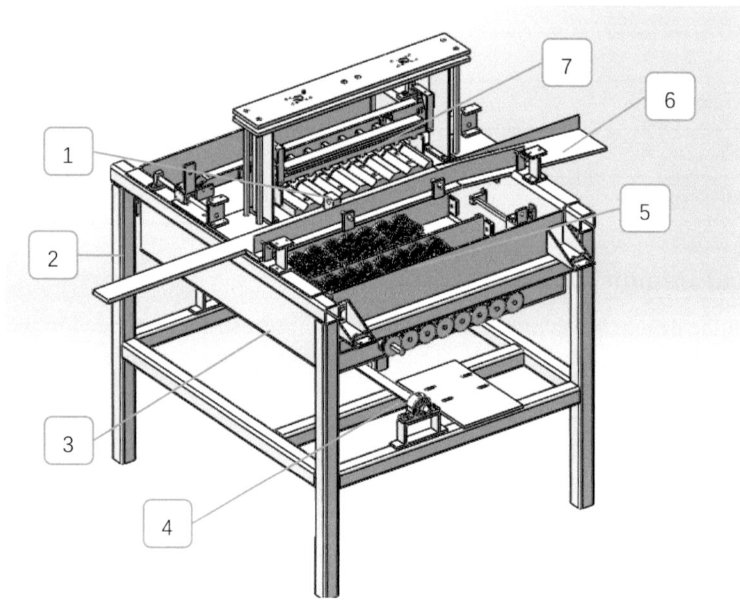

Fig. 4.93 Cleaning Machine
1 - Worktable 2 - Frame 3 - Cleaning Tank 4 - Drive Shaft 5 - Cleaning Mechanism 6 - Feed Port 7 - Propulsion Mechanism

II. Working Principle

During operation, products are fed into the cleaning machine through the feed port. The propulsion mechanism continuously moves back and forth, pushing the products into the cleaning tank. The cleaning mechanism, powered by an electric motor, rotates at high speed and washes the products by rolling and flushing.

III. Main Structure Introduction

Propulsion Mechanism: as shown in Fig. 4.94, the propulsion mechanism is controlled by a servo motor to continuously push and pull the products placed on the worktable along the guide rail (2). The worktable (1) has multiple slots, allowing products to enter the cleaning tank in an orderly manner, preventing accumulation and blockage.

Drive Mechanism: as shown in Fig. 4.95, the drive shaft (2) is connected to the electric motor. Simultaneously, the drive shaft (2) and the rotation shaft (1) of the cleaning mechanism are connected through a flat belt for transmission. The various rotation shafts (1) of the cleaning mechanism are interlinked by gear discs, achieving high-speed rotation of all rotation shafts (1) [100].

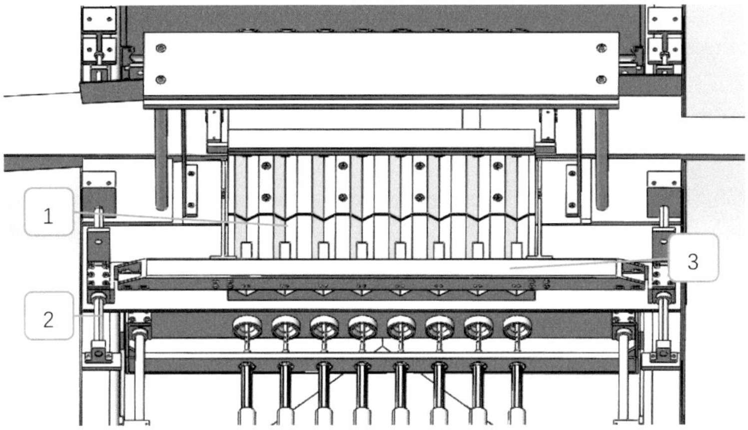

Fig. 4.94 Propulsion Mechanism
1 - Worktable 2 - Guide Rail 3 - Push Plate

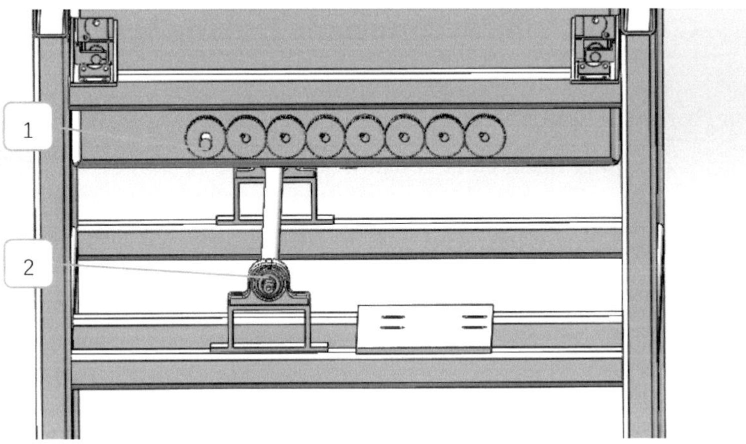

Fig. 4.95 Drive Mechanism
1 - Rotation Shaft 2 - Drive Shaft

IV. Specific Parameters (see Table 4.22)

Table 4.22 Parts Cleaning Machine Parameters

Theoretical Flow:	20-30 L/min
Electric Motor Speed:	780 r/min
Operating Temperature:	0~75°C
Cleaning Tank Size:	270 mm(H) × 1000 mm(L) × 900 mm(W)
Equipment Dimensions:	1410 mm(H) × 1915 mm(L) × 1230 mm(W)

V. Mechanical Design Highlights

The parts cleaning machine features a worktable and multiple slots set on the worktable. Through these slots and the propulsion mechanism, products can be quickly and orderly pushed into the cleaning tank. Multiple rotation shafts of the cleaning mechanism are synchronously driven by the same electric motor, completing the cleaning process. After cleaning, the products can be continuously and automatically removed, achieving good cleaning results and improving production efficiency. The design allows one person to manage multiple machines [101].

VI. Precautions

(1) Before using the cleaning machine, ensure a reliable grounding device.
(2) Strictly prohibit carrying open flames or performing operations with fire on the work site (when using volatile flammable liquids such as alcohol, acetone, gasoline, etc., cover them, and heating function should be strictly avoided).
(3) The cleaning solution temperature should not exceed 60°C, as higher temperatures may affect the cleaning effect. Generally, room temperature is sufficient.
(4) Do not start the machine when there is no cleaning solution in the tank to avoid damage due to empty operation.
(5) If the cleaning machine is not used for an extended period, drain the cleaning solution from the tank and wipe the body clean.
(6) Place the cleaning machine in a dry, clean room, strictly avoiding use in damp, polluted, or direct sunlight-exposed areas.

Case 69: Rotary Automatic Drilling Machine

I. Case Description

The rotary automatic drilling machine is designed for drilling operations on workpieces in the industrial processing field. The overall structure is shown in Fig. 4.96, consisting mainly of a pneumatic automatic drilling machine (1), automatic turntable feeding mechanism (2), and frame (3). This machine can efficiently complete drilling work on all circularly distributed holes on disc components with a single clamping. It offers advantages such as low manufacturing costs and easy maintenance.

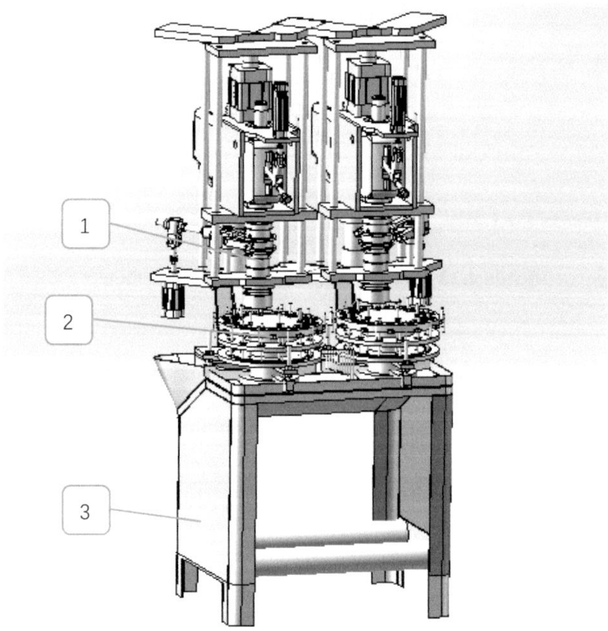

Fig. 4.96 Rotary Automatic Drilling Machine
1-Pneumatic Automatic Drilling Machine 2-Automatic Turntable Feeding Mechanism 3-Frame

II. Working Principle

When the rotary automatic drilling machine is in operation, the operator mounts the workpiece on the workpiece drilling sleeve mechanism of the automatic turntable feeding mechanism. The workpiece drilling sleeve mechanism is then installed in the automatic turntable feeding mechanism. After starting the machine, the pneumatic automatic drilling machines begin drilling. The two automatic turntable feeding mechanisms and pneumatic automatic drilling machines work alternately, significantly improving processing efficiency. The drilling head advancing cylinder and lifting cylinder of the pneumatic automatic drilling machine move synchronously. When the lifting seat is controlled to descend, the drilling head also descends synchronously, performing drilling on the workpiece [102].

III. Main Structure Introduction

The pneumatic automatic drilling machine, as shown in Fig. 4.97, is driven by an electric motor (5). When the workpiece is installed, the pneumatic automatic drilling machine's drilling head (3) rotates. The drilling head advancing cylinder (4) and lifting cylinder (1) of the pneumatic automatic drilling machine move synchronously. The advancing cylinder controls the position of the drilling head, and the lifting cylinder controls the lifting and lowering of the drilling head. When the lifting seat (2) controlled by the lifting cylinder descends, the drilling head (3) descends synchronously, performing drilling on the workpiece.

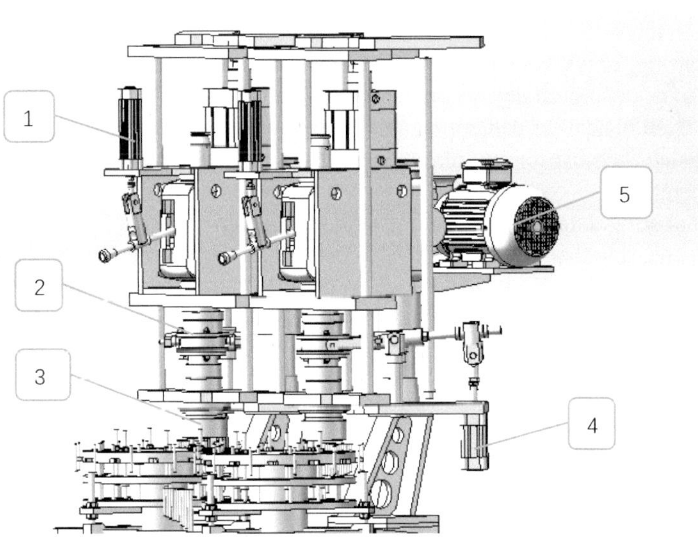

Fig. 4.97 Pneumatic Automatic Drilling Machine
1, 4 - Cylinder 2 - Lifting Seat 3 - Drilling Head 5 - Electric Motor

IV. Specific Parameters (see Table 4.23)

Table 4.23 Rotary Automatic Drilling Machine Parameters

Positioning Accuracy:	0.02 mm
Repeated Positioning Accuracy:	0.01 mm
Minimum Setting Unit:	0.01 mm

Contd.

Operating Rate:	≥93%
Yield Rate:	≥98%
Operating Temperature:	−10~55°C
Allowed Humidity:	10%~95%
Equipment Dimensions:	2050 mm(H) × 1000 mm(L) × 940 mm(W)

V. Mechanical Design Highlights

Traditional manual drilling methods using bench drills are inefficient, lack drilling precision, and have low drilling yield rates. Especially when drilling in different positions on irregular workpieces, changing fixtures is required for each hole, further reducing efficiency and yield rates. The rotary automatic drilling machine fixes the workpiece through the turntable feeding mechanism, allowing pneumatic automatic drilling machines to complete drilling work on all circularly distributed holes on disc components in a single clamping. This significantly improves drilling efficiency and accuracy.

VI. Precautions

(1) Operators must understand the machine's performance, usage, and operating precautions before operation.
(2) Operators must wear appropriate uniforms, not wear gloves, and non-operators are not allowed to approach the working machine.
(3) Before starting, check whether all parts of the machine are correctly positioned and whether there are tools or foreign objects on the worktable.
(4) Lock the fixed plate after debugging and calibrating the machine's position.
(5) When leaving the post for any reason, operators must turn off the machine to avoid chatting during operation.
(6) In case of abnormal machine operation, stop the machine immediately and have a professional inspect it, ensuring that the power is disconnected during inspection.
(7) When installing the drill bit, make sure to confirm the rotation direction of the drill bit and tighten the drill bit fixing screws.

Chapter 5

Testing Instruments and Equipment

Case 70: CCD Inspection Automatic Packaging Machine

I. Case Description

The CCD Inspection Automatic Packaging Machine (CCD stands for Charge-Coupled Device) is designed for inspecting the flatness, contact point height, needle absence, and other characteristics of various products. Through automated inspection, defective items are rejected, while the accepted ones are conveyed to the packaging unit for automatic packaging. The overall structure is illustrated in Fig. 5.1, with an enlarged view provided in Fig 5.2. The main components include a display screen (1), packaging mechanism (2), recycling bin (3), machine base (4), first inspection unit (5), fixture (6), feed inlet (7), product handling mechanism (8), and second inspection unit (9). This machine integrates feeding, inspection, and packaging, effectively enhancing the efficiency of the entire process.

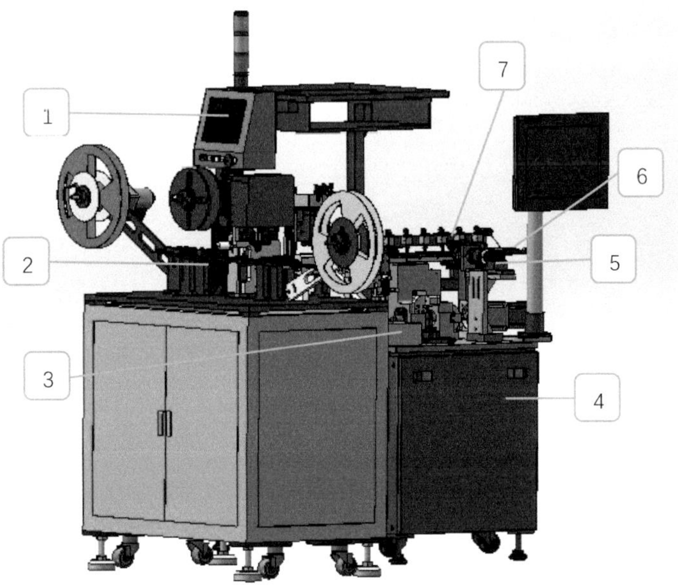

Fig. 5.1 CCD Inspection Automatic Packaging Machine – Overall Structure
1-Display Screen 2-Packaging Mechanism 3-Recycling Bin 4-Machine Base 5-First Inspection Unit 6-Fixture 7-Feed Inlet

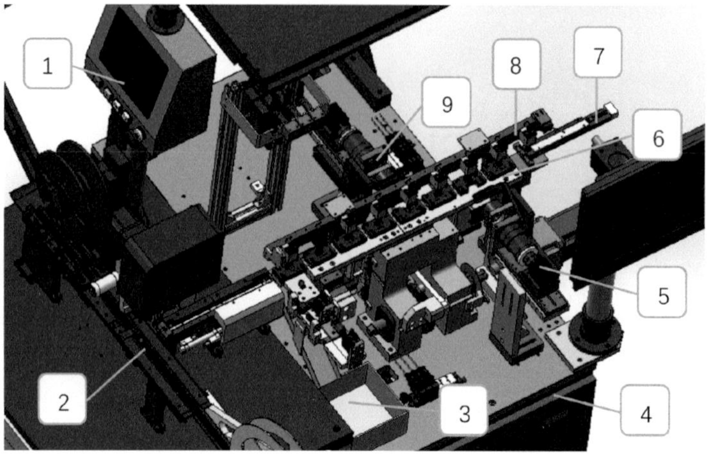

Fig. 5.2 Enlarged View of CCD Inspection Automatic Packaging Machine
1-Display Screen 2-Packaging Mechanism 3-Recycling Bin 4-Machine Base 5-First Inspection Unit 6-Fixture 7-Feed Inlet 8-Product Handling Mechanism 9-Second Inspection Unit

II. Working Principle

When the CCD Inspection Automatic Packaging Machine is activated, the products to be inspected are introduced through the feed inlet. The product handling mechanism, equipped with suction cups, transports the products to the fixture, where the first and second inspection units evaluate the products. Following inspection, defective items are discarded into the recycling bin, while acceptable products are conveyed to the packaging unit for automated packaging, completing the automatic inspection and packaging process.

III. Main Structure Introduction

Suction Cups (Fig. 5.3): The suction cups (2) are connected to the vacuum pump (1) via a pipeline. During operation, the vacuum pump (1) creates a negative pressure between the suction cup (2) and the object's surface, securely attaching the suction cup (2) to the object's surface.

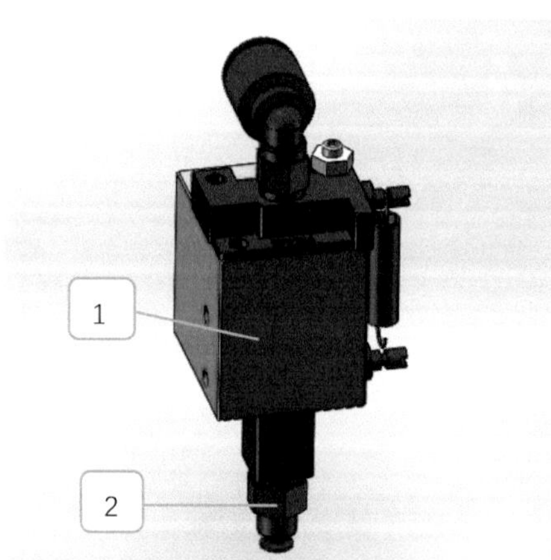

Fig. 5.3 Suction Cups
1-Vacuum Pump 2-Nozzle

Product Handling Mechanism (Fig. 5.4): Driven by a servo motor (2), the product handling mechanism utilizes a cam mechanism (3) to control the vertical movement. The suction cups are mounted on the handling frame (1), continuously transporting the products from the feed inlet to the inspection points, where the first and second inspection units assess the products.

Fig. 5.4 Product Handling Mechanism with Rotary Direction Mechanism
1 - Handling Frame 2 - Electric Motor 3 - Cam Mechanism

Product Packaging Mechanism (Fig. 5.5): Approved products are conveyed by the product handling mechanism to the product packaging mechanism, driven by a servo motor (3). The packaging mechanism extracts plastic film from the film disk (2) and covers the products on the film guide (4), completing the film packaging process.

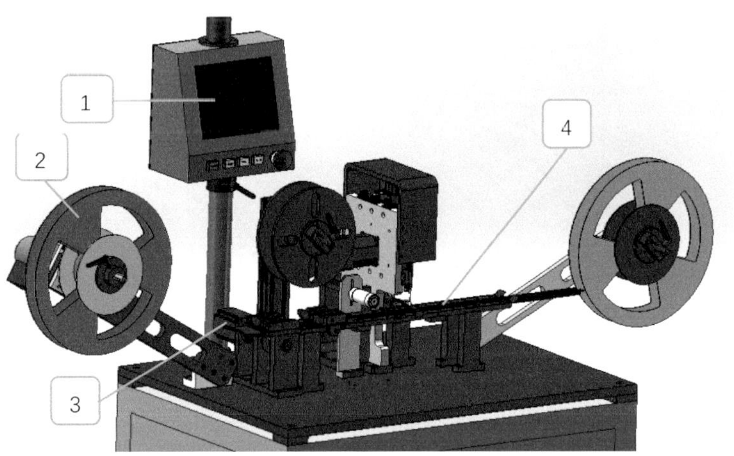

Fig. 5.5 Product Packaging Mechanism
1 - Display Screen 2 - Film Disc 3 - Servo Motor 4 - Film Guide

IV. Specific Parameters (see Table 5.1)

Table 5.1 CCD Inspection Automatic Packaging Machine Parameters

Air Pressure:	0.5~0.7 Mpa
Equipment Capacity:	800-1000 Pcs/h
Ambient Temperature:	0~55°C
Allowed Humidity:	20%~75%
Equipment Dimensions:	1780 mm(H) × 1480 mm(L) × 1485 mm(W)

V. Mechanical Design Highlights

To achieve a more compact structure, a dedicated cam mechanism is designed for the product handling mechanism, enabling smooth and accurate transportation of electronic components with a single motor. Additionally, the equipment employs the CCD inspection principle, incorporating embedded central control and industrial-grade high-speed image transmission control technology. The optical detection system is established based on CCD/CMOS and DSP/FPGA image recognition and processing technologies. Fuzzy control with adaptive adjustment allows the system to automatically identify product categories and make adaptive adjustments, realizing the automatic sorting function for products.

Case 71: USB Testing Machine

I. Case Description

Universal Serial Bus (USB) is a type of serial bus standard and a technical specification for input/output interfaces. It is widely used in information and communication products such as personal computers and mobile devices, and has expanded to other fields such as photography equipment, digital TVs (set-top boxes), and gaming consoles. The USB model is shown in Fig. 5.6.

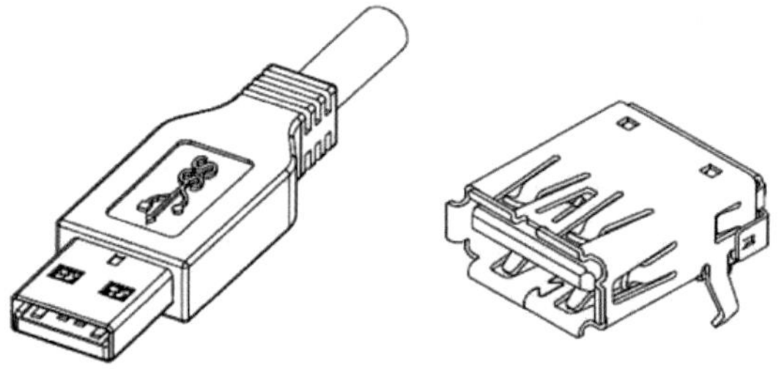

Fig. 5.6 USB Interface Model

The USB testing machine is primarily used for the testing of USB components. The overall structure is depicted in Fig. 5.7, with an enlarged view shown in Fig. 5.8. It consists of testing mechanism (1), positioning mechanism (2), touchscreen (3), frame (4), feeder (5), bending mechanism (6), retrieval mechanism (7), and mainframe (8). This device significantly improves production efficiency, product yield, reduces scrap rates, and lowers production costs.

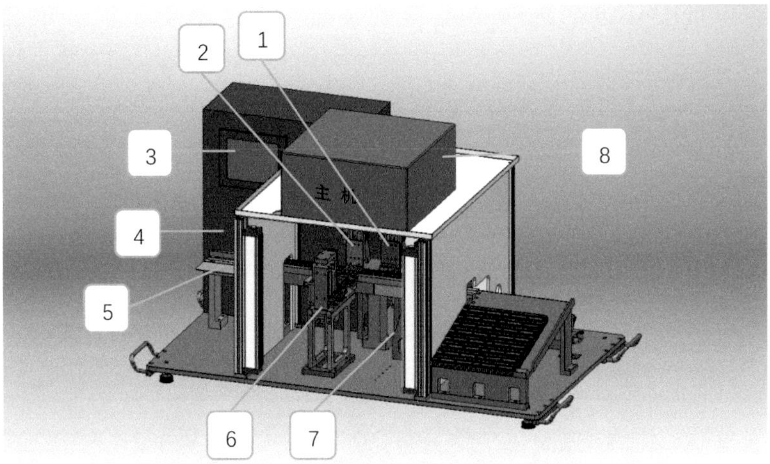

Fig. 5.7 USB Testing Machine
1-Testing Mechanism 2-Positioning Mechanism 3-Touchscreen 4-Frame 5-Feeder 6-Bending Mechanism 7-Retrieval Mechanism 8-Mainframe

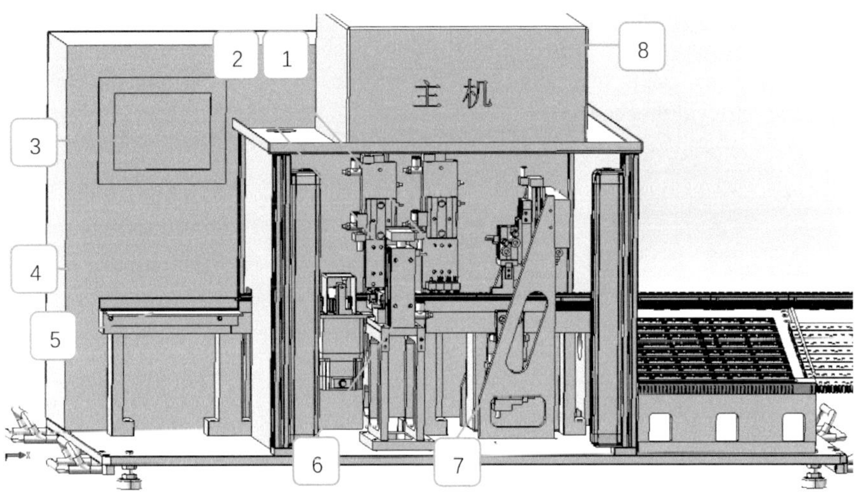

Fig. 5.8 Enlarged View of USB Testing Machine
1-Testing Mechanism 2-Positioning Mechanism 3-Touchscreen 4-Frame 5-Feeder 6-Bending Mechanism 7-Retrieval Mechanism 8-Mainframe

II. Working Principle

The USB testing machine operates by introducing the USB components through the feeding mechanism. The retrieval mechanism collects and organizes USB components, which then undergo processing and testing sequentially by the positioning mechanism, bending mechanism, and testing mechanism to improve the yield of USB components [103].

III. Main Structure Introduction

Positioning Mechanism and Bending Mechanism: When the USB testing machine is operational, the positioning mechanism (1) presses down to secure the USB component, and simultaneously, the bending mechanism (3) activates, pushing the bending head (2) to fix the USB component with the USB shell.

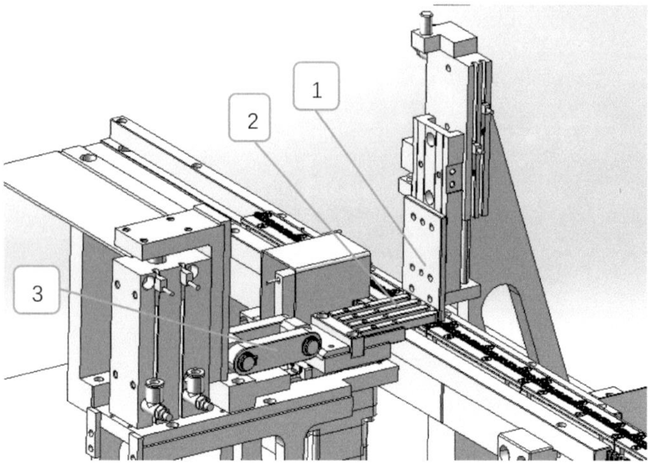

Fig. 5.9 Positioning Mechanism and Bending Mechanism
1 - Positioning Mechanism 2 - Bending Head 3 - Bending Mechanism

Testing Mechanism: As shown in Fig. 5.10, this mechanism tests the integrity and rebound of the USB's spring leaf using test pins (2). After testing, the USB component continues to be transported to the retrieval mechanism.

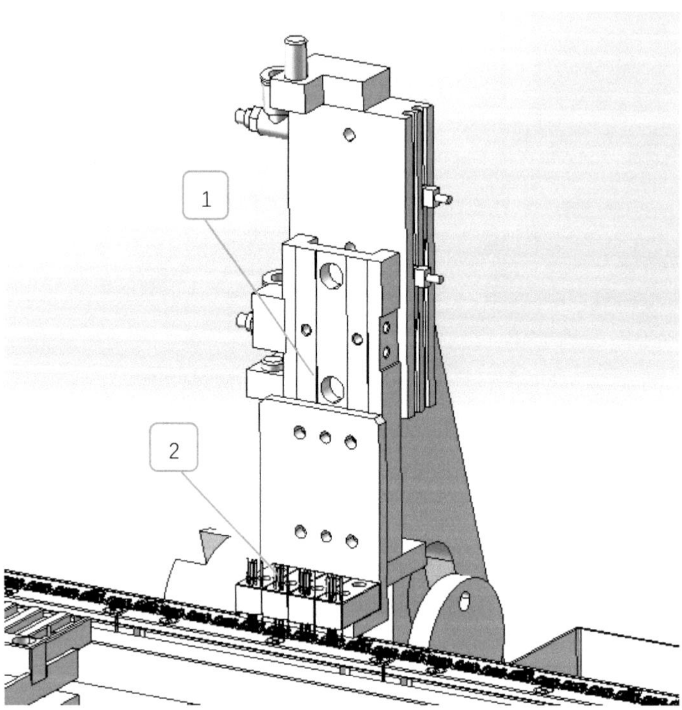

Fig. 5.10 Testing Mechanism
1 - Testing Mechanism 2 - Test Pins

The retrieval mechanism (Fig. 5.11), controlled by a cylinder (1), picks up USB components that have not passed the test and deposits them into the recycling bin (2).

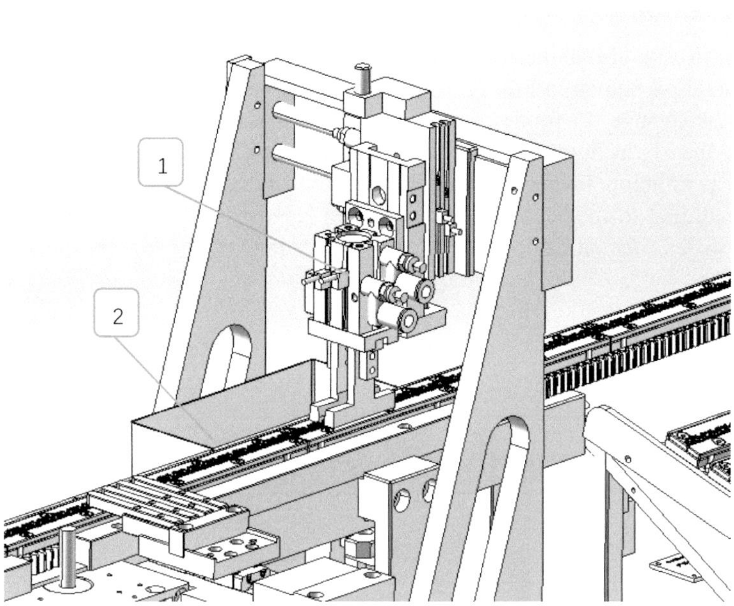

Fig. 5.11 Retrieval Mechanism
1 - Cylinder 2 - Recycling Bin

IV. Specific Parameters (see Table 5.2)

Table 5.2 USB Testing Machine Parameters

Temperature Range:	−10~40°C
Environmental Humidity:	30%~85%
USB Storage Temperature:	−20~70°C
USB Storage Humidity:	20%~90%
Device Dimensions:	585 mm(H) × 1555 mm(L) × 600 mm(W)
USB Plug Pin:	Front gold-plated 1U/2U/3U, rear tin-plated, nickel-plated iron shell, salt spray tested for 24/48/60 hours.

V. Mechanical Design Highlights

Despite its small size, USB has many components and precise structures. In the past, producing USB interfaces required over a dozen manual laborers on one production line, resulting in low efficiency and high defect rates. This USB testing machine uses structural design to replace manual labor in processing and testing USB components, significantly reducing labor and improving product quality.

Case 72: Air Tightness Testing Machine

I. Case Description

The Air Tightness Testing Machine is an automatic testing machine for the airtightness of products. The overall structure is shown in Fig. 5.12, and a detailed view is provided in Fig. 5.13. It mainly consists of a pressure gauge (1), frame (2), power switch (3), parameter setting dial (4), pressurization mechanism (5), and alarm light (6). The air tightness testing machine can automatically perform operations such as inflation, balance, detection, judgment, exhaust, display, alarm, and information transmission according to the set test parameters and standards. This eliminates the influence of human factors in leak detection, achieving standardization, efficiency, and automation in leak detection work.

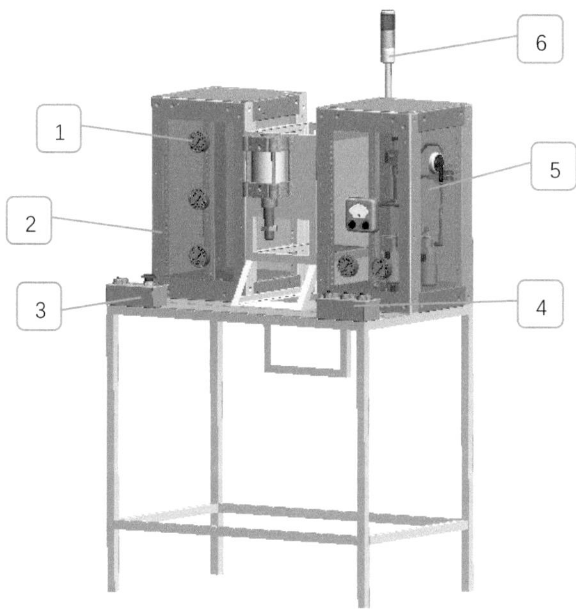

Fig. 5.12 Air Tightness Testing Machine
1 - Pressure Gauge 2 - Frame 3 - Power Switch 4 - Parameter Setting Disk 5 - Pressurization Mechanism 6 - Alarm Light

II. Working Principle

As a common means of leakage detection, pressure gauges or pressure sensors directly measure the pressure changes inside the tested workpiece, deducing the leakage amount. However, when the test environment is complex, precision requirements are high, or the test pressure is large and the leakage is small, this method places high demands on pressure gauges or sensors, making the testing cost quite expensive and sometimes even impossible. This testing machine uses high-precision electronic pressure sensors. Stable detection gas is provided by the air source according to the detection requirements, and it is pressurized to the tested workpiece through the equipment. When the internal pressure of the workpiece reaches the test pressure, the gas enters the detection phase after a period of stability. When there is no leakage in the tested workpiece, the test pressure remains basically unchanged. If there is a leak, the internal gas pressure gradually decreases as the gas leaks from the tested workpiece. The pressure sensor outputs the corresponding pressure changes in real-

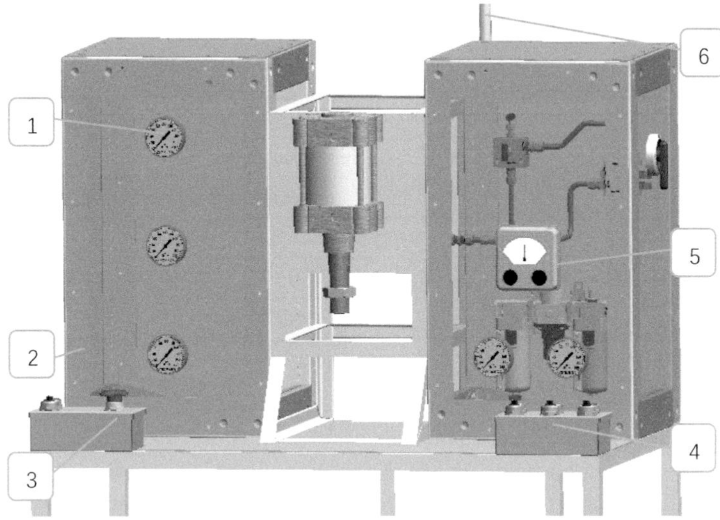

Fig. 5.13 Detailed View of the Testing Machine
1 - Pressure Gauge 2 - Frame 3 - Power Switch 4 - Parameter Setting Disk 5 - Pressurization Mechanism 6 - Alarm Light

time. According to pre-set parameters, the system automatically determines whether there is a leak. The entire testing process is fully automated, eliminating the influence of human factors [104].

III. Main Structure Introduction

The pressurization mechanism, as shown in Fig. 5.14, is pressurized according to the parameter settings. It is supplied with gas by the air source (3) through the gas supply pipeline (2) for pressurizing the airtightness testing mechanism.

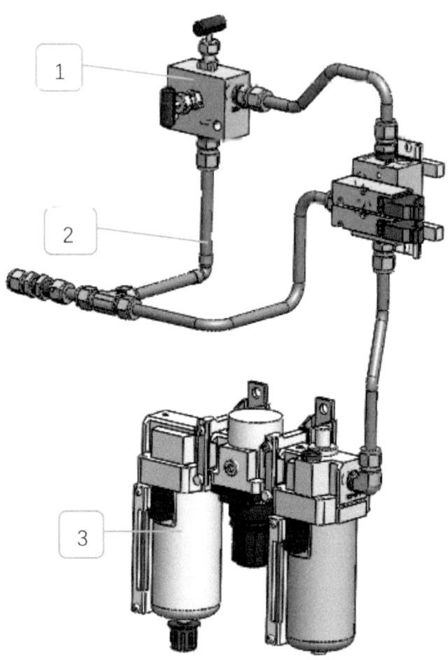

Fig. 5.14 Pressurization Mechanism
1 - Valve 2 - Gas Supply Pipeline 3 - Air Source

IV. Specific Parameters (see Table 5.3)

Table 5.3 Air Tightness Testing Machine Parameters

Pressurization Range:	0~210 MPa
Test Medium:	Air, Nitrogen, etc.
Operating Temperature:	Room temperature or customized
Pressure Accuracy:	±0.5% F.S.
Control Accuracy:	±1% F.S.
Control Mode:	Automatic/Manual

V. Mechanical Design Highlights

The airtightness testing mechanism can provide various functions for airtightness testing, including gas pressure endurance test, air tightness testing, air pressure endurance burst test, air pressure pulse test, gas filling, high-pressure gas distribution, high pressure, low-pressure environment simulation, and other non-standard gas pressure equipment [105].

Case 73: Automatic Detection Machine for Battery Electrodes

I. Case Description

The automatic detection machine for battery electrodes is primarily used for the inspection of battery electrodes. The overall structure is illustrated in Fig. 5.15, consisting mainly of the battery electrode detection mechanism (1), battery assembly mechanism (2), conveyor mechanism (3), sorting mechanism (4), and infrared detection mechanism (5). The electrode detection mechanism (1) includes mechanisms for detecting positive and negative electrodes as well as the battery casing. The operation of the automatic detection machine for battery electrodes is simple, with high detection accuracy, significantly improving the efficiency of electrode detection and thereby enhancing overall production efficiency.

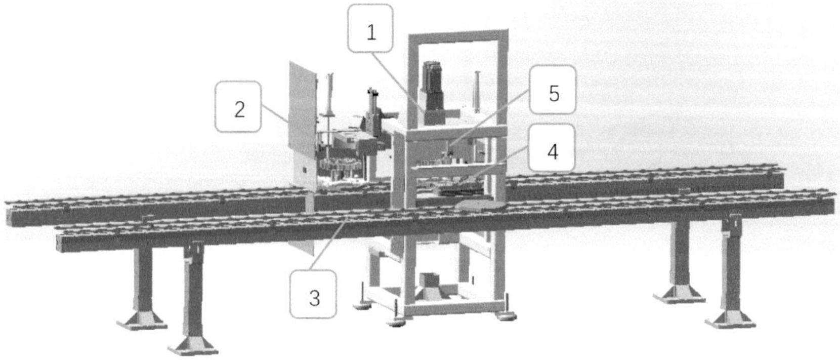

Fig. 5.15 Automatic Detection Machine for Battery Electrodes
1 - Battery Electrode Detection Mechanism 2 - Battery Assembly Mechanism 3 - Conveyor Mechanism 4 - Sorting Mechanism 5 - Infrared Detection Mechanism

II. Working Principle

When the automatic detection machine for battery electrodes is operational, the battery assembly mechanism positions the battery on the inspection platform. The conveyor mechanism then transports it to the electrode detection mechanism. The infrared detection mechanism identifies the battery, and the electrode detection mechanism automatically conducts electrode inspections. Batteries with electrode errors are sorted by the sorting mechanism into a recycling bin, while batteries with correct electrodes are output, achieving automated electrode detection for batteries [106].

III. Main Structure Introduction

The battery electrode detection mechanism, shown in Fig. 5.16, transports the battery mounted on the inspection platform to the electrode detection mechanism via a conveyor belt. The rotating disk (2), driven by cylinder (5), connects to the inspection platform (4). The electric motor (1) starts, rotating the inspection platform (4), and the infrared detection mechanism (3) inspects the battery electrodes. Simultaneously, the sorting mechanism identifies and sorts batteries with electrode errors.

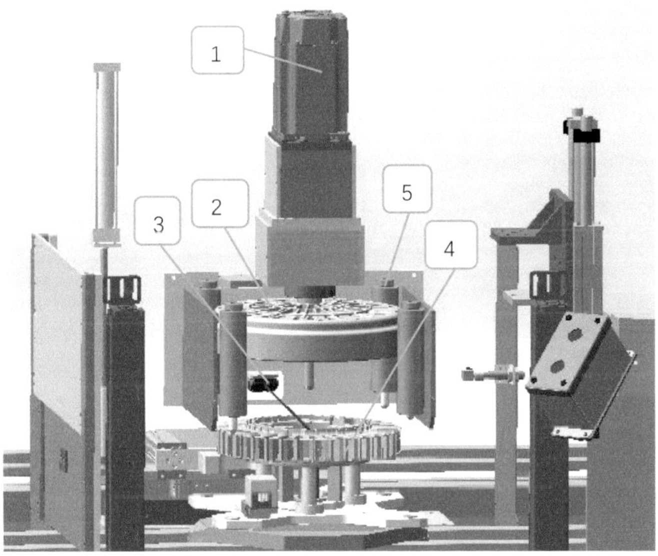

Fig. 5.16 Battery Electrode Detection Mechanism
1 - Electric Motor 2 - Rotating Disk 3 - Infrared Detection Mechanism 4 - Inspection Platform 5 - Cylinder

The battery assembly mechanism, as shown in Fig. 5.17, transports the battery-mounted inspection platform (4) from the conveyor belt to the area beneath the battery test assembly. At this point, cylinder (1) is activated, causing the pressing plate (2) to lower and simultaneously rotating the inspection platform (4). The fallen battery is corrected by the spring-loaded blade (3), facilitating the battery electrode inspection process by the battery electrode detection mechanism.

The sorting mechanism, illustrated in Fig. 5.18, consists mainly of a fixture (1) and a cylinder (2). When the battery electrode detection mechanism detects a battery with electrode errors, cylinder (2) activates, and the fixture (1) picks up the battery with electrode errors, transferring it to the recycling bin.

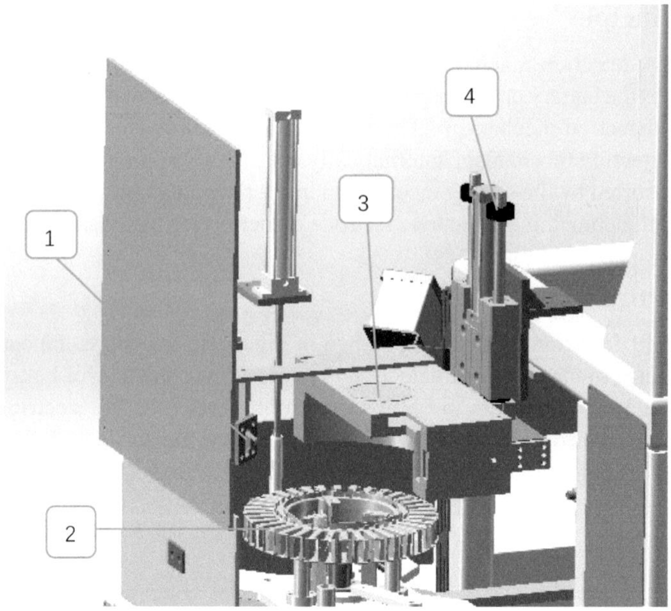

Fig. 5.17 Battery Assembly Mechanism
1 - Cylinder 2 - Pressing Plate 3 - Spring-loaded Blade 4 - Inspection Platform

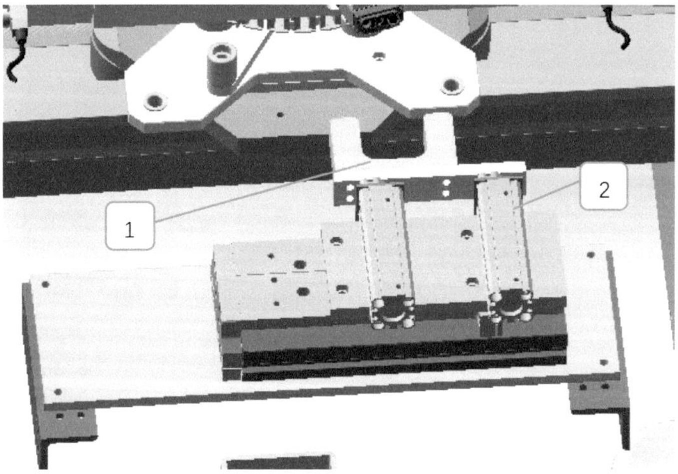

Fig. 5.18 Sorting Mechanism
1 - Fixture 2 - Cylinder

IV. Specific Parameters (see Table 5.4)

Table 5.4 Parameters for the Automatic Detection Machine for Battery Electrodes

Voltage Measurement Range:	0.00V-20V
Voltage Measurement Accuracy:	±0.2%
Working Efficiency:	55 pcs/min

Contd.

Operating Temperature:	1~45°C
Allowed Humidity:	<75%
Machine Dimensions:	1670 mm(H) × 5050 mm(L) × 1500 mm(W)

V. Mechanical Design Highlights

The automatic detection machine for battery electrodes employs AC impedance testing to obtain the AC impedance spectrum of the positive/negative electrodes, positive/reference electrode, and negative/reference electrode under different states of charge (SOC) and temperatures. It utilizes equivalent circuits for the entire battery, positive electrode, and negative electrode to fit the AC impedance spectra of positive/negative electrodes, positive/reference electrode, and negative/reference electrode, obtaining three-dimensional mapping tables related to SOC and temperature. Based on the SOC and equilibrium potential curve of the battery, it calculates the current electrode equilibrium potential. The machine can directly and conveniently measure the electrode potential of ordinary batteries, improving the accuracy of measurement results.

Case 74: Electronic Component Performance Testing Machine

I. Case Description

The Electronic Component Performance Testing Machine is designed for various performance tests on electronic components. The overall structure is shown in Fig. 5.19, with a detailed view in Fig. 5.20. It consists mainly of the handling mechanism (1), performance testing mechanism (2), sorting mechanism (3), frame (4), feeding mechanism (5), and calibration mechanism (6). This machine boasts high speed and efficiency, effectively reducing the labor intensity of employees.

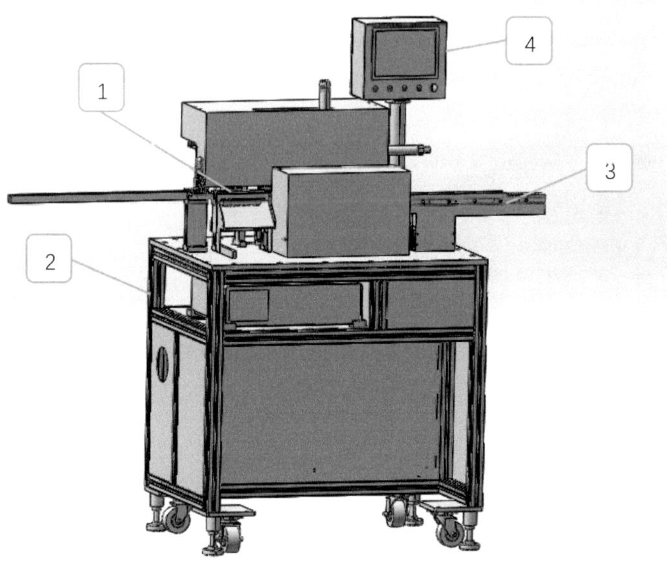

Fig. 5.19 Electronic Component Performance Testing Machine
1 - Sorting Mechanism 2 - Frame 3 - Feeding Mechanism 4 - Control Console

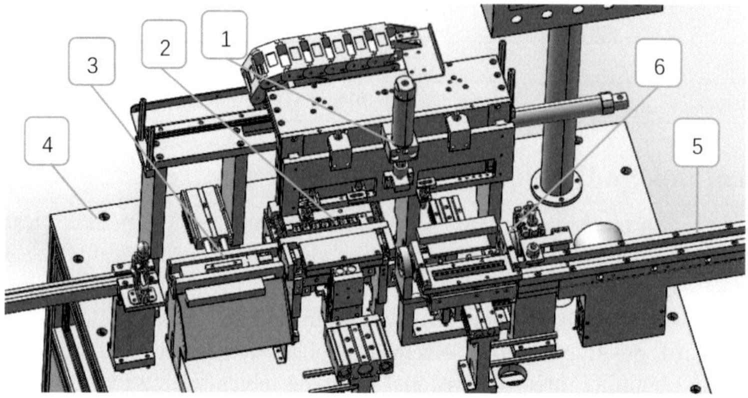

Fig. 5.20 Detailed View of Electronic Component Performance Testing Machine
1-Handling Mechanism 2-Performance Testing Mechanism 3-Sorting Mechanism 4-Frame 5-Feeding Mechanism 6-Calibration Mechanism

II. Working Principle

During operation, the feeding mechanism transports electronic components to the calibration mechanism. The calibration mechanism flips the components to a specified angle. Subsequently, the handling mechanism transfers the electronic components to the performance testing mechanism. The performance testing mechanism, driven by a cylinder (1), pushes the components into the detector (3) for testing. The cylinder (2) controls the opening and closing of the detector. The testing data is displayed on the screen. After testing, the sorting mechanism, driven by a cylinder (1), separates the defective components, while the qualified ones continue to the next process.

III. Main Structure Introduction

Feeding Mechanism (Fig. 5.21): The electronic components enter the device from the right side of the track (3). Cylinder (2) propels the components into the calibration mechanism (1).

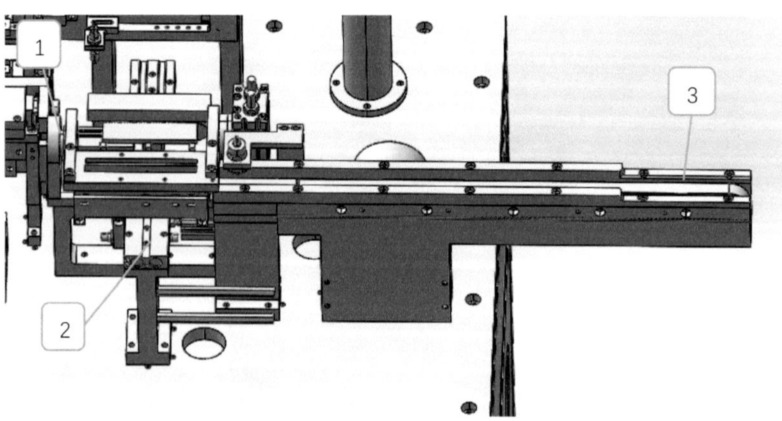

Fig. 5.21 Feeding Mechanism
1-Calibration Mechanism 2-Cylinder 3-Track

Calibration Mechanism (Fig. 5.22): When the Electronic Component Performance Testing Machine is in operation, the cylinder (4) of the feeding mechanism pushes the components into the calibration mechanism. The cylinder (3) of the calibration mechanism drives the clamp (1) to open and pick up the electronic component. The clamp (1), controlled by a servo motor (2), can rotate to adjust the orientation of the electronic component.

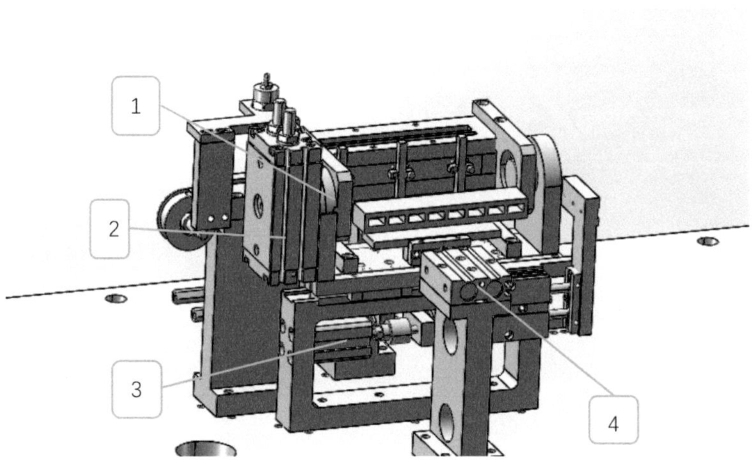

Fig. 5.22: Calibration Mechanism
1 - Clamp 2 - Servo Motor 3,4 - Cylinders

Handling Mechanism (Fig. 5.23): The suction cup (4), mounted on the installation frame (2), is connected to the installation frame (2) via a chain. This ensures the stability of the handling mechanism during operation. When the Electronic Component Performance Testing Machine is working, the cylinder (6) of the handling mechanism drives the suction cup (4) downward. The suction cup (4) picks up the electronic component that has been calibrated by the calibration mechanism. Meanwhile, the installation frame is driven by the cylinder (5) and moves horizontally along the track (3), transporting the calibrated electronic component to the performance testing mechanism (2).

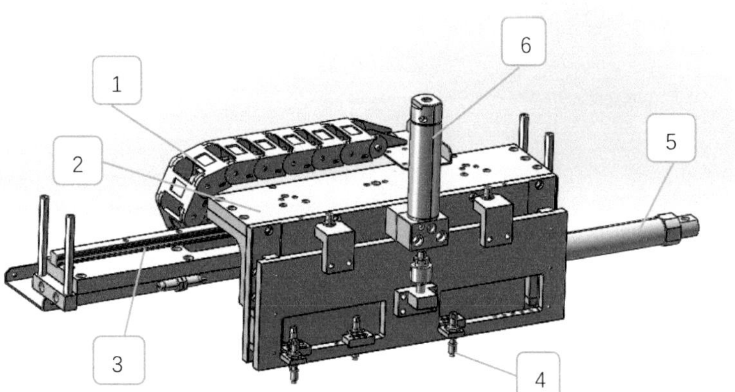

Fig. 5.23 Handling Mechanism
1 - Rack 2 - Installation Frame 3 - Track 4 - Suction Cup 5,6 - Cylinders

Performance Testing Mechanism (Fig. 5.24): After calibration, the electronic component is transferred from the handling mechanism to the performance testing mechanism. The cylinder (1) of the performance testing mechanism pushes the electronic component into the detector (3) for testing. The cylinder (2) controls the opening and closing of the detector. The test data is displayed on the screen [107].

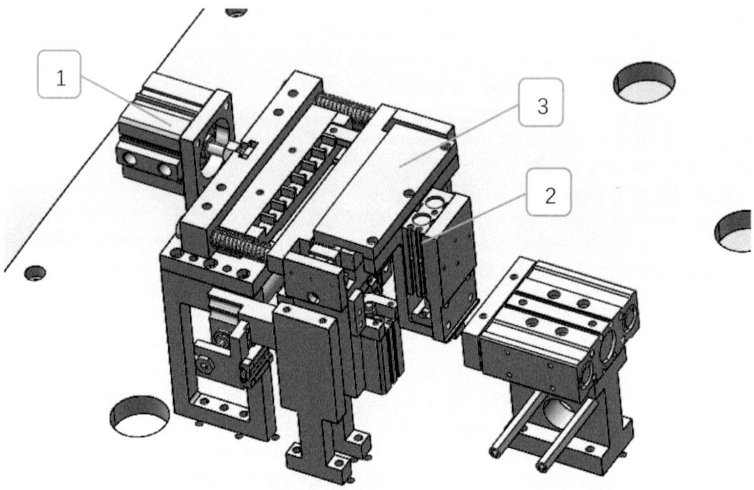

Fig. 5.24 Performance Testing Mechanism
1,2 - Cylinder 3 - Detector

Sorting Mechanism (Fig. 5.25): Driven by cylinder (1), the sorting mechanism selects defective electronic components, which are then sent to the recycling bin. Qualified electronic components continue to move along the guide rail (2).

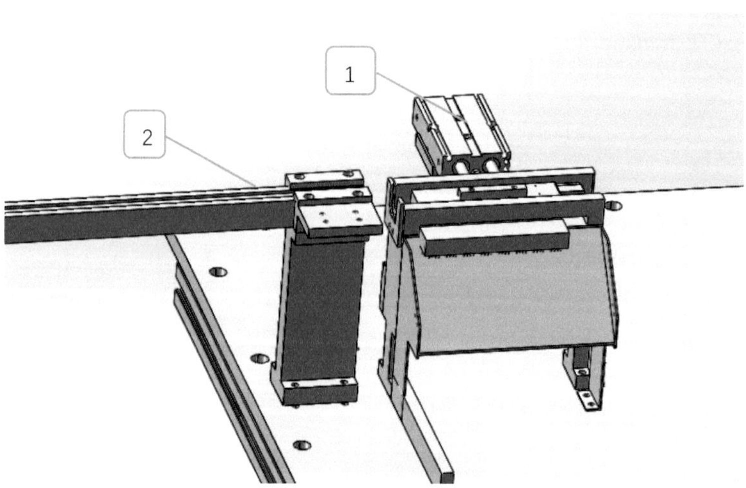

Fig. 5.25 Sorting Mechanism
1 - Cylinder 2 - Guide Rail

IV. Specific Parameters (see Table 5.4)

Table 5.5 Electronic Component Performance Testing Machine Parameters

Voltage Measurement Accuracy:	±0.2%
Operating Efficiency:	1000-2000 pcs/h
Operating Temperature:	1~45°C
Allowable Humidity:	~80%
Equipment Dimensions:	1470 mm(H) × 1620 mm(L) × 600 mm(W)

V. Mechanical Design Highlights

The Electronic Component Performance Testing Machine streamlines the testing process for various electronic components, significantly improving efficiency. Utilizing high-precision fixtures ensures testing quality. The inclusion of multiple stations for both qualified and defective components enhances the unloading efficiency and effectively prevents the accumulation of products at the unloading port [108].

VI. Testing Indicators

Source Table Section

(1) Voltage Source Output Specifications: Minimum range: 200 mV, programming resolution: 5 μV, output accuracy: 0.02% + 37 μV; Maximum range: 200 V, programming resolution: 5 mV, output accuracy: 0.02% + 50 mV.
(2) Current Source Output Specifications: Minimum range: 100 nA, programming resolution: 2 pA, accuracy: 0.06% + 100 pA. Maximum range: 1.5 A, programming resolution: 50 μA, accuracy: 0.06% + 4 mA.
(3) Voltage Measurement Specifications: Minimum range: 200 mV, resolution: 100 nV, accuracy: 0.015% + 225 μV. Maximum range: 200 V, resolution: 100 μV, accuracy: 0.015% + 50 mV.
(4) Current Measurement Specifications: Minimum range: 100 nA, resolution: 100 fA, accuracy: 0.06% + 100 pA. Maximum range: 1.5 A, resolution: 1 μA, accuracy: 0.05% + 3.5 mA.

Multi-Channel Data Acquisition Equipment

(1) Voltage Measurement: Minimum range: 100 mV, resolution: 0.01 μV, accuracy: 10 ppmreading + 9 ppm100 mV. Maximum range: 300 V, resolution: 100 μV, accuracy: 10 ppmreading + 9 ppm 300 V.
(2) Resistance Measurement: Minimum range: 1 Ω, resolution: 0.1 μΩ, accuracy: 15 ppmreading + 80 ppm 1 Ω. Maximum range: 100 MΩ, resolution: 10Ω, accuracy: 800 ppmreading + 30 ppm 100 MΩ.

Case 75: Resistance Chip Inspection Machine

I. Case Description

The Resistance Chip Inspection Machine is designed for the rapid testing and sorting of resistance chips. The overall structure, as shown in Fig. 5.26, consists of the Resistance Chip Inspection Mechanism (1), Sieve Rod Mechanism (2), Material End Box (3), Base (4), Electrical Control Box (5), and Vibration Conveyor (6). The structures are rigidly connected, providing simplicity in assembly, reliability, durability, and easy maintenance.

Fig. 5.26 Resistance Chip Inspection Machine
1 - Resistance Chip Inspection Mechanism 2 - Sieve Rod Mechanism 3 - Material End Box 4 - Base 5 - Electrical Control Box 6 - Vibration Conveyor

II. Working Principle

When operational, the Vibration Conveyor transports resistance chips to the Resistance Chip Inspection Mechanism, where probes are used to test the chips. After testing, a cylinder propels a guide block to transport the resistance chips to the Material End Box. Positioned above the Material End Box is the Sieve Rod Mechanism, equipped with two rotating electric motors and rotating blocks, effectively classifying the resistance chips [109].

III. Main Structure Introduction

The Vibration Conveyor, depicted in Fig. 5.27, features a hopper (3) with a pulse electromagnet beneath, causing vertical vibrations. Tilted spring blades induce oscillations around the vertical axis, lifting parts within the hopper along a spiral track for organized delivery to the Resistance Chip Inspection Mechanism.

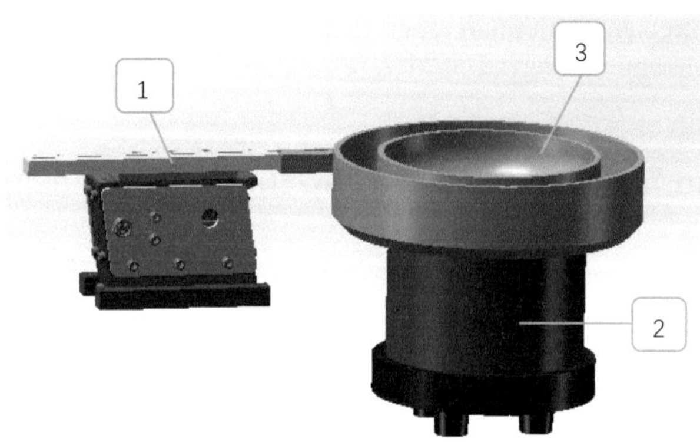

Fig. 5.27 Vibration Conveyor,
1 - Guide Rail 2 - Vibration Base 3 - Hopper

The Resistance Chip Inspection Mechanism, shown in Fig. 5.28, operates with a cylinder (3) pushing the resistance chips. Probes (2) and a detector (5) examine the resistance data. The detector (5), driven by cylinder (1), descends to connect with the resistance chip. After testing, the Sieve Rod Mechanism (4) classifies the resistance chips.

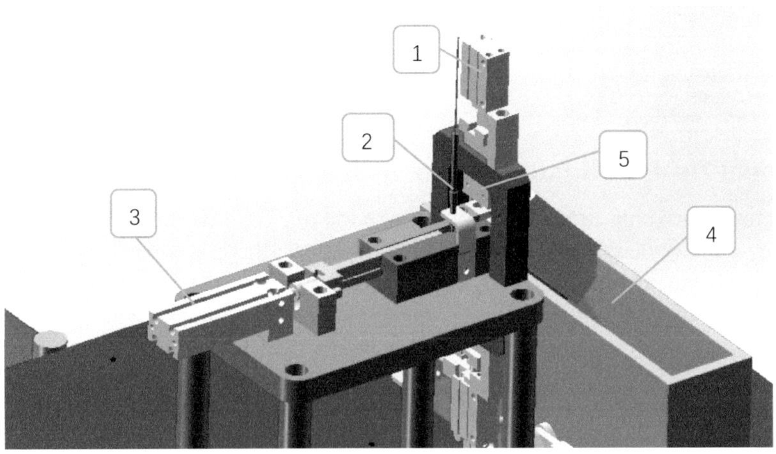

Fig. 5.28 Resistance Chip Inspection Mechanism
1, 3-Cylinder 2-Probes 4-Sieve Rod Mechanism 5-Detector

The Sieve Rod Mechanism structure, depicted in Fig. 5.29, includes two rotating electric motors and rotating blocks (1, 7), facilitating effective classification of resistance chips through the rotation of the sieve rods (3, 5).

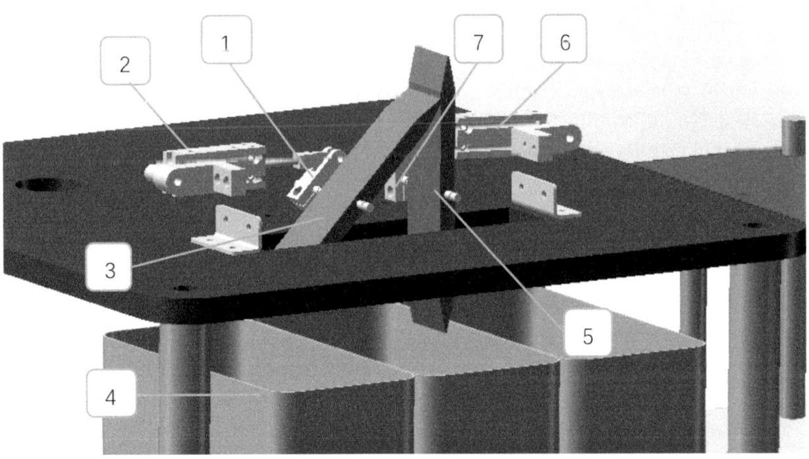

Fig. 5.29 Sieving Mechanism
1, 7-Rotary Block 2, 6-Cylinder 3, 5-Sieving Rod 4-Material Box

IV. Specific Parameters (see Table 5.6)

Table 5.6 Parameters of Resistance Chip Detection Machine

Error Rate:	≤0.2%
Operating Efficiency:	65.90 pcs/min
Temperature Requirements:	1~55°C
Humidity:	<75%
Equipment Dimensions:	1150 mm(H) × 750 mm(L) × 450 mm(W)

V. Mechanical Design Highlights

Many current resistance chip inspection machines in the market lack sufficient automation, exhibit a single mode of conveyance, and lack stability. Therefore, there is a need for a highly efficient and automated resistance chip inspection machine. The machine places a material end box under the rectangular hole of the working platform on the base and is equipped with two rotating electric motors. The motors' output shafts are connected to one end of rotating blocks, and the central hole of the rotating blocks has a connecting rod. The end of the connecting rod passes through the groove of the sieve rod for effective chip sorting. Additionally, the resistance chip inspection machine features simple assembly, high reliability, and ease of maintenance due to rigid connections between structures.

Case 76: Buckle Terminal Sorting Machine

I. Case Description

Terminals, as shown in Fig. 5.30, are components that connect batteries to external conductors. In electrical engineering, terminals mostly refer to wiring terminals, also known as terminal blocks, with types including single-hole, double-hole, plug-in, hook, etc., and made from materials such as copper-plated silver, copper-plated zinc, copper, aluminum, iron, etc. Their main function is to transmit electrical signals or conduct electricity. Generally, during terminal processing, the finished terminals need to be sorted to distinguish between good and bad. The buckle terminal sorting machine is suitable for automatic sorting of buckle terminals. The overall structure is shown in Fig. 5.31, with an enlarged view shown in Fig. 5.32. It mainly consists of a controller (1), cam mechanism (2), terminal clamp (3), detection mechanism (4), frame (5), and vibration conveyor (6), etc. The buckle terminal sorting machine has the advantages of high detection efficiency, high accuracy, and simple operation.

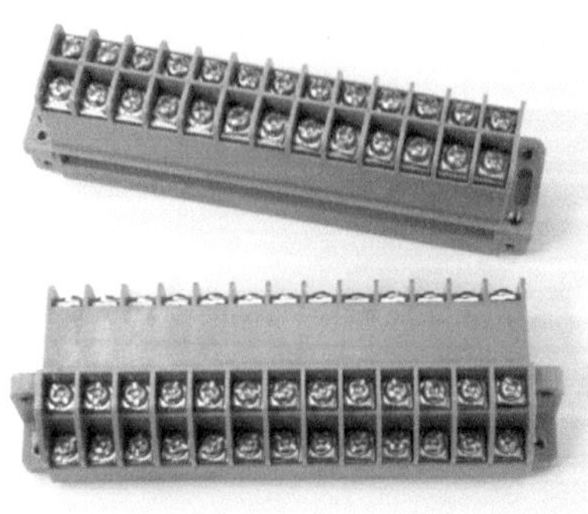

Fig. 5.30 Terminals

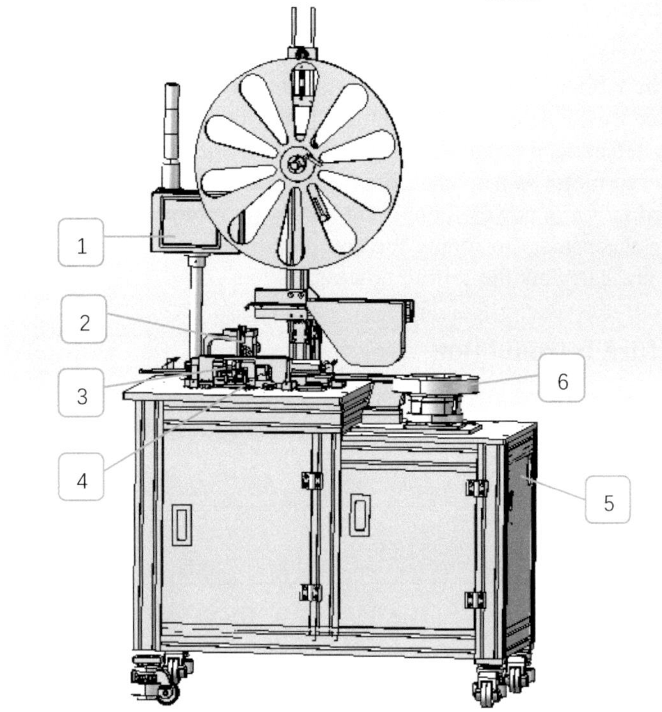

Fig. 5.31 Buckle Terminal Sorting Machine
1 - Controller 2 - Cam Mechanism 3 - Terminal Clamp 4 - Detection Mechanism 5 - Frame 6 - Vibration Conveyor

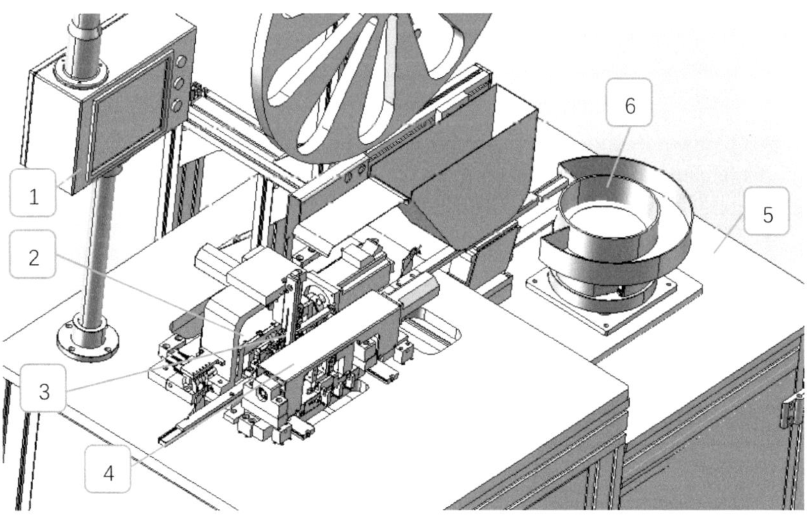

Fig. 5.32 Enlarged View of Buckle Terminal Sorting Machine
1 - Controller 2 - Cam Mechanism 3 - Terminal Clamp 4 - Detection Mechanism 5 - Frame 6 - Vibration Conveyor

II. Working Principle

When the buckle terminal sorting machine is in operation, buckle terminals to be tested are fed into the machine from the vibration conveyor. They are transported to the detection point by the guide rail and then detected by the detection mechanism. If the terminal is good, it is placed in the good terminal box; if it is defective, it is placed in the recycle bin. The terminal clamp is driven by a servo motor. When the servo motor is activated, the terminal clamp automatically adjusts its position. Every time the servo motor is powered on, it automatically returns to the origin. The servo motor then drives the cam mechanism to rotate, thereby controlling the terminal clamp to sequentially grasp the terminals and carry out the sorting process.

III. Main Structure Introduction

The terminal clamp, as shown in Fig. 5.33, is driven by a servo motor (2) to move vertically from top to bottom, used to send the tested buckle terminals into the corresponding storage box.

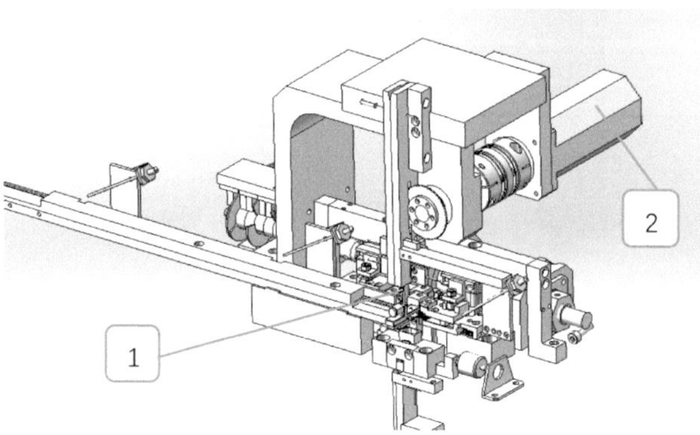

Fig. 5.33 Terminal Clamp
1 - Terminal Clamp 2 - Servo Motor

The cam mechanism, as shown in Fig. 5.34, is driven by a servo motor (1) to rotate, controlling the horizontal displacement of the terminal clamp (2). It works with the servo motor to sequentially place the tested buckle terminals into the corresponding storage box.

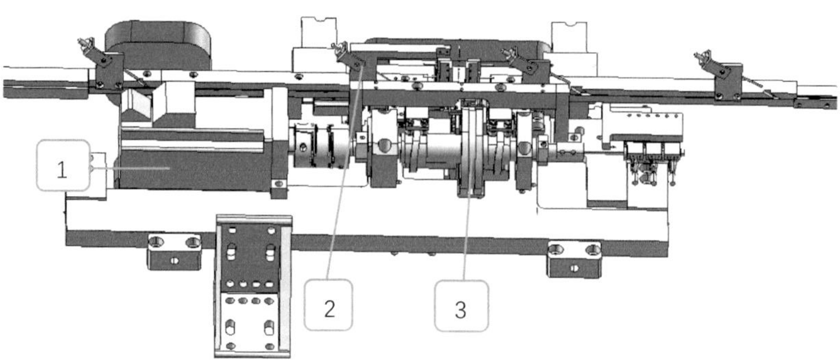

Fig. 5.34 Cam Mechanism
1 - Servo Motor 2 - Terminal Clamp 3 - Cam

The detection mechanism, as shown in Fig. 5.35, is driven by a servo motor (2) to control the detection head (1) for testing the buckle terminals. There is a fixed installation of a visual lens inside the detection head (1) for testing, which can determine the opening size of the terminal. If the terminal is good, it is placed in the good terminal box; if it is defective, it is placed in the defective terminal box.

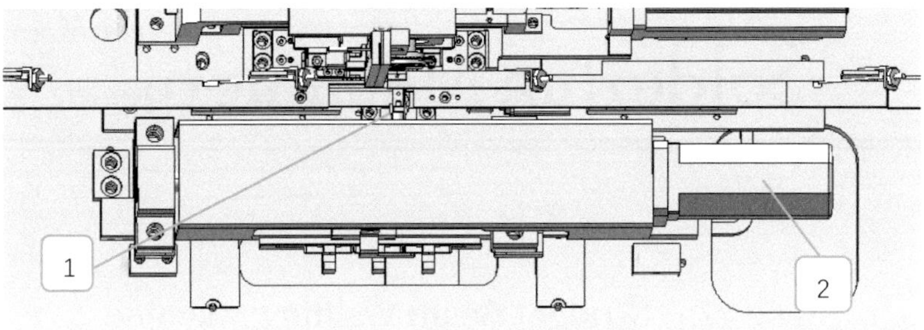

Fig. 5.35 Detection Mechanism
1 - Detection Head 2 - Servo Motor

IV. Specific Parameters (Refer to Table 5.7)

Table 5.7 Buckle Terminal Sorting Machine Parameters

Efficiency:	2500-3000 pieces/hour
Power:	1.2 kW
Utilization Rate:	≥93%
Ambient Temperature:	0~45°C
Allowed Humidity:	30%~95%
Equipment Dimensions:	1930 mm(H) × 1120 mm(L) × 690 mm(W)

V. Precautions

(1) In case of abnormal operation, turn off the power and promptly call maintenance personnel for inspection and adjustment.
(2) Only designated personnel are allowed to adjust or dismantle machine parts.
(3) Before exiting during work hours, the power to the motor must be turned off.
(4) When replacing terminals, the power must be disconnected first, and then replacement operations can be performed.
(5) Before starting the equipment, ensure that the output and circuit connections are firm.

Chapter 6

Soldering Machinery

Case 77: Straight Seam Welding Machine

I. Case Description

The straight seam welding machine is an automated welding device that completes linear weld seams automatically through automation and mechanization. The overall structure is shown in Fig. 6.1, consisting mainly of components such as cylinders (1), crossbeams (2), welding heads (3), core shaft mechanisms (4), main units (5), servo motors (6), rotary mechanisms (7), and welding gun fine-tuning mechanisms (8). The crossbeam includes a rack and crossbeam driving components. The straight seam welding machine can significantly replace manual labor, greatly improve production efficiency and welding quality, reduce labor costs, and improve the working environment for welders.

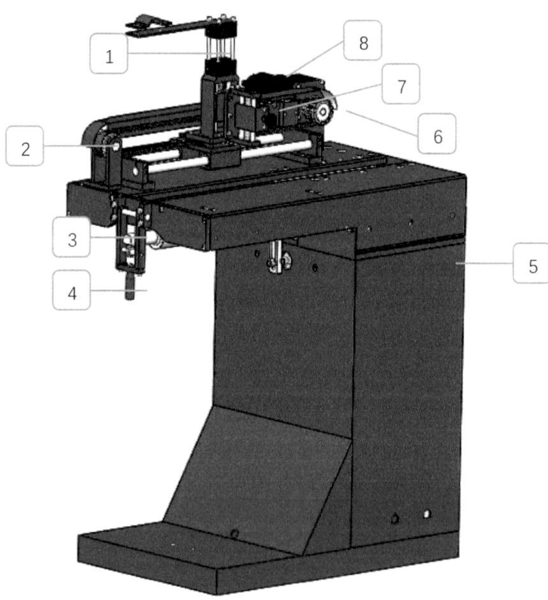

Fig. 6.1 Straight Seam Welding Machine
1 - Cylinder 2 - Crossbeam 3 - Welding Head 4 - Core Shaft Mechanism 5 - Main Unit 6 - Servo Motor 7 - Rotary Mechanism 8 - Welding Gun Fine-Tuning Mechanism

II. Working Principle

When the straight seam welding machine is operational, the welding area of the workpiece is pressed, aligned with the welding main beam and welding head, and the welding gun height is adjusted by the cylinder. The welding gun fine-tuning mechanism finely adjusts the welding gun to improve welding accuracy, aligning the tungsten electrode with the weld seam. The servo motor drives the welding head to make a horizontal displacement, completing the straight seam welding of the workpiece in one pass.

III. Main Structure Introduction

The welding gun fine-tuning mechanism, as shown in Fig. 6.2, finely adjusts the welding head through top supports (1) and (2), greatly improving welding accuracy.

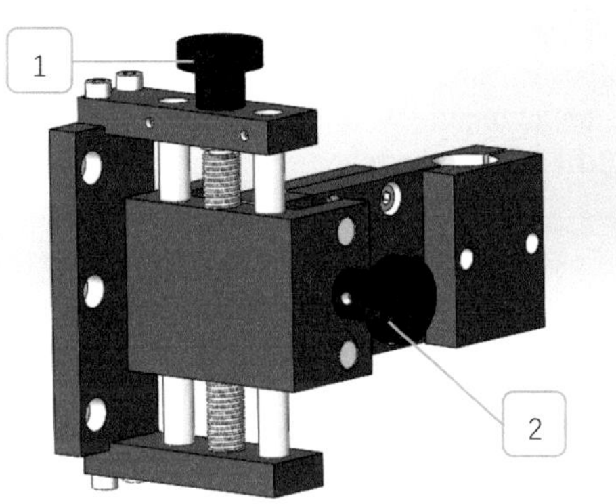

Fig. 6.2 Welding Gun Fine-Tuning Mechanism
1, 2 - Top Supports

IV. Specific Parameters (Refer to Table 6.1)

Table 6.1 Straight Seam Welding Machine Parameters

Welding Thickness Range:	0.5~6 mm
Effective Length of Welded Workpiece:	<300 mm
Minimum Diameter of Welded Workpiece:	60 mm
Welding Gun Movement Speed Range:	100~1600 mm/min
Welding Gun Lifting Stroke:	200 mm
Permissible Humidity:	<75%
Air Source Pressure:	0.4~0.7 MPa
Equipment Dimensions:	1230 mm(H) × 490 mm(L) × 790 mm(W)

V. Mechanical Design Highlights

(1) The cantilever structure is adopted, and the two cantilever beams undergo stress relief annealing after welding to ensure long-term deformation resistance.

(2) Pneumatic clamping structure, arranged closely along both sides of the straight seam, ensures uniform clamping of the butt weld throughout the entire welding length range. The spacing

between the left and right pressure fingers is adjustable to adapt to the welding of different workpieces.

(3) Depending on the thickness of the workpiece, pneumatic or cylinder types can be used to ensure sufficient clamping force to prevent thermal deformation during welding.

(4) The welding core shaft is embedded with a copper-made mold to provide back gas protection for the weld seam. Different welding process slots are designed according to the processing of cylindrical or flat workpieces to achieve single-side welding and double-side forming.

(5) The spacing between the welding core shaft and the pressure fingers is adjustable to meet the welding requirements of different workpieces.

(6) The welding gun movement adopts a stepper motor drive, gear-rack transmission, and the rail surface is processed by grinding, ensuring smooth movement and stable and reliable welding.

VI. Precautions

Straight seam automatic welding machines are now widely used in steel structures, shipbuilding, automotive, household appliances, and some architectural decoration manufacturing industries. During the use of straight seam automatic welding machines, different supporting products need to be configured according to different products.

(1) For welding workpieces with a length of more than 1500 mm, due to equipment errors or workpiece errors, it is easy to cause poor weld seam formation or weld seam deviation. Therefore, it is necessary to configure tracking systems or swing systems to overcome defects such as edge biting or incomplete penetration caused by the bending of the weld seam, in order to truly achieve automatic welding and achieve efficiency and welding quality.

(2) For welding the cylinder body with a steel plate thickness of more than 2 mm, due to the large gap of the weld seam, the parent material of the workpiece itself cannot fill the weld seam during melting, resulting in welding defects. In this case, an automatic wire feeding mechanism needs to be added to increase the amount of metal melting during welding to fill the weld seam and achieve high-quality welding.

Case 78: Electrical Component Inspection, Tray Alignment, and Sealing Integrated Machine

I. Case Description

The integrated machine for inspecting electrical components, aligning trays, and sealing can simultaneously perform sealing, inspection, and tray alignment of electrical components. The overall structure is shown in Fig. 6.3, and the top view is shown in Fig. 6.4. It mainly consists of a support frame (1), control box (2), suspension (3), sealing mechanism (4), support plate (5), inspection mechanism (6), and tray alignment mechanism (7). This integrated machine has the advantages of a high efficiency and stable operation.

II. Working Principle

When the integrated machine is in operation, a batch of electrical components is input into the sealing mechanism for sealing. After sealing, the components are input into the first inspection mechanism for detecting any anomalies such as virtual welding or burrs during welding. Components that pass the inspection are then conveyed to the tray alignment mechanism for packaging [110].

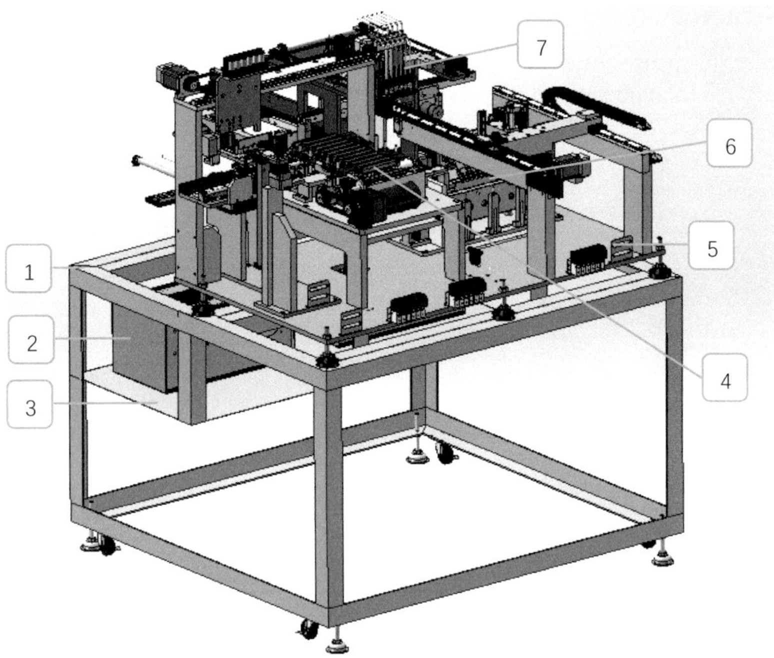

Fig. 6.3 Electrical Component Inspection, Tray Alignment, and Sealing Integrated Machine
1 - Support Frame 2 - Control Box 3 - Suspension 4 - Sealing Mechanism 5 - Support Plate 6 - Inspection Mechanism 7 - Tray Alignment Mechanism

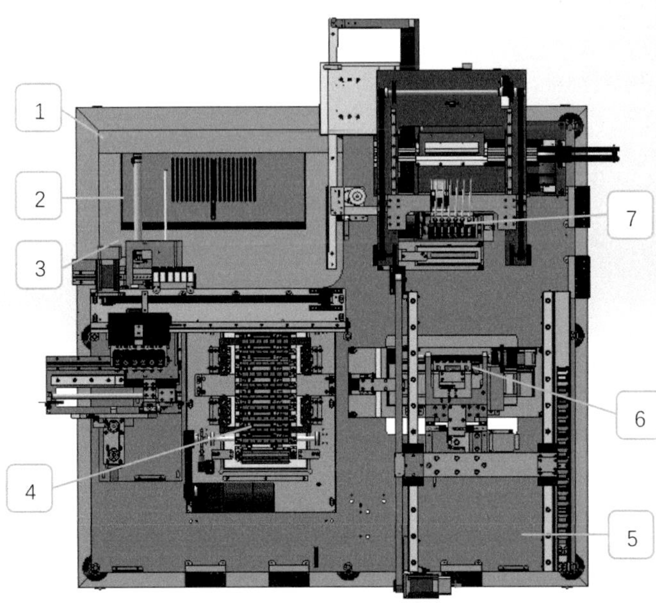

Fig. 6.4 Top View of the Integrated Machine
1 - Sealing Mechanism 2 - Support Frame 3 - Control Box 4 - Suspension 5 - Support Plate 6 - Inspection Mechanism 7 - Tray Alignment Mechanism

III. Main Structure Introduction

The sealing mechanism, as shown in Fig. 6.5, transports electrical components to be sealed from the inlet (2) through the rotation support frame (5), driven by the servo motor (3).

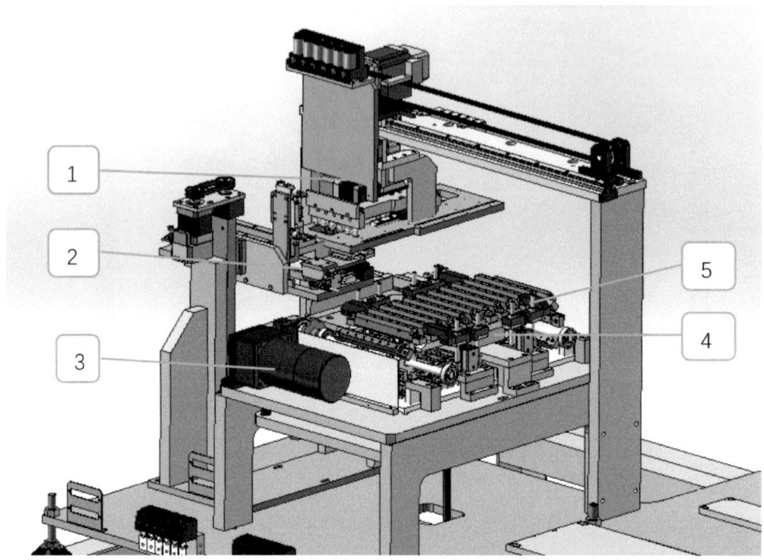

Fig. 6.5 Sealing Mechanism
1 - Transport Mechanism 2 - Inlet 3 - Servo Motor 4 - Sealing Mechanism 5 - Rotation Support Frame

The inspection mechanism, as shown in Fig. 6.6, positions the sealed electrical components under the inspection microscope (2). The inspection microscope (2), driven by a servo motor (1), horizontally moves to inspect the sealed area, and components with good inspection results are sent to the tray alignment mechanism.

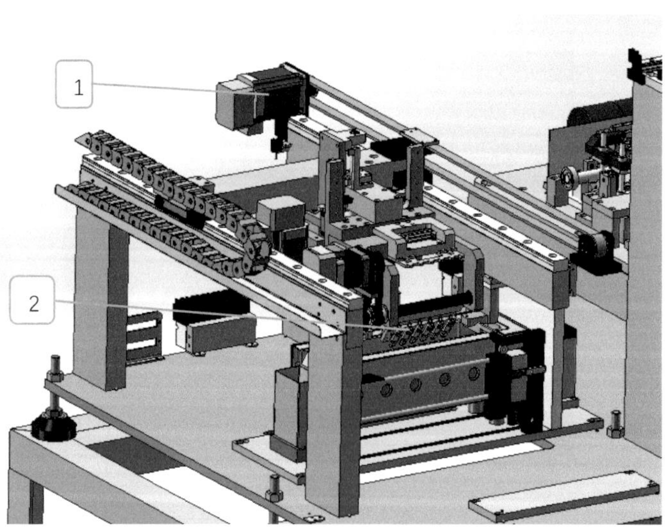

Fig. 6.6 Inspection Mechanism
1 - Servo Motor 2 - Inspection Microscope

The tray alignment mechanism, as shown in Fig. 6.7, is driven by the servo motor (3). The suction mechanism (1) picks up electrical components with good inspection results and places them on the tray track (2), which then conveys the components to the tray.

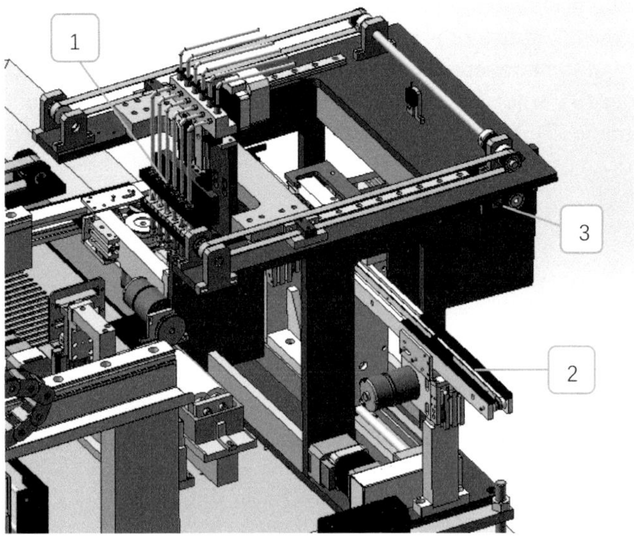

Fig. 6.7 Tray Alignment Mechanism
1 - Suction Mechanism 2 - Tray Track 3 - Servo Motor

IV. Specific Parameters (Refer to Table 6.2)

Table 6.2 Electrical Component Inspection, Tray Alignment, and Sealing Integrated Machine Parameters

Working Efficiency:	1200-1500 pcs/h
Operating Rate:	≥93%
Yield Rate:	≥98%
Environmental Temperature:	0~55°C
Allowable Humidity:	20%~90%
Equipment Dimensions:	1465 mm(H) × 1300 mm(L) × 1200 mm(W)

V. Mechanical Design Highlights

Due to technological limitations, the welding of electrical components is often manually done, making it challenging to guarantee welding quality, especially for the welding of electrical components with internal fuses. This process requires welding a wire with a diameter typically ranging from 0.3 to 0.8 mm to the protruding aluminum foil on the end face of the component. This demands high skills from the operator, and the quality of welding is hard to ensure. Electrical components operate in conditions of high voltage and field strength, demanding high-quality welding. Any abnormalities such as virtual welding or burrs during welding can impact the product's lifespan and, in severe cases, lead to explosions. The integrated machine adopts automated sealing technology, providing a high level of intelligence. Additionally, a visual inspection system is set up on the equipment to detect the welding situation of electrical components in real-time, ensuring product quality.

Case 79: Automatic Soldering Machine

I. Case Description

The automatic soldering machine is an automated soldering device designed to complete soldering tasks efficiently. The overall structure is shown in Fig. 6.8, with an enlarged view depicted in Fig. 6.9. It consists mainly of the horizontal displacement mechanism (1), lifting mechanism (2), angle adjustment mechanism (3), soldering mechanism (4), positioning and clamping mechanism (5), solder pool (6), and frame (7). The automatic soldering machine is characterized by high soldering accuracy and a high degree of automation.

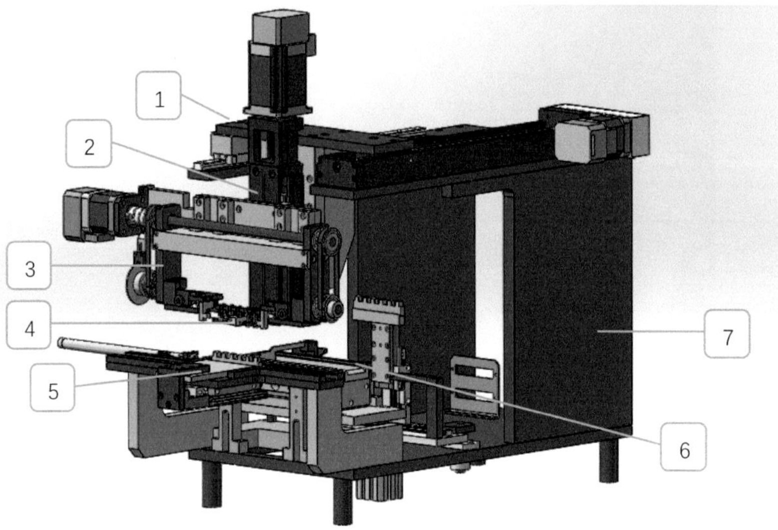

Fig. 6.8 Automatic Soldering Machine
1-Horizontal displacement mechanism 2-Lifting mechanism 3-Angle adjustment mechanism 4-Soldering mechanism 5-Positioning and clamping mechanism 6-Solder pool 7-Frame

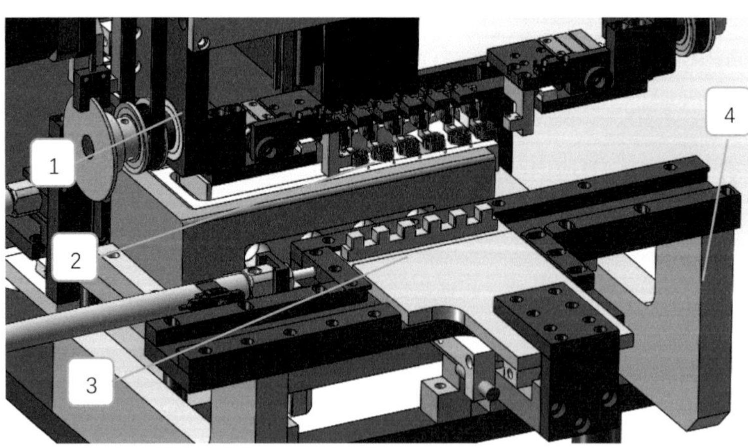

Fig. 6.9 Detailed View of the Automatic Soldering Machine
1-Angle adjustment mechanism 2-Soldering mechanism 3-Positioning and clamping mechanism 4-Frame

II. Working Principle

The automatic soldering machine fixes the product to be soldered on the processing base. The positioning and clamping mechanism locates the product, and the soldering mechanism automatically heats and melts the solder wire to solder the components. After cooling, a firm and reliable solder joint is formed, completing the automatic soldering process. When the soldering process changes, the automatic soldering machine can adjust the programming to switch between spot welding, drag welding, and other techniques [111].

III. Main Structure Introduction

The positioning and clamping mechanism, as shown in Fig. 6.10, fixes the product to be soldered on the processing base using the cylinder (2) to control the position of the product and dock with the soldering mechanism (1).

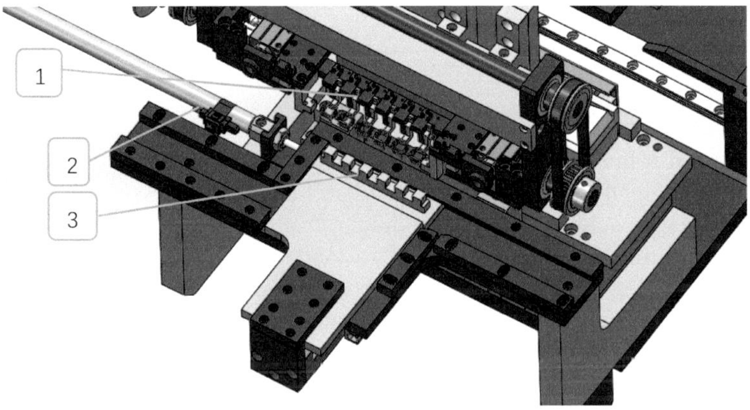

Fig. 6.10 Positioning and Clamping Mechanism
1 - Soldering mechanism 2 - Cylinder 3 - Processing base

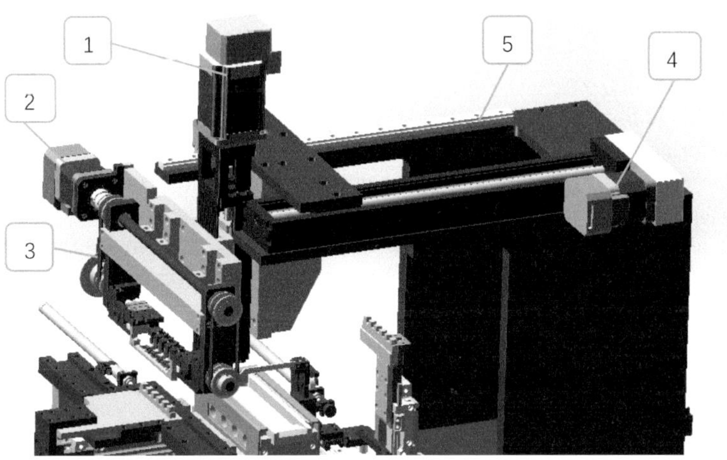

Fig. 6.11 Automatic Soldering Machine Motion System
1 - Cylinder 2,4 - Servo motors 3 - Pulley 5 - Guide rail

The motion system of the automatic soldering machine, depicted in Fig. 6.11, is controlled by cylinder (1) for vertical lifting of the soldering mechanism, servo motors (2) and (4) for controlling the soldering angle using pulleys (3), and servo motor (4) for horizontal displacement along guide rail (5). This equipment completes soldering tasks through the coordinated operation of multiple mechanisms [112].

IV. Specific Parameters (see Table 6.3)

Table 6.3 Automatic Soldering Machine Parameters

Temperature deviation:	±1°C
Soldering accuracy:	±0.02 mm
Operating rate:	≥93%
Yield rate:	≥98%
Permissible humidity:	<75%
Equipment dimensions:	535 mm(H) × 490 mm(L) × 650 mm(W)

V. Mechanical Design Highlights

(1) The adoption of a multi-axis linkage manipulator ensures the quality and stability of the automatic soldering machine technology.
(2) It supports both spot welding and drag welding, with all process parameters adjustable to adapt to various high-difficulty operations and micro-soldering techniques.
(3) Rapid heating and cooling ensure efficient operation, with the temperature from room temperature to the set temperature of 300°C requiring only 10 seconds, and reheating within 0.1 seconds (temperature difference within 10°C) with a temperature deviation of ±1°C.
(4) Additional nitrogen supply fixtures ensure the best soldering quality.
(5) Can be paired with automated conveyors for fully automated operations.
(6) Optional CCD for full-process monitoring.

VI. Precautions

During soldering operations, ensure the following:
(1) Avoid oversaturation of solder joints, maintaining them at three or four points in the middle to prevent short circuits.
(2) The solder should form a small semicircle with the component pin to ensure neat soldering and optimal performance.
(3) The voltage drop during soldering should not exceed 5% of the initial voltage, and proper protective gear such as gloves and goggles should be worn during operations.
(4) After each use, do not clean the soldering iron tip on a sponge; instead, place the solder on the soldering iron tip into a tin collection box to maintain the tip's temperature and extend its lifespan.
(5) Clean the soldering machine daily, replenish solder, and then immediately cut off the power supply to prolong the machine's lifespan.

Case 80: Test Tinning Machine

I. Case Description

The test tinning machine is capable of automatically cutting and tinning the leads of stator coils. The overall structure is shown in Fig. 6.12, with an enlarged view in Fig. 6.9. The main components include the display screen (1), handling mechanism (2), lead cutting mechanism (3), frame (4), discharge chute (5), fixture (6), and solder pot (7). The machine offers high production efficiency and stable product quality.

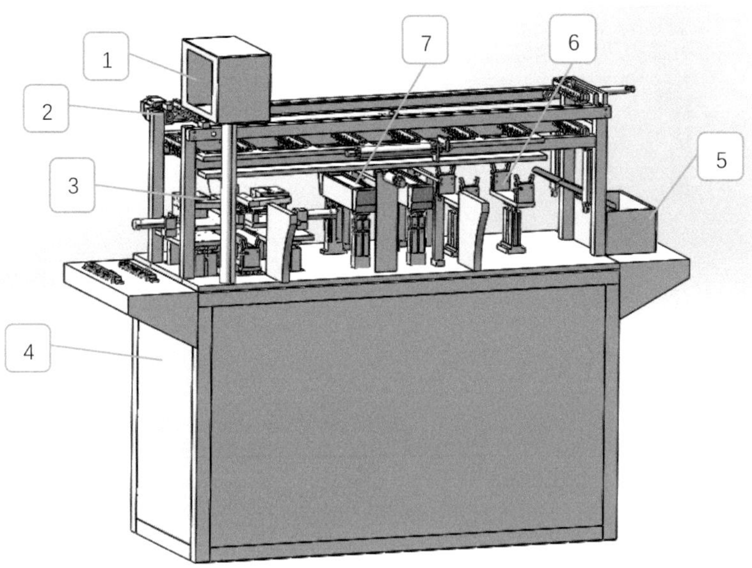

Fig. 6.12 Testing and Tinning Machine
1 - Display Screen 2 - Handling Mechanism 3 - Lead cutting Mechanism 4 - Frame 5 - Discharge Chute 6 - Fixture 7 - Solder Pot

II. Working Principle

When the test tinning machine operates, the power switch is turned on as required. The temperature is set between 255–265 degrees (higher in winter, lower in summer). Suitable solder bars are added to the solder pot in the soldering furnace, where they are heated and melted. The lead cutting mechanism is adjusted to the specified position in terms of height and width, ensuring that the width and smoothness of the handling mechanism match the PCB. The handling mechanism transports the workpieces to be soldered above the lead cutting mechanism. The excess parts of the coils are cut off by the lead cutting mechanism. After cutting, the portion to be tinned is immersed in the solder pot of the soldering furnace. Due to capillarity, solder adheres to the part. The workpiece is then removed and cooled, completing the tinning process.

III. Main Structure Introduction

Handling Mechanism (Fig. 6.13): Consists of cylinders (1, 3) controlling the horizontal displacement and vertical movement of the product frame (2). The synchronous belt (5) above the product frame is driven by a servo motor (4), transporting the product fixtures in batches for recycling.

Lead Cutting Mechanism (Fig. 6.14): Mainly controlled by cylinders (2, 3) to cut off the excess feet of the coils.

Soldering Furnace (Fig. 6.15): Equipped with a solder pot (1). Coils completed after the lead cutting process are transported by the handling mechanism to the solder pot (2) for tinning. The solder pot (1) is connected to cylinder (2), slightly shaking to evenly heat the solder inside.

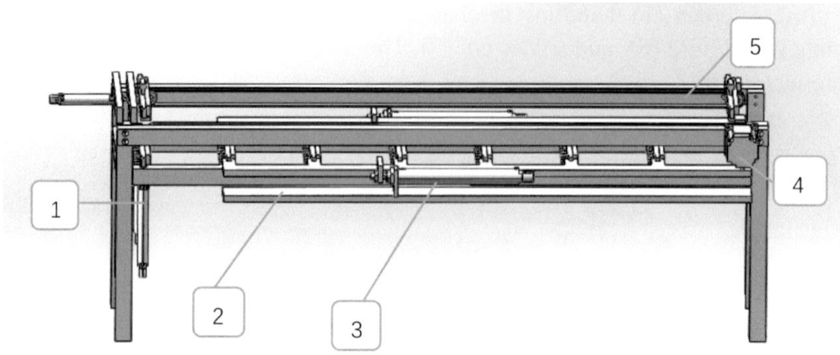

Fig. 6.13 Conveying Mechanism
1, 3 - Cylinder 2 - Product Frame 4 - Servo Motor 5 - Synchronous Belt

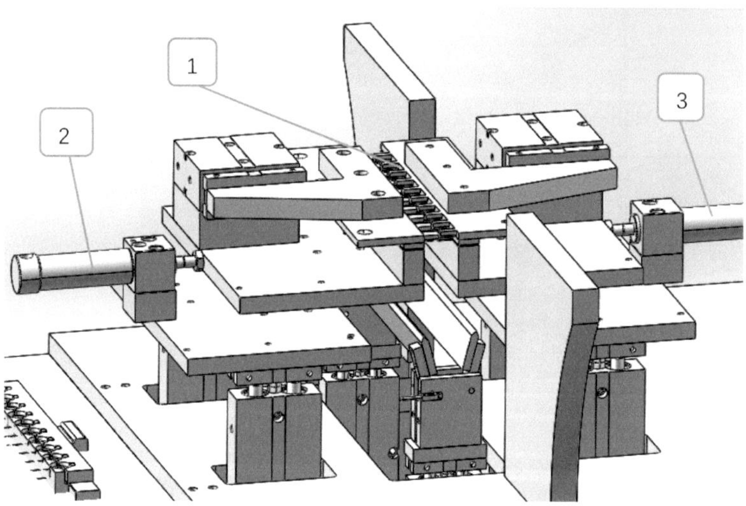

Fig. 6.14 Lead Cutting Mechanism
1 - Cutting Blade 2,3 - Cylinders

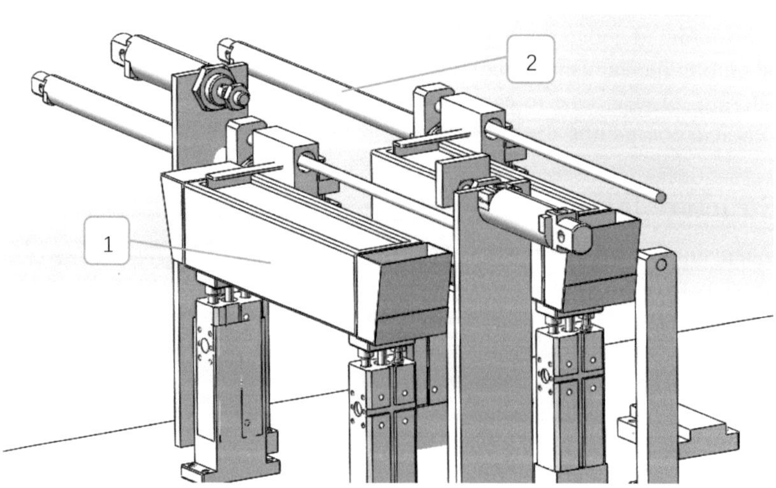

Fig. 6.15 Soldering Furnace
1 - Solder Pot 2 - Cylinder

IV. Specific Parameters (see Table 6.4)

Table 6.4 Parameters of the Testing Tin-Dipping Machine

Processing efficiency:	1000-1500 pcs/h
Operating rate:	≥93%
Yield:	≥98%
Permissible humidity:	10%~75%
Equipment dimensions:	1400 mm(H) × 1600 mm(L) × 500 mm(W)

V. Mechanical Design Highlights

(1) The handling mechanism has an input product frame and synchronous belt, synchronizing tinning operations with shelf recovery, enhancing efficiency, and minimizing production space.
(2) Both the handling mechanism and the soldering furnace have components like cylinders and grippers to swing the product frame, allowing the solder to bond quickly with the legs.

VI. Precautions

(1) Poorly soldered circuits must be resoldered, and a second resoldering should be performed after cooling.
(2) Avoid touching the soldering furnace during operation to prevent burns.
(3) Flux and diluents are flammable; store and use them away from open flames.
(4) If not used for an extended period, recover the flux and seal it.
(5) Ensure ventilation during soldering to prevent air pollution. Operators should wear work clothes and masks.
(6) Exercise caution during solder changes to avoid burns.
(7) Regularly inspect the heating wire to prevent aging and leakage.
(8) Check the solder level frequently, ensuring it is not more than 20mm below the top of the cylinder.

Case 81: Large Diameter Pipe Internal Welding Machine

I. Case Description

The Large Diameter Pipe Internal Welding Machine is an efficient automatic welding machine suitable for root welding inside large-diameter pipes. The overall structure is shown in Fig. 6.16, consisting mainly of a gas tank (1), drive mechanism (2), frame (3), welding unit (4), clamping mechanism (5), and synchronous automatic positioning and centering mechanism (6). This machine is characterized by simple operation and easy control, enabling efficient and high-quality welding of circumferential seams inside pipes.

II. Working Principle

To achieve higher welding speed and precision, the ends of the two pipes to be welded are aligned and securely fixed in place. The clamping mechanism secures the vertical pipes, and the synchronous drive mechanism controls the welding unit to perform argon arc welding on the pipes. Argon arc welding uses a high-voltage breakdown method, employing a tungsten electrode and inert gas

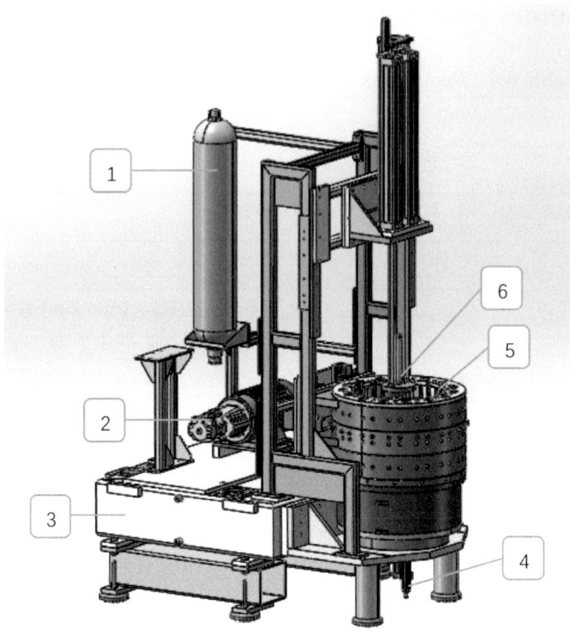

Fig. 6.16 Large Diameter Pipe Internal Welding Machine
1 - Gas tank 2 - Drive mechanism 3 - Frame 4 - Welding unit 5 - Clamping mechanism 6 - Synchronous automatic positioning and centering mechanism

(argon) for arc welding. The non-melting electrode is made of industrial tungsten, and the inert gas (argon) serves as a protective welding method. High-frequency high-voltage is applied between the electrode needle (tungsten needle) and the workpiece, breaking through the argon gas to make it conductive. Then, a continuous current is supplied to ensure a stable arc.

III. Main Structure Introduction

The clamping mechanism (Fig. 6.17) uses a multi-petal structure and a pressure plate (1) to fix the two straight pipes, ensuring the verticality of the pipes and the stability of the welding.

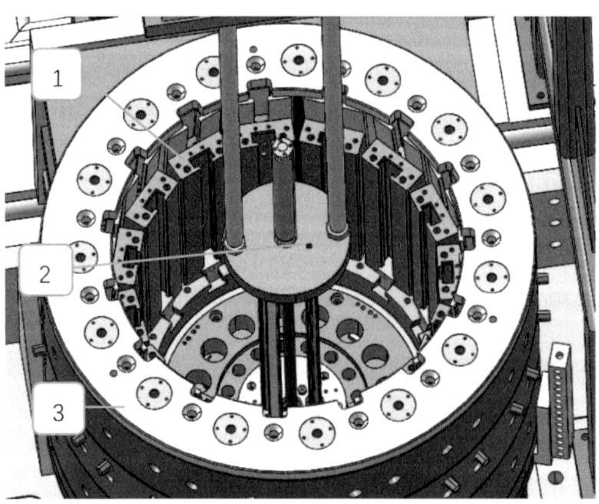

Fig. 6.17 Clamping mechanism
1 - Pressure plate 2 - Synchronous automatic positioning and centering mechanism 3 - Vertical pipe

The synchronous positioning and centering mechanism (Fig. 6.18), driven by cylinder (1), controls the vertical displacement of the welding unit (4) after aligning and fixing the straight pipes. The welding unit (4) then performs welding inside the pipes.

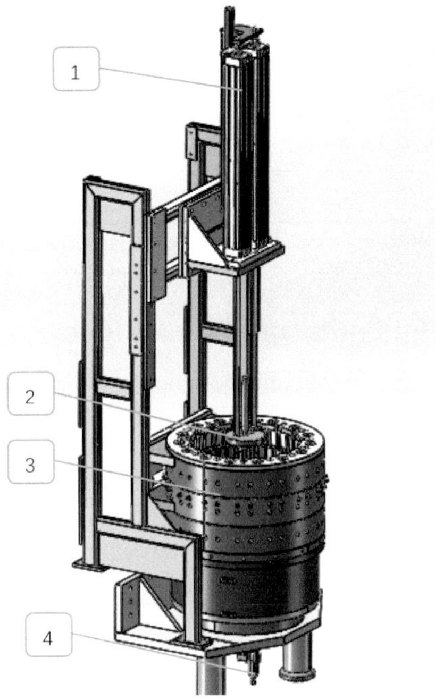

Fig. 6.18 Synchronous automatic positioning and centering mechanism
1 - Cylinder 2 - Synchronous automatic positioning and centering mechanism 3 - Tensioning mechanism 4 - Welding unit

The protective gas supply system (Fig. 6.19) is used to provide protective gas for welding.

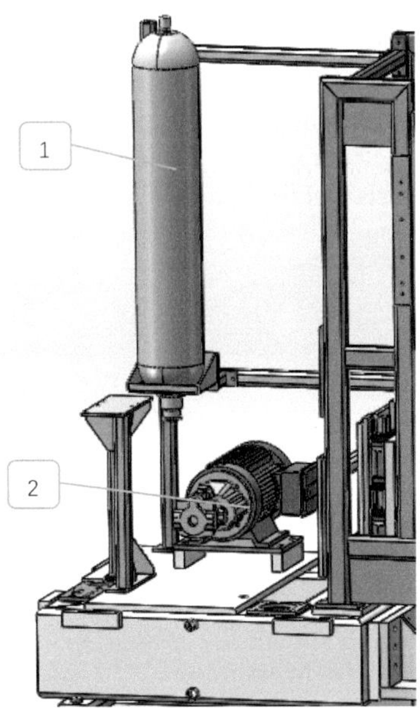

Fig. 6.19 Protective Gas Supply System
1 - Gas tank 2 - Drive mechanism

IV. Specific Parameters (see Table 6.5)

Table 6.5 Parameters of the Large Diameter Pipe Internal Welding Machine

Rated power:	2.0 kW
Welding method:	Argon arc/gas protection
Applicable pipe diameter:	915-1016 mm
Maximum tensioning force:	1400 kN
Welding speed:	0~1000 mm/min
Equipment dimensions:	3170 mm(H) × 2000 mm(L) × 1350 mm(W)

V. Mechanical Design Highlights and Precautions

(1) Pre-supply of Argon Gas: Argon gas should be supplied before welding, completing the welding and then stopping the gas supply. Argon gas, an easily breakdown inert gas, should be filled between the workpiece and the electrode needle, aiding in arc ignition.
(2) Post-Welding Gas Supply: Maintaining gas supply after welding helps prevent rapid cooling and oxidation of the workpiece, ensuring good welding results.
(3) High Voltage for Arc Initiation: Argon arc welding requires high voltage during arc initiation, with the high voltage disappearing after ignition.
(4) Electromagnetic Interference: Argon arc initiation involves high-frequency components that can cause serious interference with the entire circuit. Therefore, the circuit should have excellent anti-interference capability.

Case 82: Large Chassis Welding Clamp

I. Case Description

The large chassis welding clamp is mainly used for spot welding at points located in the middle of workpieces, such as the welding of the floor to the main beams and the reinforcement beams between the floor and the chassis of a vehicle. The overall structure is shown in Fig. 6.20, primarily composed of a cylinder (1), hanging component (2), clamp arm (3), electrode arm (4), electrode (5), and transformer (6). The large chassis welding clamp features fast movement speed, good reliability, high safety, and accurate positioning.

II. Working Principle

The clamp arms (3) of the large chassis welding clamp have a pair of electrodes at their extended ends, with the cylinder and the connected clamp arms controlling the separation and closure of the two electrodes. During welding, the high voltage and large current through electrode (5) instantaneously melt the metal, thereby welding the components together.

III. Main Structure Introduction

The electrode, as shown in Fig. 6.21, positions the product to be welded between the two electrodes when the large chassis welding clamp is in operation.

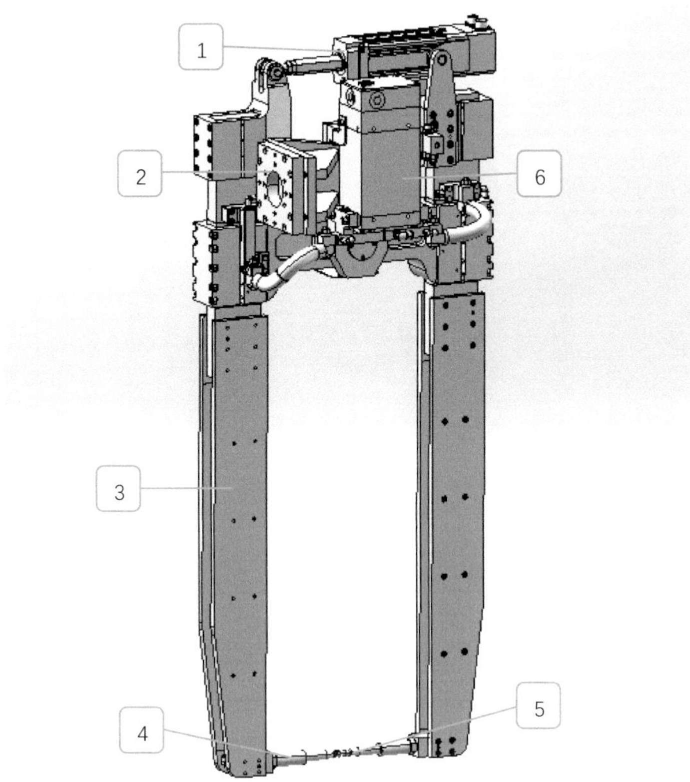

Fig. 6.20 Large Chassis Welding Clamp
1 - Cylinder 2 - Hanging component 3 - Clamp arm 4 - Electrode arm 5 - Electrode 6 - Transformer

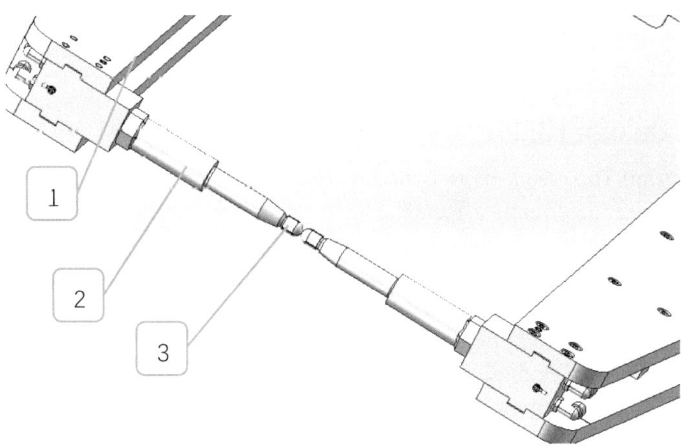

Fig. 6.21 Electrode
1 - Clamp arm 2 - Electrode arm 3 - Electrode

The motion structure of the large chassis welding clamp, as shown in Fig. 6.22, is controlled by cylinder (1), which opens and closes the two clamp arms.

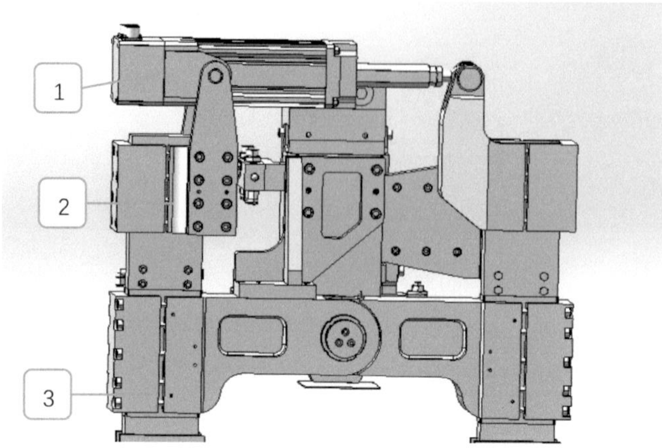

Fig. 6.22 Motion Structure
1 - Cylinder 2 - Hanging component 3 - Clamp arm

IV. Specific Parameters (see Table 6.6)

Table 6.6 Parameters of the Large Chassis Welding Clamp

Door width:	480 mm
Throat depth:	1250 mm
Concentricity of welding clamp electrodes:	Error < 0.5 mm
Insulation resistance:	Between upper and lower electrodes > 3 MΩ
Electrode pressure:	200-250 kg
Compressed air pressure:	0.4 MPa–0.6 MPa
Allowable humidity:	<75%
Equipment dimensions:	1880 mm(H) × 810 mm(L) × 430 mm(W)

V. Mechanical Design Highlights and Precautions

(1) Electrode Section: The electrode is a crucial component of the pneumatic welding clamp. The quality of the electrode directly affects the welding process, production efficiency, and welding quality. The electrode section consists of an electrode cap, electrode stem, and electrode joint.

(2) Clamp Body Section: The clamp body connects the upper and lower electrode arms of the welding clamp and transmits the cylinder pressure. The clamp body typically comprises upper and lower clamp bodies and hinge shafts. The upper and lower electrode arms are fixed in the upper and lower clamp bodies, respectively. They rotate around the hinge shaft for relative motion, using the lever principle to transfer the force applied by the cylinder to the upper and lower electrodes, thus generating welding pressure.

(3) Cylinder Assembly: The cylinder is the kinetic energy device for the welding clamp's welding pressure. The cylinder is connected to the upper and lower electrode arms through fixed sleeves, hinge plates, and connecting plates. It transmits pressure through the lever mechanism formed by the electrode arms and clamp body to apply pressure through the electrodes to the workpiece, meeting the pressure requirements for spot welding.

(4) Hanging Component: The hanging ring is primarily used to balance the welding clamp. The welding clamp is connected to the balancer through the hanging component. The balancer, by

its action, counteracts the self-weight of the welding clamp, making it seem as if the welding clamp is in a weightless state. During operation, workers do not have to exert effort to overcome the self-weight of the welding clamp, making it easy to move the welding clamp to the welding position for spot welding operations.

Case 83: Large Reclaimer Positioning Welding Machine

I. Case Description

The large reclaimer positioning welding machine is suitable for submerged arc welding of large pipe components. The overall structure is shown in Fig. 6.23, mainly composed of a reclaimer trough (1), L-shaped positioning machine housing (2), horizontal rotating positioning component (3), lifting platform bracket (4), platform walking mechanism (5), roller component (6), etc. The horizontal rotating positioning component (3) includes a rotating workpiece holder, rotating support, and a rotating motor. The roller component includes rollers and a positioning motor. The machine has the advantages of simple structure, fast movement, good reliability, high safety, and accurate positioning.

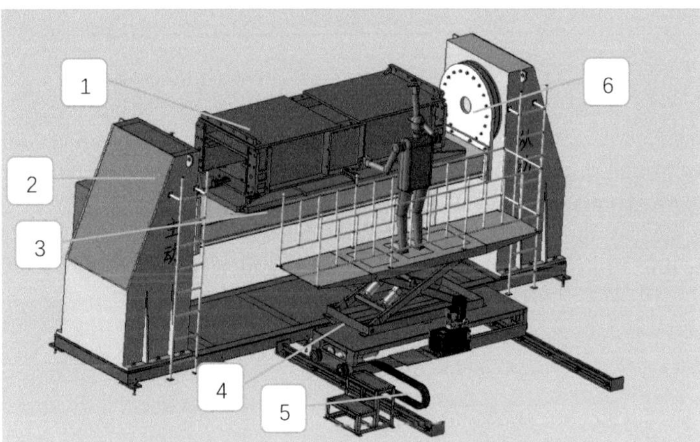

Fig. 6.23 Large Reclaimer Positioning Welding Machine
1-Reclaimer trough 2-L-shaped positioning machine housing 3-Horizontal rotating positioning component 4-Lifting platform bracket 5-Platform walking mechanism 6-Roller component

II. Working Principle

The large reclaimer positioning welding machine uses a heavy-duty walking mechanism for displacement, and the workpiece holder can freely rotate for welding. The drive mechanism uses a variable frequency motor to provide driving force. Operators stand on the platform walking mechanism to perform welding. The workpiece is placed on the reclaimer trough, which has a rotating support and a rotating workpiece holder beneath it, capable of rotating in the horizontal plane. Roller components on both sides of the rotating trough allow the workpiece to rotate around the axis while the operator controls the welding gun's wielding status, achieving the purpose of welding large pipe components without blind spots.

III. Main Structure Introduction

The L-shaped positioning machine housing component, as shown in Fig. 6.24, has multiple support blocks at the bottom to ensure stable operation. Inside the housing, the roller component is assembled. The roller (1) is driven by the roller motor (3), allowing the workpiece to rotate around the axis.

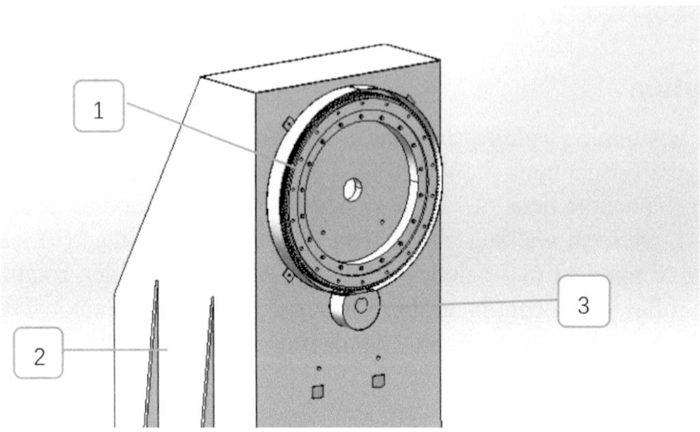

Fig. 6.24 L-shaped positioning machine housing component
1 - Roller 2 - L-shaped positioning machine housing 3 - Roller motor

The reclaimer trough and horizontal rotating positioning component, as shown in Fig. 6.25, include a rotating workpiece holder (2) and a rotating support (3). The rotating support (3) is equipped with an electric motor. During operation, the workpiece is placed inside the reclaimer trough (1), and the operator controls the workpiece's horizontal rotation.

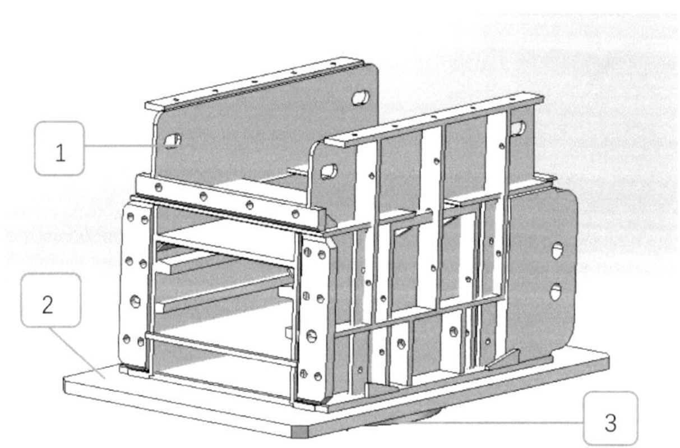

Fig. 6.25 Reclaimer trough and horizontal rotating positioning component
1 - Reclaimer trough 2 - Rotating workpiece holder 3 - Rotating support

The lifting platform and platform walking mechanism, as shown in Fig. 6.26, allow the operator to weld the workpiece on the lifting platform 1. The lifting platform 1 is equipped with a safety guardrail 6 to prevent the operator from accidentally falling. The lifting platform is driven by an electric motor 5, using a cross mechanical structure. When the platform ascends with both feet closed, and descends when open. The platform walking mechanism 4 is driven by the platform walking motor 3, enabling horizontal displacement of the lifting platform.

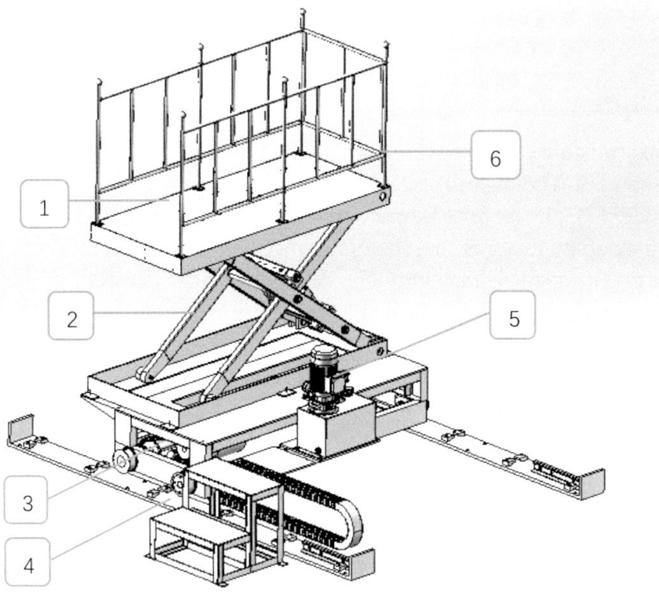

Fig. 6.26 Lifting platform and platform walking mechanism
1 - Lifting platform 2 - Lifting platform bracket 3 - Lifting platform motor 4 - Safety guardrail 5 - Platform walking mechanism 6 - Platform walking motor

IV. Specific Parameters (see Table 6.7)

Table 6.7 Parameters of the Large Reclaimer Positioning Welding Machine

Yield rate:	≥95%
Operating pressure range:	4-6 kg/cm^2
Welding depth:	2.0 mm
Maximum welding thickness:	10 mm
Power:	100 kVA
Input power:	500 kW

V. Mechanical Design Highlights

The large reclaimer positioning welding machine supports the rotating workpiece holder through the rotating support to rotate the workpiece, achieving welding positioning. The drive mechanism uses a variable frequency motor to provide driving force and includes at least one set of gear pairs. The welding head adjustment mechanism can adjust the welding head position in "XYZ" three directions, and the vertical plane's "XY" direction translation adjustment is completed by a linkage welding head adjustment mechanism based on the parallelogram principle. Since the workpiece holder rotates around the structure's center of gravity, the required torque and driving power are significantly reduced. The servo-driven and linkage welding head adjustment mechanisms have a simple structure, good followability of the welding head to the weld seam, and maintain the welding gun and the current welding point in a ship-shaped welding state. This design is particularly suitable for the automatic submerged arc welding of large pipe components.

Case 84: Spot Welding Machine

I. Case Description

The spot welding machine is used to form weld spots between the contact surfaces of two overlapping workpieces. The overall structure is shown in Fig. 6.27, consisting mainly of a cylinder (1), electrode holder (2), electrodes (3), pressure positioning mechanism (4, 5), and main frame (6). The spot welding machine features short heating time, minimal deformation and stress, low welding cost, high production efficiency, and simplicity of operation. It is easily mechanized and automated, and in large-scale production, it can be integrated into assembly lines with other manufacturing processes.

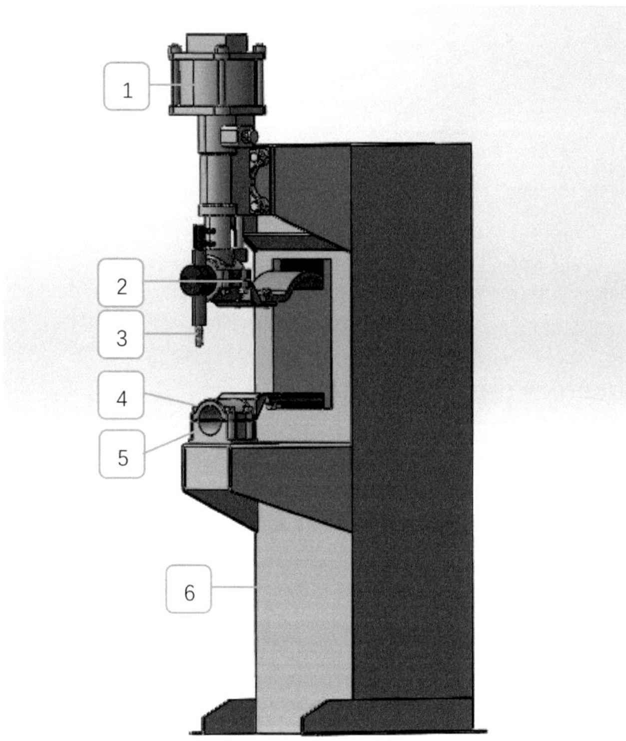

Fig. 6.27 Spot Welding Machine
1 - Cylinder 2 - Electrode holder 3 - Electrode 4, 5 - Pressure positioning mechanism 6 - Frame

II. Working Principle

When the spot welding machine operates, the surface of the welded piece is cleaned, assembled accurately, and fed into the pressure positioning mechanism. Pressure is applied to ensure good contact on the surface of the welded piece. After applying power, the electrode moves downward, heating the contact surface of the two workpieces, causing localized melting and forming a weld nugget. After power is cut off, pressure is maintained to allow the nugget to cool and solidify, forming a weld spot. Release the pressure, remove the workpiece, and complete the spot welding process.

III. Main Structure Introduction

The welding electrode part, as shown in Fig. 6.28, involves placing the cleaned welded piece in the pressure positioning mechanism (4), securing it with screws, and ensuring good alignment. After fixing the workpiece, power is supplied to the equipment, and the cylinder (1) is controlled to move the electrode (2) downward. This causes the contact surface of the workpiece to heat up, melt locally, and form a weld nugget [113].

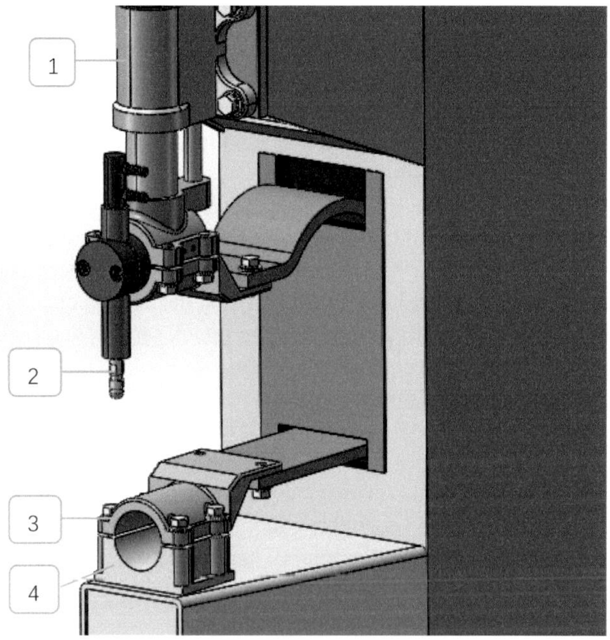

Fig. 6.28 Welding Electrode Part
1 - Cylinder 2 - Electrode 3, 4 - Pressure positioning mechanism

IV. Specific Parameters (see Table 6.8)

Table 6.8 Spot Welding Machine Parameters

Maximum Power:	10 kVA
Primary Current:	5~40 A
Secondary Short Circuit Current:	1800 A
Secondary No-load Voltage:	5.5 V
Maximum Cylinder Stroke:	25 mm
Electrode Diameter:	3.00 mm
Minimum Electrode Distance:	0.1 mm
Equipment Dimensions:	1640 mm(H) × 840 mm(L) × 400 mm(W)

V. Mechanical Design Highlights

The spot welding machine uses the principle of double-sided double-point overcurrent welding. During operation, the two electrodes press the workpiece, creating a certain contact resistance between the two layers of metal under the pressure of the two electrodes. The welding current flows through one electrode to another, forming an instantaneously heated joint at the two contact resistance points. The welding current then flows through the other electrode along the two workpieces to this electrode, completing the circuit. This process does not damage the internal structure of the welded workpiece.

VI. Precautions

(1) Steel welded parts must be cleaned of dirt, oil, oxide, and rust before welding. For hot-rolled steel, it is best to acid wash, sandblast, or use a grinding wheel to remove oxide. Uncleaned welded parts can be spot welded, but it significantly reduces the electrode's lifespan and decreases the production efficiency and quality of spot welding.

(2) For on-site use, there should be a shed that is rainproof, moisture-proof, and sun-proof, equipped with corresponding firefighting equipment.

(3) Within a 10 m range of the welding site, flammable and explosive items such as oil, wood, oxygen cylinders, and acetylene generators should not be stacked.

(4) The connection between the secondary tapping and the copper plate should be tightened, and the terminal should have washers. Before closing the circuit, check the terminal nuts, bolts, and other components in detail to ensure they are intact, secure, and undamaged. Protective covers are required at the terminal connection points.

(5) Before use, check and confirm the correct connection of primary and secondary lines, the input voltage complies with the welding machine's nameplate requirements, and understand the types and applicable ranges of spot welding machine welding currents. After powering on, contact with the energized part of the primary circuit is strictly prohibited. Protective covers must be installed at the primary and secondary junctions.

(6) When moving the spot welding machine, the power must be cut off, and the machine should not be moved by dragging the cable. In case of a sudden power outage during welding, cut off the power immediately.

(7) When welding copper, aluminum, zinc, tin, lead, and other non-ferrous metals, it must be done in a well-ventilated area. Welding personnel should wear a gas mask or respirator filter.

(8) When multiple spot welding machines are used together, they should be connected to a three-phase power network to balance the three-phase load. The grounding devices of multiple machines should be separately connected from the grounding pole and should not be connected in series.

(9) Welding is strictly prohibited on operating pressure pipelines, containers containing flammable and explosive materials, and stressed components.

(10) When preheating welded parts, a baffle should be set up to isolate the radiant heat emitted by the preheated welded parts.

Case 85: Inductor Automatic Soldering and Testing Machine

I. Case Description

The inductor automatic soldering and testing machine can achieve automatic soldering and electrical performance testing of inductor components. The overall structure is shown in Fig. 6.29, and a detailed enlarged view is shown in Fig. 6.30. It mainly consists of a display screen (1), cooling box (2), frame (3), feeding mechanism (4), inductor clamping plate (5), automatic foot trimming mechanism (6), grabbing robot (7), and inductor testing mechanism (8). The machine not only improves the processing accuracy of inductor coil pins, ensuring the accuracy of product processing but also realizes the automation of inductor production, increasing product qualification rate and production efficiency.

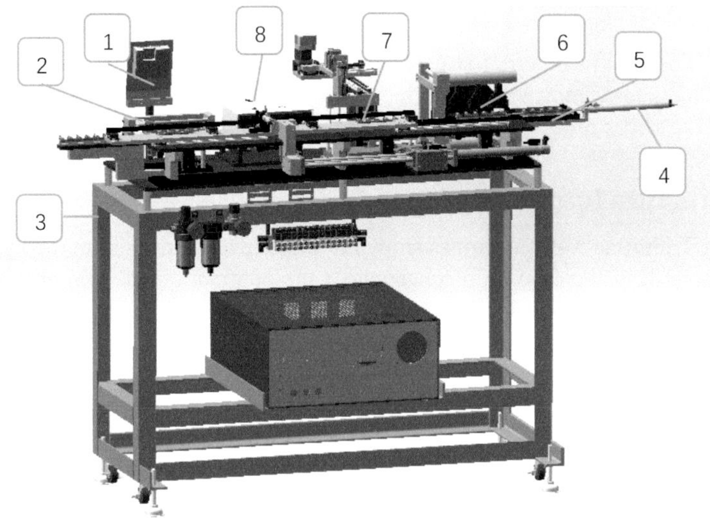

Fig. 6.29 Inductor Automatic Soldering and Testing Machine
1 - Display screen 2 - Cooling box 3 - Frame 4 - Feeding mechanism 5 - Inductor clamping plate 6 - Automatic foot trimming mechanism 7 - Automatic soldering mechanism 8 - Inductor testing mechanism

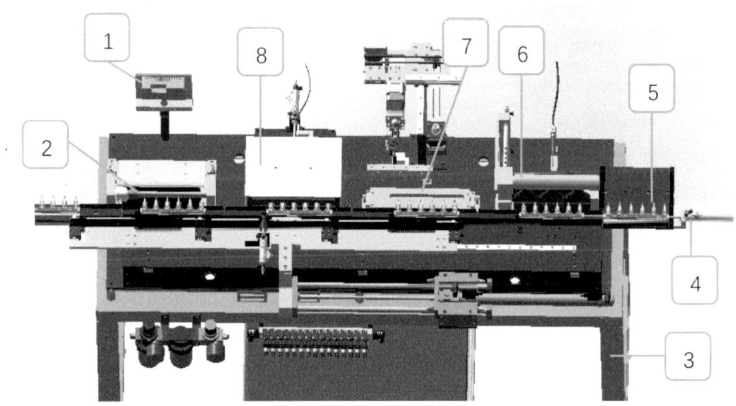

Fig. 6.30 Detailed Enlarged View of Inductor Automatic Soldering and Testing Machine
1 - Display screen 2 - Cooling box 3 - Frame 4 - Feeding mechanism 5 - Inductor clamping plate 6 - Automatic foot trimming mechanism 7 - Automatic soldering mechanism 8 - Inductor testing mechanism

II. Working Principle

The workpiece to be soldered is placed on the inductor clamping plate. When the inductor automatic soldering and testing machine operates, the inductor clamping plate is fed into the feeding mechanism from the feeding mechanism, and the automatic foot trimming mechanism performs initial processing on the workpiece, removing the oxide layer on the surface of the workpiece pins. The workpiece is then transferred to the automatic soldering mechanism, where the robot automatically dips into the solder and accurately positions itself above the inductor clamping plate to perform soldering on the workpiece pins. After soldering, the inductor clamping plate is transferred to the inductor testing mechanism for testing. The test data can be viewed on the display screen, and products with unsatisfactory soldering are filtered out. After testing, the inductor clamping plate is transported to the cooling box for cooling. After the surface of the workpiece is cooled, it is output from the discharge port. The hollow internal structure of the inductor clamping plate can accommodate multiple sets of inductors, improving soldering efficiency [114]. After passing the test, the output is completed, and the entire process of automatic soldering and inductor testing is finished.

III. Main Structure Introduction

Automatic Foot Trimming Mechanism, as shown in Fig. 6.31, operates with the inductor clamping plate (1) transported by the feeding mechanism to the automatic foot trimming mechanism and stops. The automatic foot trimming mechanism is activated, and six foot grinding mechanisms vibrate to grind the surface of the workpiece pins, removing the oxide layer.

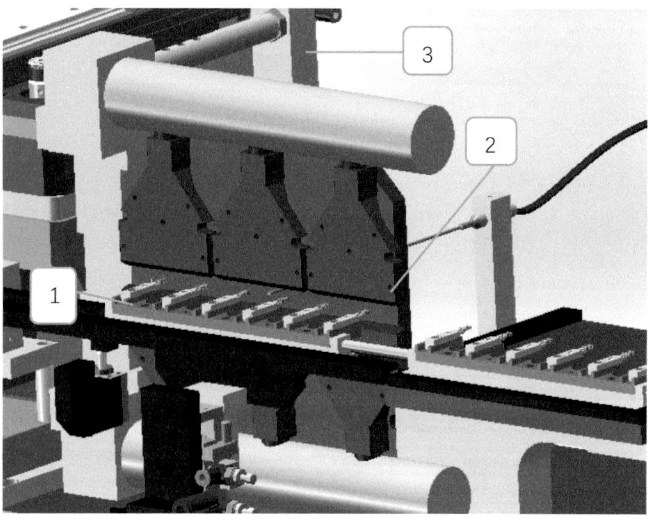

Fig. 6.31 Automatic Foot Trimming Mechanism
1 - Inductor clamping plate 2 - Foot grinding mechanism 3 - Installation bracket

Automatic Soldering Mechanism, as shown in Fig. 6.32, operates when the feeding mechanism transports the workpiece to the automatic soldering mechanism. At this point, the robot starts, controls the soldering iron to dip into the solder, and then accurately positions itself above the inductor clamping plate to solder the workpiece pins.

Inductor Testing Mechanism, as shown in Fig. 6.33, operates based on the principle of sampling the voltage and current of the tested capacitor branch with a sampling circuit. The output ends of the sampling circuit are connected to the amplification circuit. The phase difference signal is output by the phase discriminator after the voltage signal and current signal from the amplification circuit are subjected to A/D conversion. The phase difference signal is then processed by the processor to obtain the capacitance value of the tested capacitor. The use of a movable current sampling unit eliminates the need to disconnect the connecting wires to directly measure the capacitance value.

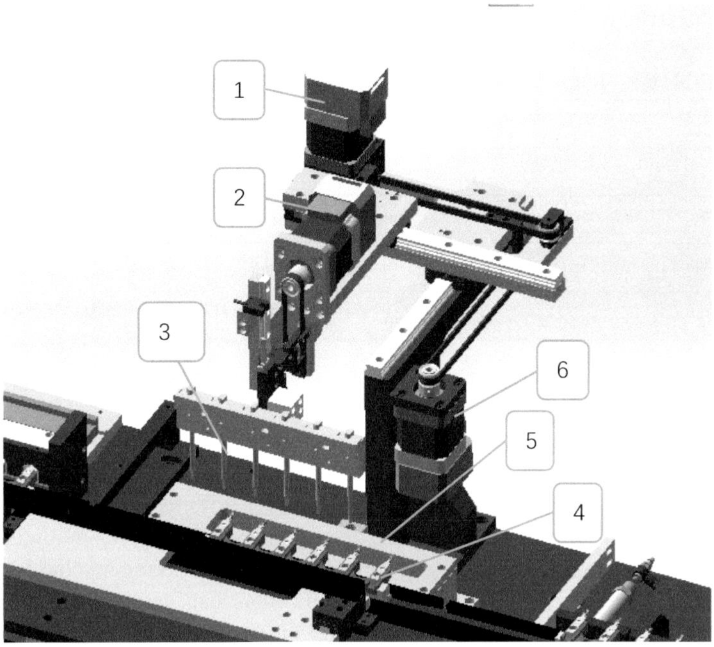

Fig. 6.32 Automatic Soldering Mechanism
1, 2, 6 - Electric motor 3 - Soldering iron 4 - Inductor clamping plate 5 - Solder tank

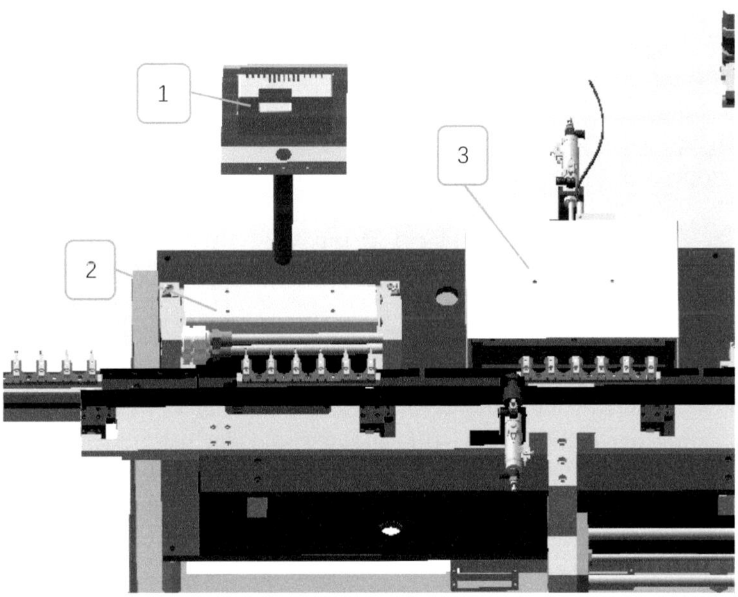

Fig. 6.33 Inductor Testing Mechanism
1 - Display screen 2 - Cooling box 3 - Inductor testing mechanism

IV. Specific Parameters (see Table 6.9)

Table 6.9 Inductor Automatic Soldering and Testing Machine Parameters

Inductance Value:	1.5 mH ± 15%
DC Resistance:	2Ω MAX
Dielectric Strength:	500 VAC5s
Insulation Resistance:	500 V DC
Equipment Dimensions:	1230 mm(H) × 1790 mm(L) × 650 mm(W)

V. Mechanical Design Highlights

Currently, in the manufacturing process of inductors, the welding between the leads and lead-out wires of the inductor is manually carried out by immersing the inductor in a soldering furnace. This manual operation is labor-intensive, and the welding quality is difficult to guarantee, resulting in low efficiency. The inductor automatic soldering and testing machine achieves automatic soldering of the pins by dipping them into the solder bath using the automatic soldering mechanism. Additionally, the hollow internal structure of the inductor clamping plate can accommodate multiple sets of inductors, allowing for efficient replacement and increasing soldering efficiency while reducing the labor intensity of operators.

VI. Precautions

(1) Do not touch the soldering furnace during the operation to avoid burns. Do not let water or oily substances fall into the soldering furnace.
(2) Ensure ventilation during soldering operations to prevent air pollution. Operators should wear work clothes and masks.
(3) When changing solder, pay attention to the safety of the operators to avoid burns.
(4) Regularly inspect the lead wires at the heating point to prevent aging and leakage.

Case 86: Dual-Feed Wire Welding Machine

I. Case Description

The dual-feed wire welding machine is designed for welding wires. The overall structure, as shown in Fig. 6.34 with an enlarged view in Fig. 6.35, consists of various components such as a vibratory conveyor (1), welding mechanism (2), wire transfer mechanism (3), wire feeding mechanism (4), motor (5), frame (6), recovery bin (7), and shelf (8). The machine features a symmetrical design with mechanisms on both sides driven by multiple cams, significantly enhancing production efficiency and reducing costs.

II. Working Principle

Raw wire material is placed on the shelf. During operation, the vibratory conveyor introduces the flux into the welding mechanism. The wire is fed into the wire transfer mechanism through the wire feeding mechanism. After passing through the wire transfer mechanism's winding pulley, the wire is inserted into the welding machine for welding. Upon completion, the continuously driven cam system transports the welded wire into the recovery bin, completing the wire welding process [115].

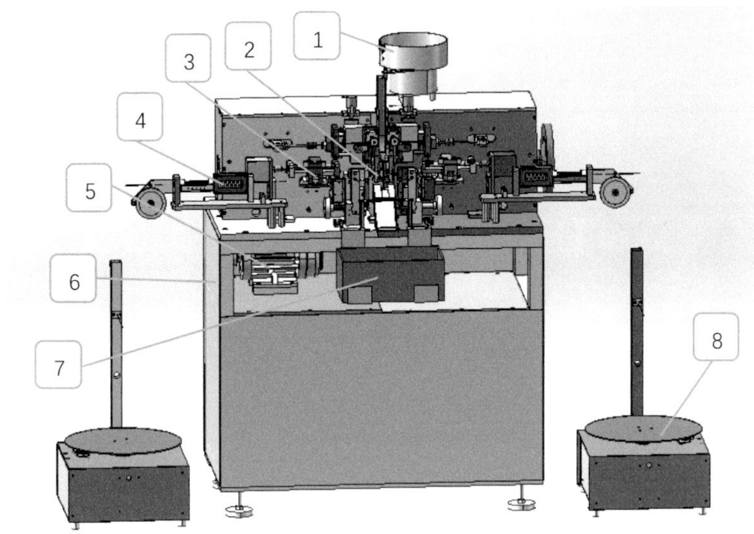

Fig. 6.34 Dual-Feed Wire Welding Machine
1-Vibratory Conveyor, 2-Welding Mechanism, 3-Wire Transfer Mechanism, 4-Wire Feeding Mechanism, 5-Motor, 6-Frame, 7-Recovery Bin, 8-Shelf

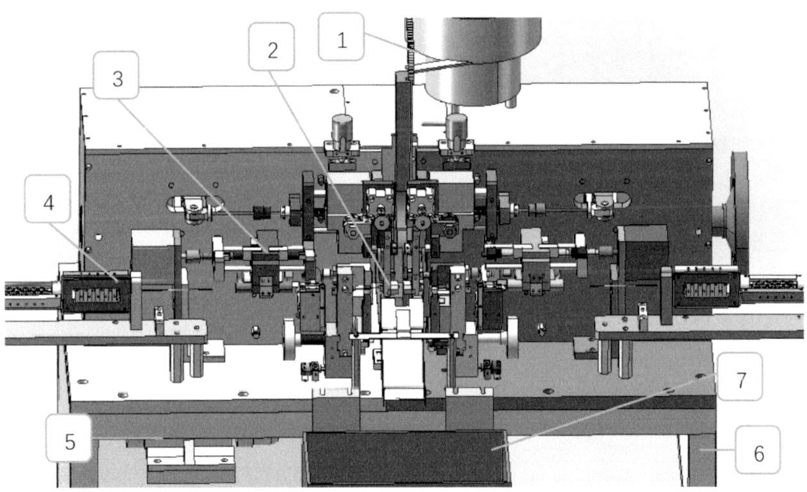

Fig. 6.35 Enlarged View of Dual-Feed Wire Welding Machine
1-Vibratory Conveyor, 2-Welding Mechanism, 3-Wire Transfer Mechanism, 4-Wire Feeding Mechanism, 5-Motor, 6-Frame, 7-Recovery Bin

III. Main Structure Introduction

Cam System (Fig. 6.36): Driven by an electric motor through a pulley (4), the cam system controls the horizontal displacement of the welding mechanisms on both sides (cams 3 and 5), the wire transfer mechanism (cams 2 and 6), and the vertical displacement of the welding mechanism (cams 1 and 7). The optimized design of cam curves integrates multiple mechanisms, significantly reducing production costs.

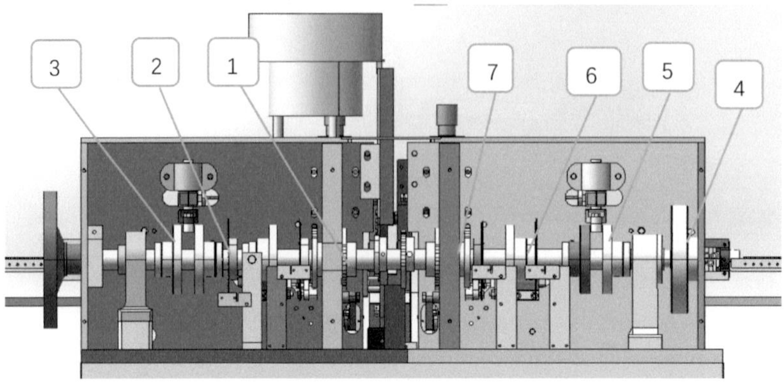

Fig. 6.36 Cam System
3, 5 - Horizontal Displacement Cams, 2, 6 - Transfer Displacement Cams, 1, 7 - Vertical Displacement Cams, 4 - Pulley

Wire Feeding Mechanism (Fig. 6.37): The wire is guided into the wire feeding mechanism through guide wheel (3), passing through both horizontal (2) and vertical (1) wire straightening mechanisms before entering the wire transfer mechanism.

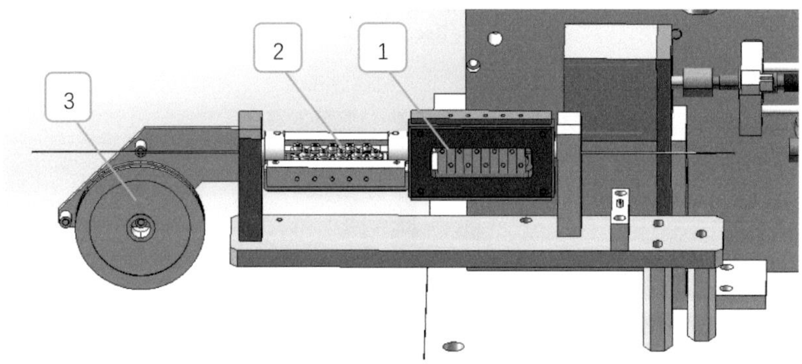

Fig. 6.37 Wire Feeding Mechanism
1 - Vertical Wire Straightening Mechanism, 2 - Horizontal Wire Straightening Mechanism, 3 - Guide Wheel

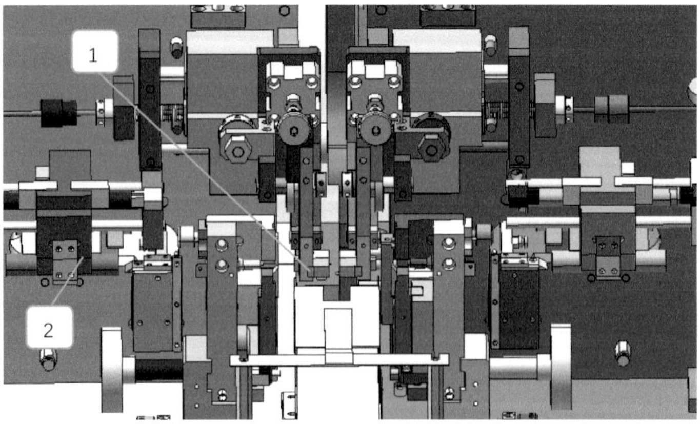

Fig. 6.38 Welding Mechanism
1 - Welding Mechanism, 2 - Wire Transfer Mechanism

Welding Mechanism (Fig. 6.38): The wire is introduced into the welding mechanism from the wire transfer mechanism (2), and flux is introduced from the vibratory conveyor. Controlled by the cam system, the welding mechanisms on both sides (1) make contact and perform electric welding, completing the wire welding operation.

IV. Specific Parameters (see Table 6.10)

Table 6.10 Parameters of Dual-Feed Wire Welding Machine

Yield Rate:	≥98%
Allowed Humidity:	<75%
Drive Type:	Stepper Motor
Equipment Dimensions:	1300 mm(H) × 1400 mm(L) × 500 mm(W)

V. Mechanical Design Highlights and Precautions

(1) The symmetrical design and multi-cam drive significantly improve production efficiency and reduce costs.
(2) Limit the temperature rise during welding to A-level 60 degrees or B-level 80 degrees; stop welding and cool down if exceeded.
(3) In case of automatic stop device failure during welding, immediately stop the machine, cut power, and perform maintenance.
(4) Ensure a dry and well-ventilated working environment. When moving the equipment, cut off the power and avoid using dragging methods. In the event of a sudden power outage during welding, cut off the power.
(5) During welding, do not adjust the current. Adjustments must be made when welding is stopped, using the adjustment handle, avoiding rapid or excessive adjustments to prevent damage to the regulator.
(6) When removing weld slag, wear safety glasses. Ensure the head is away from the direction of flying weld slag to avoid injury. Do not strike the weld slag towards people present.

Case 87: Robot Automatic Welding Machine

I. Case Description

The robot automatic welding machine is suitable for batch automatic welding of electronic products. The overall structure is shown in Fig. 6.39 and consists mainly of detection mechanism (1), five-axis robot (2), frame (3), trolley (4), positioning mechanism (5), welding mechanism (6), and electrical control box (7). The robot automatic welding machine features high automation and high welding efficiency, among other advantages.

II. Working Principle

When the robot automatic welding machine operates, the trolley transports the gas cylinder to the appropriate position and connects it with the welding mechanism. The workpiece to be welded is placed in the welding mechanism, and the positioning mechanism is adjusted to ensure a tight fit. The control box controls the welding mechanism to weld the workpiece. After welding, the five-axis robot picks up the workpiece and transfers it to the inspection mechanism for testing. Qualified products are then moved to the next process [116].

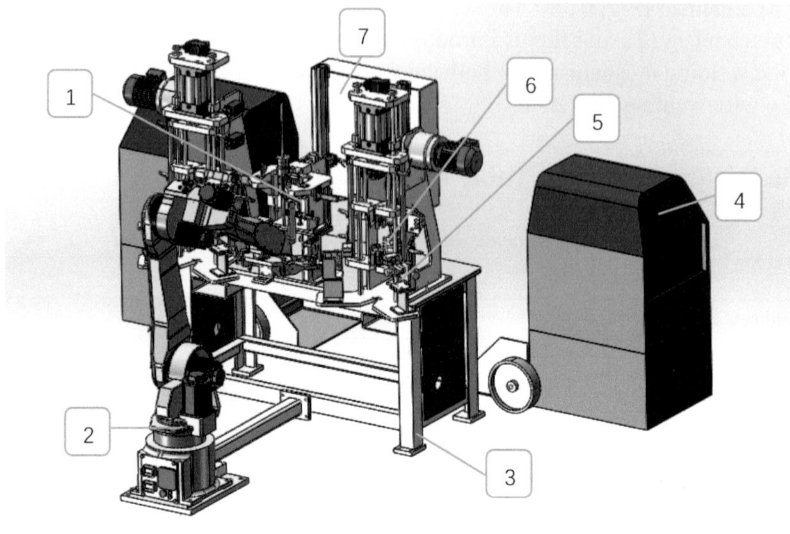

Fig. 6.39 Robot Automatic Welding Machine
1-Detection Mechanism 2-Five-Axis Robot 3-Frame 4-Trolley 5-Positioning Mechanism 6-Welding Mechanism 7-Electrical Control Box

III. Main Structure Introduction

Five-Axis Robot: Shown in Fig. 6.40, driven and controlled by a servo motor (1), used for gripping and welding tasks.

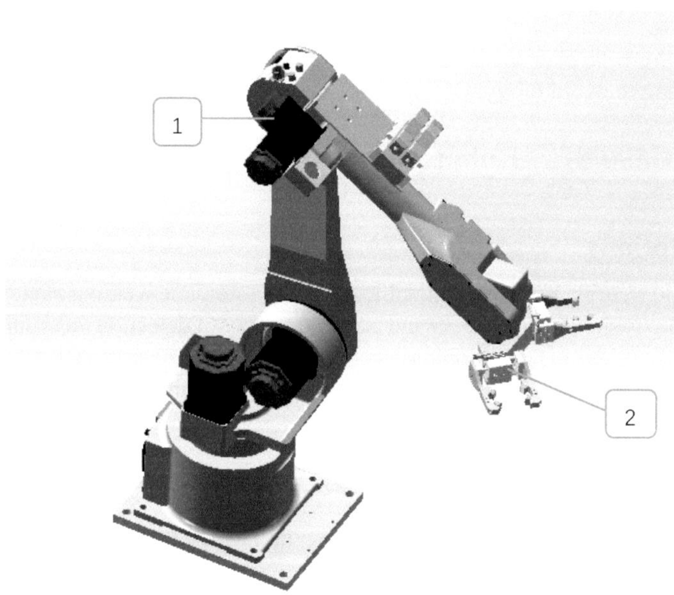

Fig. 6.40 Five-Axis Robot
1-Servo Motor 2-Mechanical Arm

Positioning Mechanism and Welding Mechanism: Illustrated in Fig. 6.41, the five-axis robot places the product on the base (2) of the positioning mechanism. The servo motor (4) drives the upper die (1) to press down, positioning and tightly fitting the workpiece. Once positioned, the argon arc welding machine (3) in the welding mechanism carries out the welding task.

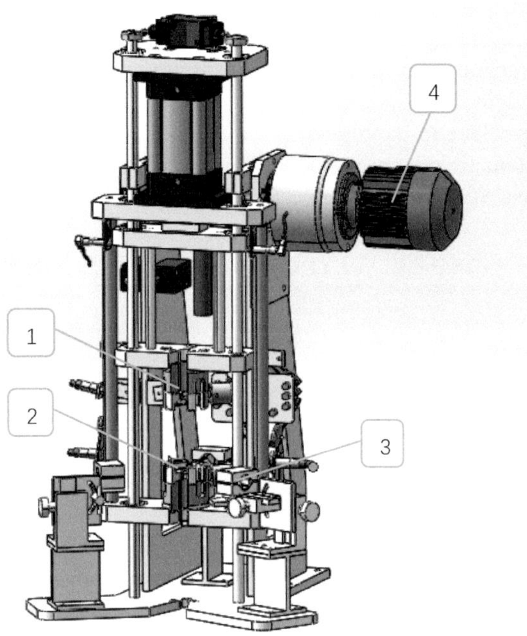

Fig. 6.41 Positioning Mechanism and Welding Mechanism
1 - Upper Die 2 - Base 3 - Welding Mechanism 4 - Servo Motor

IV. Specific Parameters (see Table 6.11)

Table 6.11 Robot Automatic Welding Machine Parameters

Input Power	1.5 kW
Welding Workpiece Thickness	0.5~6 mm
Minimum Welding Workpiece Diameter	60 mm
Effective Length of Welded Workpiece	<300 mm
Welding Depth	1.5 mm
Equipment Dimensions	1800 mm(H) × 1970 mm(L) × 2070 mm(W)

V. Mechanical Design Highlights

Industrial robots are versatile, programmable automatic control devices with multiple programmable axes used in industrial automation. The Robot Automatic Welding Machine boasts high automation and exceptional welding efficiency. To enhance its efficiency, a dedicated gas cylinder transport trolley is included, significantly improving the overall productivity [117].

VI. Precautions

(1) Teaching Operation: Before the robot performs automatic welding, operators must teach the robot the welding gun trajectory and set welding conditions. This is not suitable for welding products with multiple varieties and small quantities.
(2) Ensure Workpiece Accuracy: The robot's trajectory accuracy is ±0.1mm, requiring precise workpiece accuracy within the radius of the welding wire.
(3) Setting Welding Conditions: The setting of welding conditions depends on the technical level of the teaching operator, including welding programs, gun posture, angles, current, voltage, speed, etc.
(4) Safety Awareness: Due to the high-speed movement of the robot, no one is allowed near the robot during automatic operation (safety barriers must be installed). Operators must undergo specialized education in labor safety [118].

Case 88: Precision Hot Melt Machine

I. Case Description

The precision hot melt machine is designed for the thermal fusion welding of plastic products, as illustrated in Fig. 6.42. The primary components include a cylinder (1), a heater lifting mechanism (2), a hot melt device (3), a hot melt base (4), a frame (5), and a human-machine interface (6). This machine is characterized by its simple and precise operation, as well as high processing efficiency.

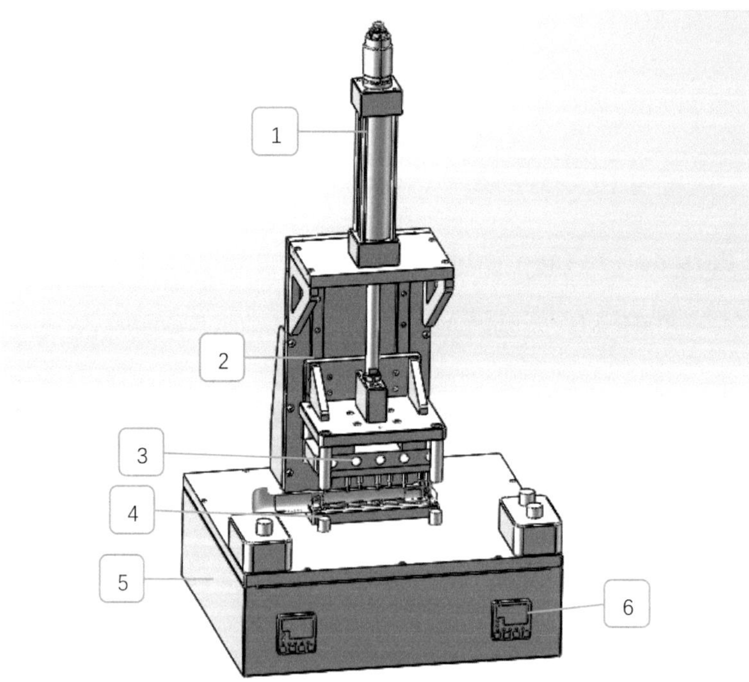

Fig. 6.42 Precision Hot Melt Machine
1 - Cylinder 2 - Heater Lifting Mechanism 3 - Hot Melt Device 4 - Hot Melt Base 5 - Frame 6 - Human-Machine Interface

II. Working Principle

The precision hot melt machine secures multiple heating heads onto a heater. The displacement of the heater is controlled by the heater lifting mechanism, allowing the surface of the product's heated area to come into contact with the heating surface. This results in surface melting, followed by the bonding and solidification of the heated surface. The precision hot melt machine can accurately control temperature changes and, when coupled with mechanical devices, achieves product fusion. It is particularly suitable for the thermal fusion welding of large, irregularly shaped plastic products, such as plastic welding, embedded nuts, and riveting [119].

III. Main Structure Introduction

The hot melt device, as shown in Fig. 6.43, is driven by a cylinder (1) to exert downward pressure and heat the product's end face.

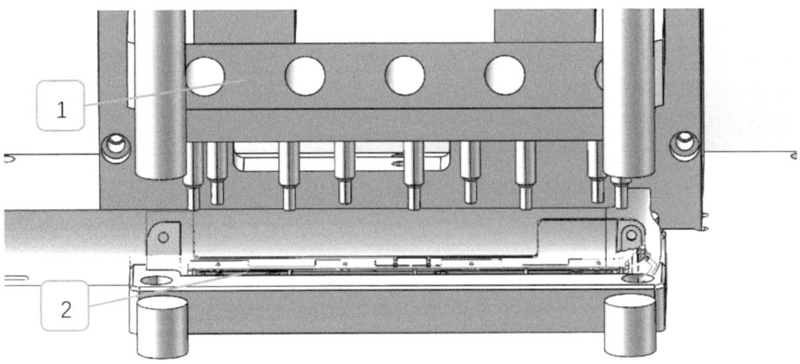

Fig. 6.43 Hot Melt Device
1 - Cylinder 2 - Hot Melt Base

The heater lifting mechanism, depicted in Fig. 6.44, is controlled by a cylinder (1) to regulate the lifting of the hot melt device (3).

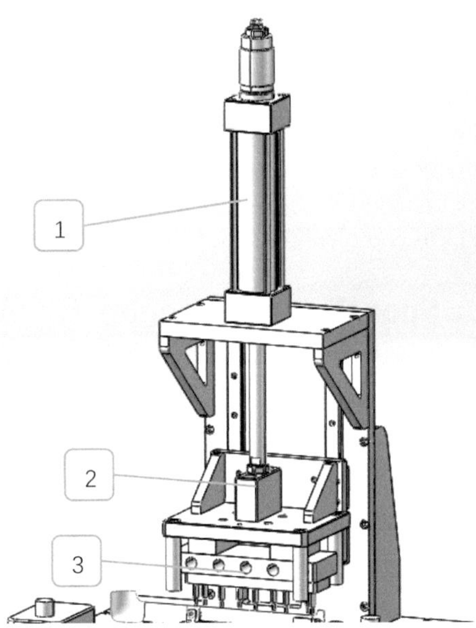

Fig. 6.44 Heater Lifting Mechanism
1 - Cylinder 2 - Heater Lifting Mechanism 3 - Hot Melt Device

IV. Specific Parameters (see Table 6.12)

Table 6.12 Precision Hot Melt Machine Parameters

Usage Area:	400×225 mm
Heating Area:	250×160 mm
Maximum Power:	1600 W
Permissible Humidity:	<75%
Equipment Dimensions:	750 mm(H) × 400 mm(L) × 350 mm(W)

V. Mechanical Design Highlights

(1) Equipped with electronic slow start and electronic limit, suitable for various hot melt methods such as embedding and riveting.
(2) Combines hydraulic shock absorber with buffer circuit for smooth operation.
(3) Suitable for processing large objects and simultaneous multiple-point fusion.
(4) Adopts multi-stage heating tubes and an imported temperature control system for accurate and reliable temperature control.
(5) Utilizes various surface treatments such as electrostatic spraying and hard chrome plating for an aesthetically pleasing and durable appearance.

VI. Precautions

(1) After connecting to the rated voltage power supply, turn on the heater power switch. The heater power indicator light will then illuminate.
(2) Adjust the temperature controller of the glue reservoir to 130-140°C and the roller temperature controller to 140-150°C. These temperatures are for reference only; actual temperatures should be based on the melting point of the hot melt adhesive. This completes the hot melt machine heater preheating procedure.
(3) After 30 minutes of preheating, when the hot melt adhesive in the reservoir has melted, and the motor start indicator light is on, open the motor power switch. Adjust the motor speed adjustment knob to the desired working speed. If the motor does not turn, wait until the hot melt adhesive is completely melted before starting.
(4) 5-10 minutes before shutting down, close the heater power switch to reduce the temperature of the roller and glue reservoir to prevent charring of the hot melt adhesive on the roller.
(5) After closing the heater power switch for 5-10 minutes and the temperature has lowered, adjust the motor speed adjustment knob to zero. Subsequently, close the motor power switch and turn off the input machine power to complete the shutdown procedure.

Case 89: Automatic Spot Welding Machine for Lithium Batteries

I. Case Description

The automatic spot welding machine for lithium batteries is mainly used for the nickel wire welding on the cap of lithium batteries. The overall structure is shown in Figure 6.45, and an enlarged view is shown in Fig. 6.46. It mainly consists of nickel wire tray (1), nickel wire transfer mechanism (2), spot welding machine (3), cap feeding mechanism (4), nickel wire tray (5), nickel wire transfer mechanism (6), spot welding machine (7), frame (8), rotating transfer mechanism (9), and conveyor mechanism (10). The automatic spot welding machine for lithium batteries has the advantages of a relatively simple structure, good calibration effect, even transmission of caps, accurate transfer of nickel wires, and high production efficiency.

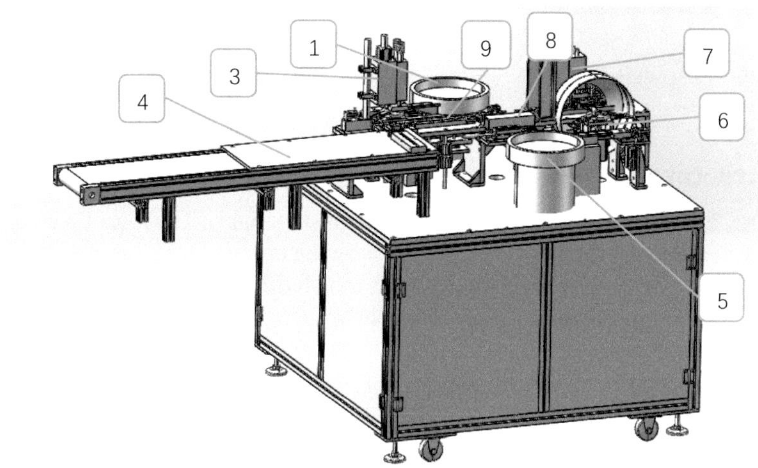

Fig. 6.45 Automatic Spot Welding Machine for Lithium Batteries
1-Nickel Wire Tray 3-Spot Welding Machine 4-Cap Feeding Mechanism 5-Nickel Wire Tray 6-Nickel Wire Transfer Mechanism 7-Spot Welding Machine 8-Rotating Transfer Mechanism 9-Conveyor Mechanism

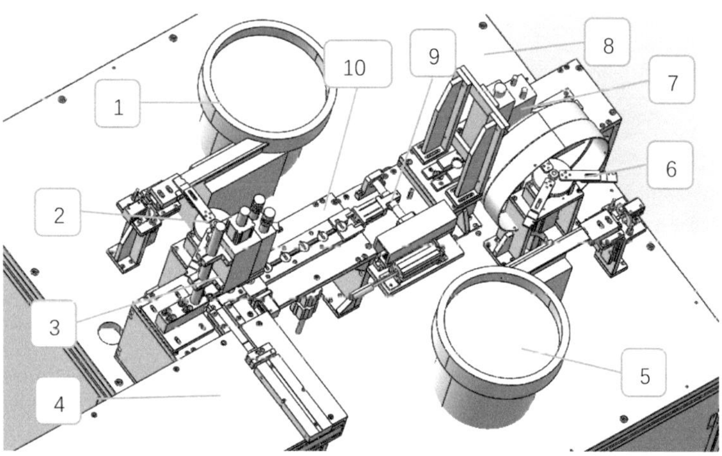

Fig. 6.46 Enlarged View of Automatic Spot Welding Machine for Lithium Batteries
1-Nickel Wire Tray 2-Nickel Wire Transfer Mechanism 3-Spot Welding Machine 4-Cap Feeding Mechanism 5-Nickel Wire Tray 6-Nickel Wire Transfer Mechanism 7-Spot Welding Machine 8-Frame 9-Rotating Transfer Mechanism 10-Conveyor Mechanism

II. Working Principle

The cap to be welded is input into the equipment from the cap feeding mechanism, and the nickel wire to be welded is input from the nickel wire tray. When the automatic spot welding machine for lithium batteries is in operation, press the start button to power on the device. The caps from the cap feeding mechanism are conveyed along the transfer track by the conveyor mechanism, and the caps are transported below the spot welding machine. Nickel wire is fed from the nickel wire tray, and after being cut by the cutting cylinder, it is transferred to the top center of the cap through the nickel wire transfer mechanism. At this point, the spot welding machine is started to perform the welding of nickel wire and the lithium battery on the cap. After welding, the caps with welded nickel wires are transported to the rotating transfer mechanism by the conveyor mechanism, and the spot welding machine performs spot welding on the reverse side of the lithium battery, completing the entire process of automatic spot welding for lithium batteries. The entire process is fully automated, with high automation, high production efficiency, and uniform and accurate transmission of caps and nickel wires.

III. Main Structure Introduction

Nickel Wire Tray and Nickel Wire Transfer Mechanism (Fig. 6.47): The nickel wire is fed from the nickel wire tray 1, and the cutting cylinder 3 driven by the cutting knife 2 moves back and forth to cut the nickel wire. After cutting, the nickel wire is transferred to the top center of the cap through the nickel wire transfer mechanism 4, and the spot welding machine 5 performs the welding of nickel wire and lithium battery.

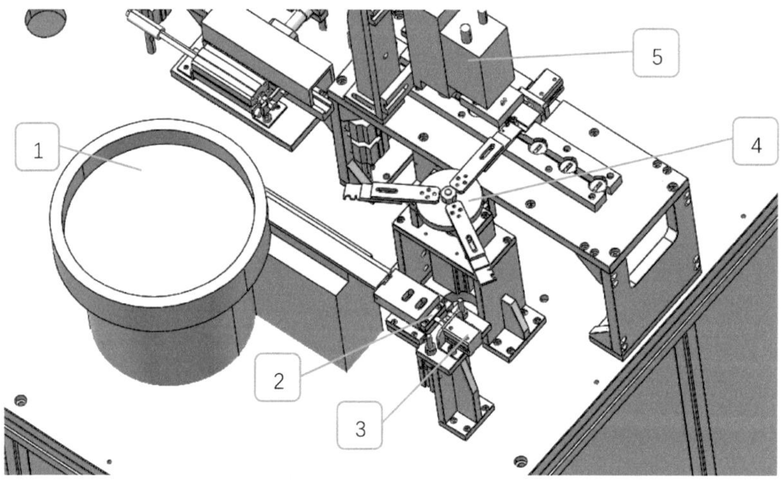

Fig. 6.47 Nickel Wire Tray and Nickel Wire Transfer Mechanism
1 - Nickel Wire Tray 2 - Cutting Knife 3 - Cutting Cylinder 4 - Nickel Wire Transfer Mechanism 5 - Spot Welding Machine

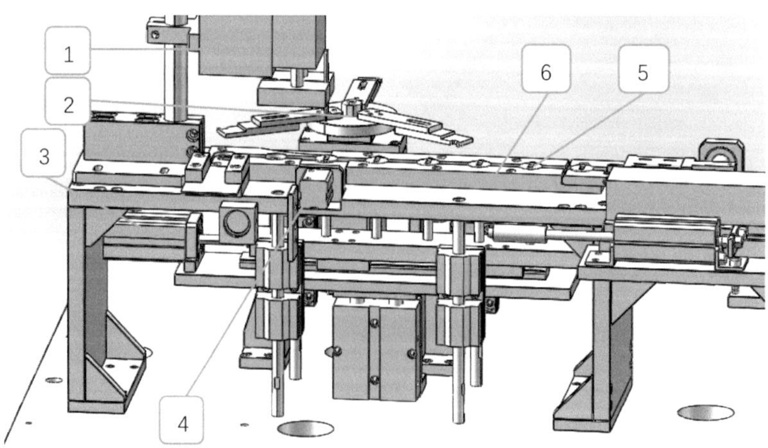

Fig. 6.48 Conveyor Mechanism
1 - Nickel Wire Tray 2 - Nickel Wire Transfer Mechanism 3, 4 - Cylinder 5 - Battery Seat 6 - Transfer Track

Conveyor Mechanism (Fig. 6.48): The conveyor mechanism, driven by cylinder 3, continuously transports lithium batteries to the next process. Cylinder 4 fixes the lithium battery to ensure that the lithium battery does not pop out of the battery seat 5 during spot welding.

Rotating Transfer Mechanism (Fig. 6.49): The rotating transfer mechanism, driven by cylinder 3, uses the fixture 1 to grab the battery from the transfer track 2, transfer it to the transfer track 4, and perform a flip.

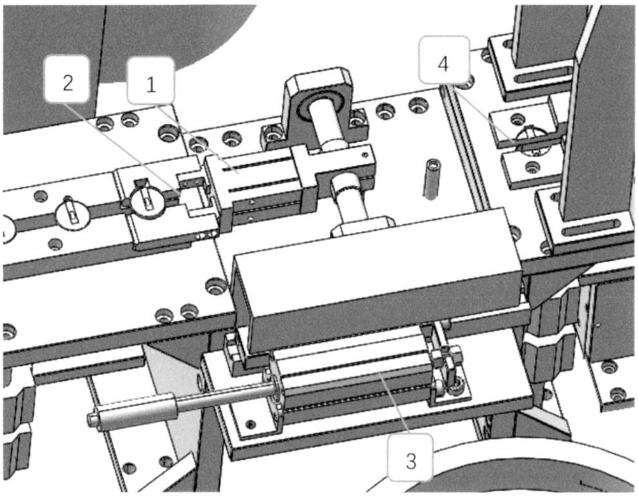

Fig. 6.49 Rotating Transfer Mechanism
1 - Fixture 2, 4 - Transfer Track 3 - Cylinder

Spot Welding Machine (Fig. 6.50): Used for welding nickel wire and caps.

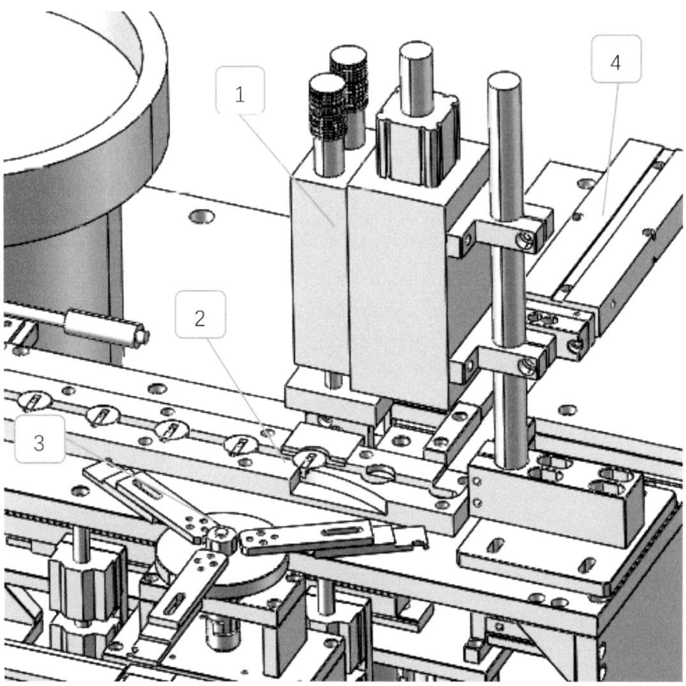

Fig. 6.50 Spot Welding Machine
1 - Spot Welding Machine 2 - Transfer Track 3 - Nickel Wire Transfer Mechanism 4 - Pushing Mechanism

IV. Specific Parameters (see Table 6.13)

Table 6.13 Automatic Spot Welding Machine for Lithium Batteries Parameters

Power:	25 kVA
Output Current:	12000 A
Welding Capacity:	3~6 mm
Operation Rate:	≥93%
Good Product Rate:	≥98%
Allowed Humidity:	<75%
Equipment Dimensions:	1275 mm(H) × 2135 mm(L) × 1150 mm(W)

V. Mechanical Design Highlights

The welding of nickel wires on lithium battery caps is generally completed by manually controlled semi-automatic equipment, with welding efficiency and quality depending on the operator's skills and proficiency. The automatic spot welding machine for lithium batteries optimizes the structure of the transportation system, designed with a relatively simple structure, good transport effect, even transmission of caps, and accurate transfer of nickel wires.

VI. Precautions

(1) Operators must undergo professional spot welding training, familiarize themselves with the performance of the spot welding machine, and be aware of usage precautions.
(2) The spot welding machine should not operate under conditions exceeding 40°C ambient temperature and 90% relative humidity.
(3) The workplace should be free of corrosive gases (vapor, salt spray, mold, dust, etc.) and flammable or explosive materials that could seriously affect the insulation performance of the welding machine.
(4) The spot welding machine casing must be reliably grounded to prevent harm to the human body from static electricity, leakage, etc.
(5) Do not place hands on the spot welding electrodes during operation to avoid safety accidents such as hand crushing.
(6) During spot welding, the welding area of the workpiece should be placed flat on the spot welding electrode for good contact, to prevent sparks from flying around due to poor contact between the workpiece and the electrode, causing injuries.
(7) In case of machine failure, the power should be cut off immediately, and relevant maintenance personnel should be notified for inspection. The machine can only be restarted after troubleshooting.
(8) Do not leave the spot welding machine with power on. Ensure that the machine has been turned off when people are not present.

Chapter 7

Lifting and Transfer Devices

Case 90: Stepper-Driven Lifting Mechanism

I. Case Description

The stepper-driven lifting mechanism is used for vertical transportation in logistics systems such as factories and automated warehouses. The overall structure, as shown in Fig. 7.1, consists of a lifting platform (1), lifting frame (2), lifting control mechanism (3), and base (4). This mechanism is known for its simplicity, robust structure, smooth lifting operation, and high load-bearing capacity.

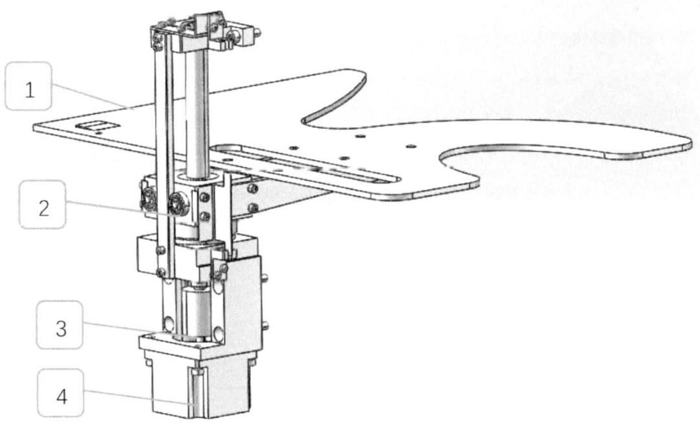

Fig. 7.1 Stepper-Driven Lifting Mechanism
1 - Lifting Platform 2 - Lifting Frame 3 - Lifting Control Mechanism 4 - Base

II. Working Principle

In practical use, the base is fixed horizontally, and the lifting control mechanism regulates the ascent or descent of the lifting frame. This achieves various platform heights to meet specific operational requirements [120]. The stepper-driven lifting mechanism features a compact design with fewer support arms compared to existing scissor-type structures, reducing space occupation, lowering production costs, and achieving greater lifting heights.

III. Main Structure Introduction

The lifting control mechanism, illustrated in Fig. 7.2, is powered by an electric motor (2). It controls the ascent and descent of the lifting frame (1). While one lifting frame ascends or descends, the remaining lifting frames remain stationary, systematically raising or lowering the stepper-driven lifting mechanism, minimizing overall space requirements, and reducing production costs.

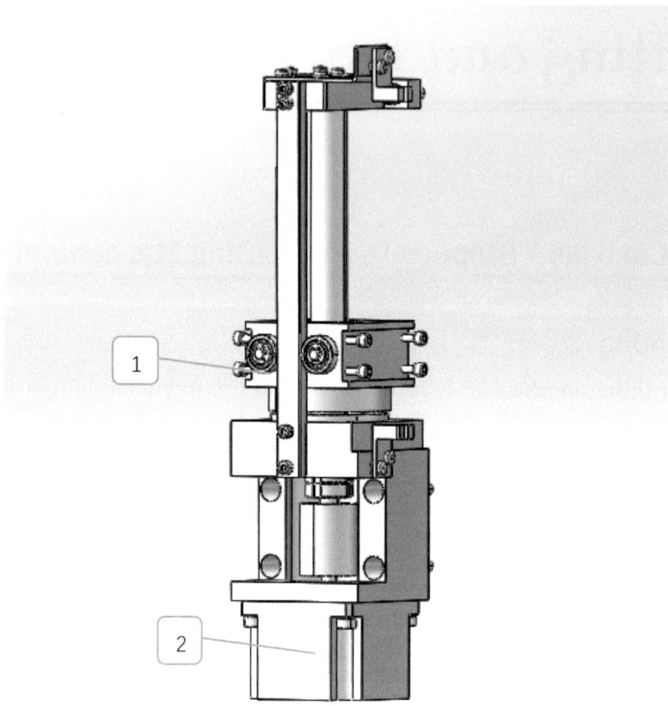

Fig. 7.2 Lifting Control Mechanism
1 - Lifting Frame 2 - Electric Motor

IV. Specific Parameters (see Table 7.1)

Table 7.1 Parameters of Stepper-Driven Lifting Mechanism

Maximum Lifting Height	2500 mm
Lifting Speed	80 mm/s
Rated Load	80 kg
Electric Motor Power	3 kW
Operating Pressure Range	4-7 kg/cm^2
Environmental Temperature	0~65°C
Allowable Humidity	10%~95%
Equipment Dimensions	272 mm(H) × 360 mm(L) × 370 mm (W)

V. Mechanical Design Highlights

The stepper-driven lifting mechanism can incorporate multiple lifting frames and lifting control mechanisms. By controlling the ascent and descent of individual lifting frames, the entire mechanism can systematically raise or lower, reducing the overall space occupied and decreasing production costs. Additionally, the number of lifting frames can be adjusted based on height requirements, minimizing stacking space and simplifying the structure [121].

Case 91: Palletizing Robot

I. Case Description

The palletizing robot is an automated palletizing equipment used to dismantle items in different packaging forms (such as bagged or boxed) from storage stack locations for loading or direct processing. The overall structure is shown in Fig. 7.3 and consists mainly of the frame (1), operating console (2), recovery frame (3), electrical control box (4), product frame (5), conveying mechanism (6), product handling mechanism (7), palletizing robot (8), and product gripping mechanism (9). The palletizing robot replaces manual work in tasks such as palletizing, handling, or feeding, achieving mechanical and automated production, liberating people from simple, heavy, repetitive labor, and hazardous, toxic, or harmful environments.

Fig. 7.3 Palletizing Robot
1-Frame 2-Operating Console 3-Recovery Frame 4-Electrical Control Box 5-Product Frame 6-Conveying Mechanism 7-Product Handling Mechanism 8-Palletizing Robot 9-Product Gripping Mechanism

II. Working Principle

The palletizing robot mainly performs gripping and placing operations. Products are input into the palletizing robot from the product frame, and the product gripping mechanism transports trays from the product frame to the conveying mechanism. The conveying mechanism transports the product under the palletizing robot, which then removes the product. Empty product pallets are conveyed to the recovery frame, and the process is repeated to complete the palletizing operation.

III. Main Structure Introduction

The product gripping mechanism, as shown in Fig. 7.4, is driven by a servo motor to horizontally move the fixture (2) within the horizontal plane. The opening and closing of the fixture (2) are controlled by the cylinder (1), continuously transporting trays from the product frame (3) to the conveying mechanism.

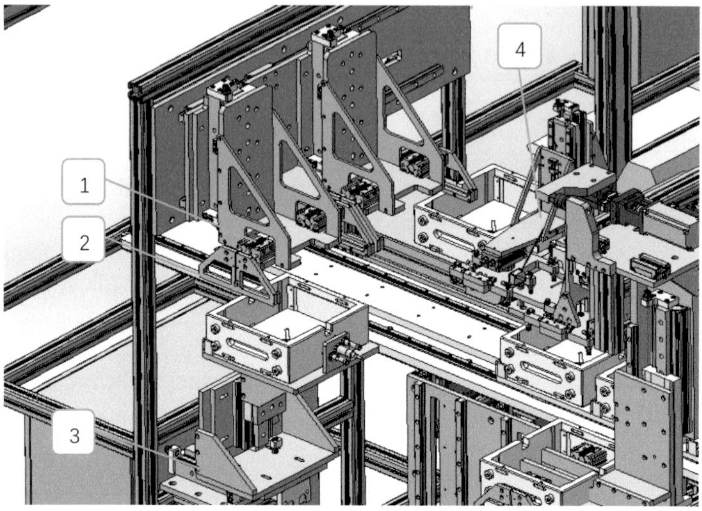

Fig. 7.4 Product Gripping Mechanism
1 - Cylinder 2 - Fixture 3 - Product Frame 4 - Palletizing Robot

The palletizing robot, as shown in Fig. 7.5, uses suction cups (1) to lift products and rotates to transport them to the tray (3). The product handling mechanism then sends them to the next process.

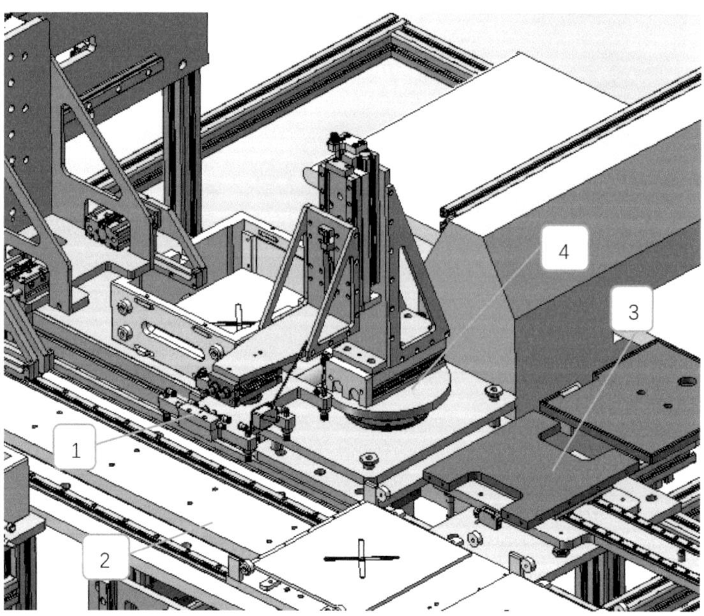

Fig. 7.5 Palletizing Robot
1 - Suction Cups 2 - Conveying Mechanism 3 - Tray 4 - Rotation Axis

The product handling mechanism, as shown in Fig. 7.6, is driven by a servo motor to horizontally move the fixture (3) within the horizontal plane, continuously transporting products from the tray to the flat belt (1).

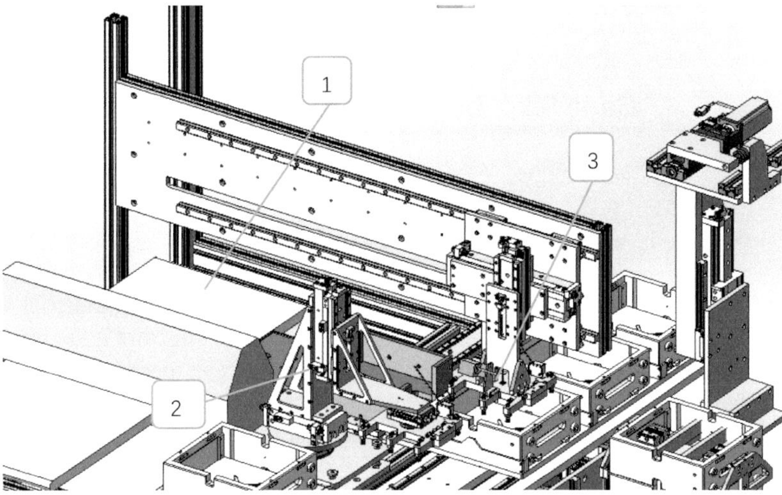

Fig. 7.6 Product Handling Mechanism
1 - Flat Belt 2 - Fixture 3 - Palletizing Robot

IV. Specific Parameters (see Table 7.2)

Table 7.2 Palletizing Robot Parameters

Total Motor Power:	1.2 kW
Weight:	1500 kg
Production Efficiency:	600-800 pcs/h
Operating Temperature:	0~45°C
Allowed Humidity:	20%~95%
Equipment Dimensions:	1700 mm(H) × 2000 mm(L) × 1700 mm(W)

V. Mechanical Design Highlights

The palletizing robot can complete palletizing for bags, blocks, boxes, and other products according to the required grouping and layers, reducing labor and lowering labor intensity. It is a high-tech product in modern industrial development, playing a significant role in improving productivity and reducing costs [122]. The automatic palletizing robot is suitable for automatic palletizing of products such as boxes, bags, appliances, sheets, tiles, etc. By replacing the gripper of the robot, it can handle palletizing for different goods, reducing customer purchasing costs. The palletizing robot has the following advantages:

(1) Simple structure and fewer components, leading to low failure rates, reliable performance, simple maintenance, and fewer stocked components.
(2) Occupies less floor space, facilitating the layout of customer production lines and allowing for larger storage areas.

(3) Strong adaptability, easy modification on the touch screen when the size, volume, shape, or tray dimensions of customer products change, without affecting normal production.
(4) Low energy consumption, significantly reducing customer operational costs.
(5) All controls can be operated on the control cabinet screen, making operation very simple.
(6) It o nly needs to position the gripping and placing points, with a simple and easy-to-understand teaching method.

Case 92: Lathe Feeding Robot

I. Case Description

The lathe feeding robot is an automation technology device that mimics certain functions of the human upper limbs, automatically transporting workpieces or gripping tools for operations according to predetermined requirements. It plays a crucial role in achieving industrial automation and promoting the development of industrial production. The overall structure is shown in Fig. 7.7, consisting mainly of a transfer mechanism (1), controller (2), frame (3), robot body (4), servo motor (5), and warning light (6). The lathe feeding robot can replace manual labor, significantly reducing the labor intensity of workers, improving working conditions, and enhancing labor productivity and automation levels [123]. Additionally, it can operate in high temperature, low temperature, deep water, radioactive, and other toxic and polluted environments, demonstrating its superiority and having broad development prospects.

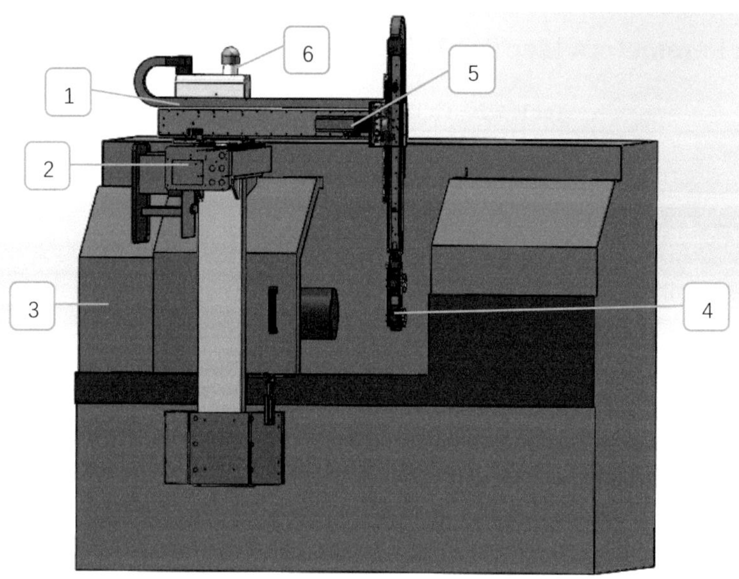

Fig. 7.7 Lathe Feeding Robot
1-Transfer Mechanism 2-Controller 3-Frame 4-Robot Body 5-Servo Motor 6-Warning Light

II. Working Principle

The main working principle of the lathe feeding robot is to simulate manual feeding. The controller and robot body collaborate to perform tasks. When the material frame is filled, the lathe feeding robot automatically comes to fetch the material. The robot body grabs the product and feeds it.

III. Main Structure Introduction

The robot body, as shown in Fig. 7.8, is mainly controlled by cylinders (3) and (4) to open and close.

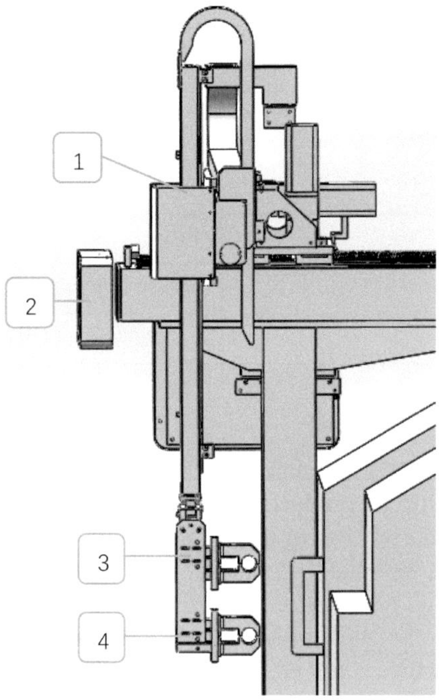

Fig. 7.8 Robot Body
1 - Transfer Mechanism 2 - Controller 3, 4 - Cylinders

The transfer mechanism, as shown in Fig. 7.9, is controlled by servo motors (1), (2), and (4) to move the robot horizontally and vertically.

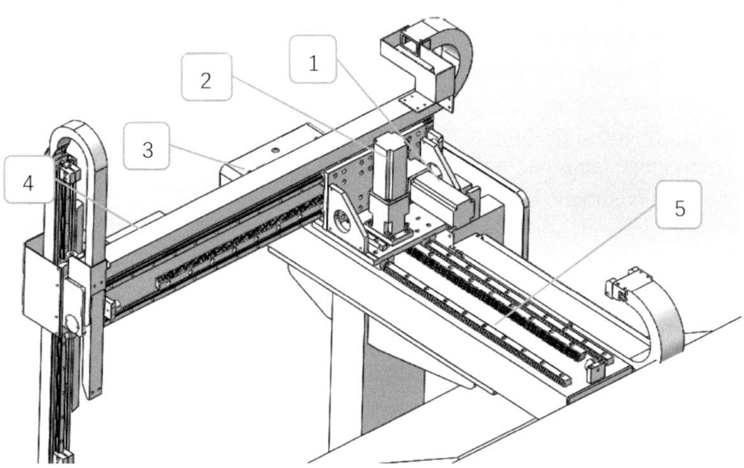

Fig. 7.9 Transfer Mechanism
1, 2, 4 - Servo Motors 3 - Controller 5 - Guide Rail

IV. Specific Parameters (see Table 7.3)

Table 7.3 Lathe Feeding Robot Parameters

Coordinate System:	Rectangular coordinate system
Arm Up-Down Movement:	500 mm
Arm Extension (Retraction):	500 mm
Operating Pressure Range:	4-7 kg/cm^2
Positioning Accuracy:	± 0.5 mm
Finger Grip Force:	392 N
Drive Mode:	Electric motor drive
Equipment Dimensions:	2175 mm(H) × 2110 mm(L) × 1500 mm(W)

V. Mechanical Design Highlights

Conventional CNC lathes generally rely on manual handling for material feeding, causing fatigue for operators. Manual clamping of products is inefficient, and manual clamping may result in misalignment, leading to product processing quality issues. The lathe feeding robot reduces the labor intensity of operators through automatic feeding, improving production efficiency by increasing the feeding speed, and reducing the occurrence of product quality incidents [124]. Additionally, a pressure sensor is installed on the guide pillar of the lathe feeding robot, transmitting the pressure received by the guide pillar to the central control unit. The central control unit determines whether the guide pillar clamps the product and adjusts the spacing between guide pillars accordingly. This prevents excessive spacing between guide pillars, causing product slippage, or insufficient spacing, resulting in product damage [125].

Case 93: Automatic Feeding Machine for Electronic Components

I. Case Description

The automatic feeding machine for electronic components is suitable for placing multiple electronic components into the machine for feeding through a product rack, improving work efficiency. The overall structure is shown in Fig. 7.10, consisting mainly of a feeding mechanism (1), display screen (2), product rack (3), material pushing mechanism (4), and base (5). The automatic feeding machine for electronic components can load multiple electronic components at once using the product rack, thereby enhancing the feeding efficiency.

II. Working Principle

When the automatic feeding machine for electronic components is in operation, operators set parameters through the human-machine interface and arrange electronic components neatly on the product rack. The material pushing mechanism pushes the electronic components from the product rack to the feeding mechanism. The feeding mechanism includes a rotating material transfer mechanism and a conveying mechanism. The conveying mechanism transports the electronic components in batches to the next process, completing the automatic feeding of multiple electronic components.

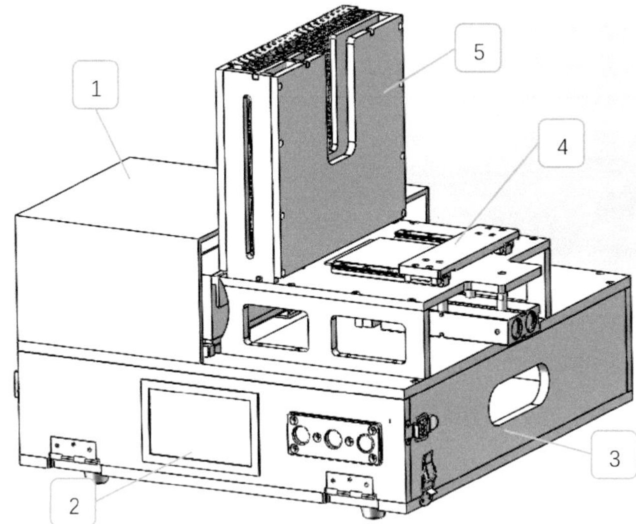

Fig. 7.10 Automatic Feeding Machine for Electronic Components
1 - Feeding Mechanism 2 - Display Screen 3 - Base 4 - Material Pushing Mechanism 5 - Product Rack

III. Main Structure Introduction

The product rack, as shown in Fig. 7.11, has electronic components (1) neatly arranged on it.

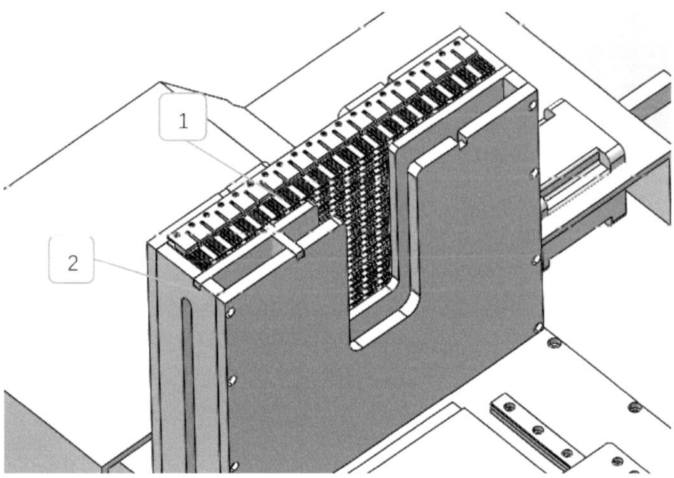

Fig. 7.11 Product Rack
1 - Electronic Components 2 - Product Rack

The material pushing mechanism, illustrated in Fig. 7.12, is driven by cylinder (4) during the operation of the automatic feeding machine for electronic components. The push plate (2) pushes the electronic components from the bottom of the product rack (1) to the rotating material transfer mechanism (5), and the stopper (3) prevents the material from being pushed too far, preventing accidents.

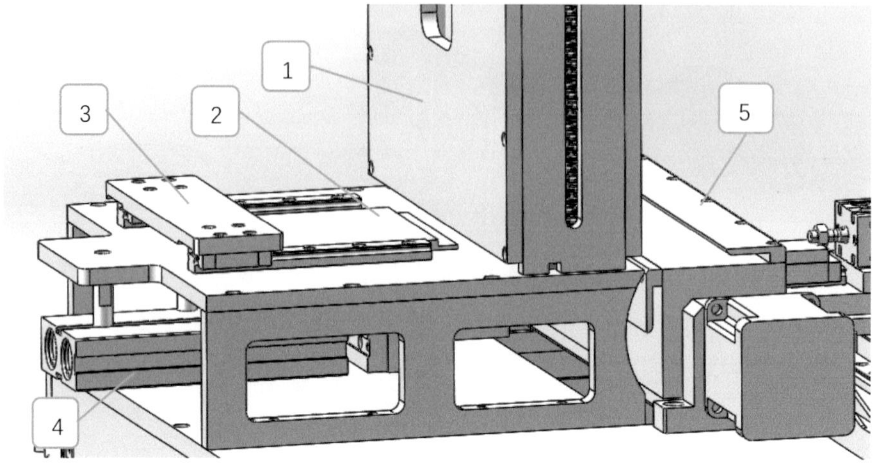

Fig. 7.12 Material Pushing Mechanism
1 - Product Rack 2 - Push Plate 3 - Stopper 4 - Cylinder 5 - Rotating Material Transfer Mechanism

The feeding mechanism, shown in Fig. 7.13, involves the material pushing mechanism pushing electronic components into the slot of the rotating material transfer mechanism (2). The rotating material transfer mechanism (2) is driven by an electric motor (3) to rotate clockwise and dock with the discharge port (4). At this point, the cylinder (7) controls the lateral pushing mechanism (8) to press against one end of the electronic component, and the electric motor (5) drives the lateral movement of the pushing mechanism (8) through the pulley (6), sending the electronic component out through the discharge port.

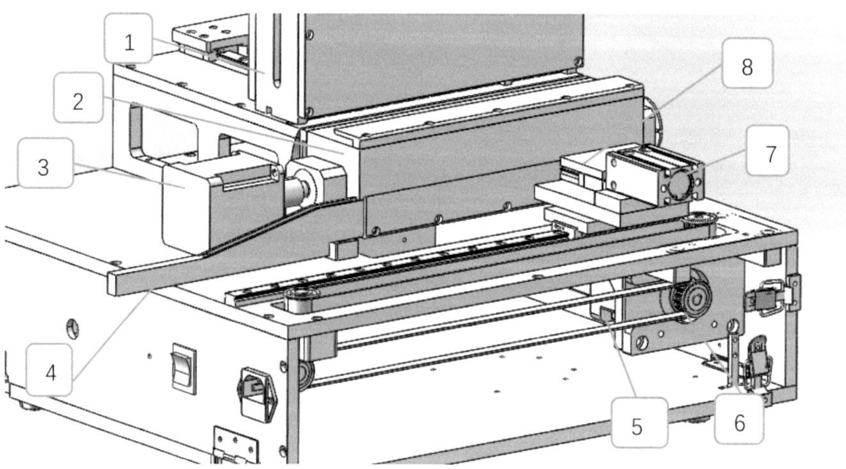

Fig. 7.13 Feeding Mechanism
1 - Product Rack 2 - Rotating Material Transfer Mechanism 3, 5 - Electric Motor 4 - Discharge Port 6 - Pulley 7 - Cylinder 8 - Lateral Pushing Mechanism

IV. Specific Parameters (see Table 7.4)

Table 7.4 Automatic Feeding Machine for Electronic Components Parameters

Working Efficiency:	1000-2000 pcs/h
Utilization Rate:	≥93%
Yield Rate:	≥98%
Ambient Temperature:	0~55°C
Permissible Humidity:	<75%
Equipment Dimensions:	410 mm(H) × 450 mm(L) × 420 mm (W)

V. Mechanical Design Highlights

Current automatic feeding machines can only load one tray at a time, resulting in lower feeding efficiency. The automatic feeding machine for electronic components transports entire columns of electronic components out of the product rack using the rotating material transfer mechanism, maintaining consistency in the orientation of electronic components, thereby improving feeding efficiency.

Case 94: Automated Suction Transfer Device

I. Case Description

The automated suction transfer device is primarily used for the suction and transfer of products. The overall structure is shown in Fig. 7.14, with an enlarged view in Fig. 7.15. It consists mainly of a cylinder (1), pushing mechanism (2), suction mechanism (3), conveyor track (4), transfer mechanism (5), and frame (6). The suction mechanism (3) includes a frame, three suction nozzles, and a vacuum pump, offering a compact structure, simple operation, and reliable performance.

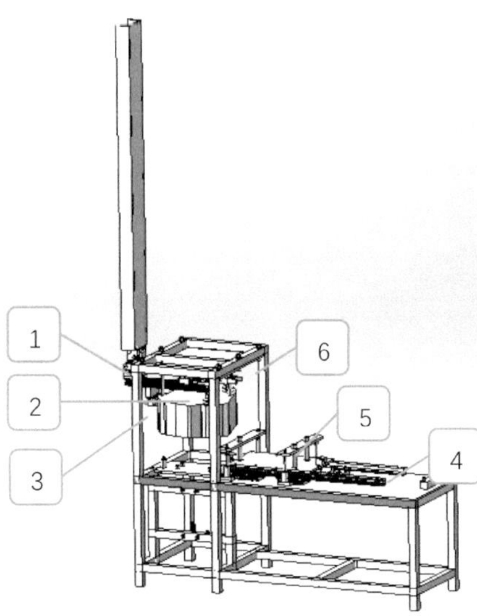

Fig. 7.14 Automated Suction Transfer Device
1 - Cylinder 2 - Pushing Mechanism 3 - Suction Mechanism 4 - Conveyor Track 5 - Transfer Mechanism 6 - Frame

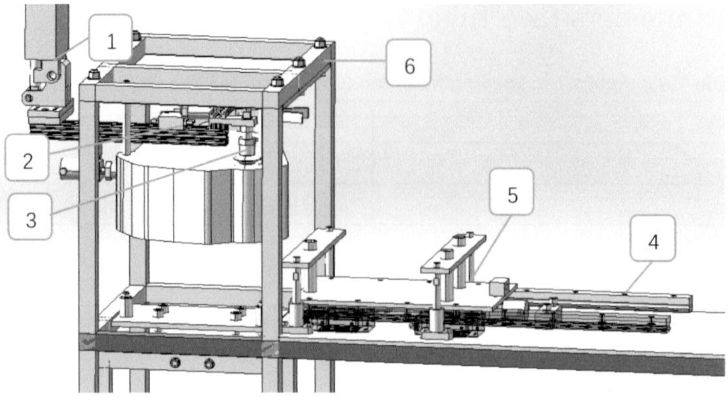

Fig. 7.15 Enlarged View of Automated Suction Transfer Device
1-Cylinder 2-Pushing Mechanism 3-Suction Mechanism 4-Conveyor Track 5-Transfer Mechanism 6-Frame

II. Working Principle

When the automated suction transfer device is in operation, the cylinder of the suction mechanism is activated, causing the suction mechanism to move down and snugly attach to the product. The suction nozzles of the suction mechanism suction the product and lift it to a slightly higher position than the transfer mechanism. The pushing mechanism then pushes the suctioned product to the transfer mechanism. The suction nozzles release, and the transfer mechanism conveys the product to the next process.

III. Main Structure Introduction

The suction mechanism and pushing mechanism, as shown in Fig. 7.16, consist mainly of a frame (2), three suction nozzles (3), and a vacuum pump (4). During the operation of the suction mechanism, the suction nozzles snugly adhere to the product surface, and the vacuum pump evacuates the air inside the suction nozzles, allowing them to securely grip the product. The pushing mechanism (5) is driven and controlled by a cylinder (6) to push the product to the transfer mechanism [126].

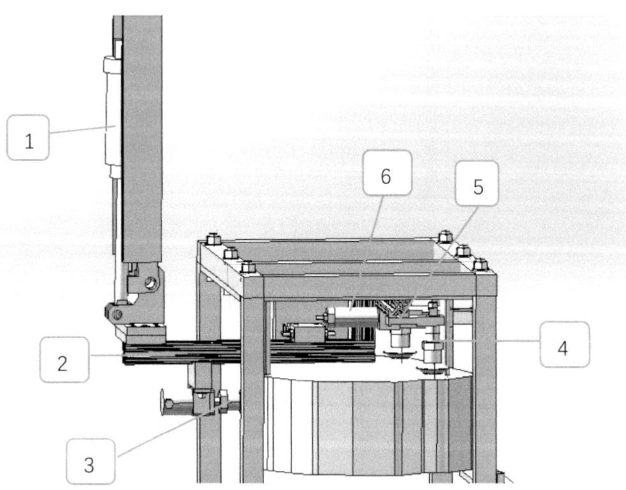

Fig. 7.16 Suction Mechanism and Pushing Mechanism
1, 6-Cylinder 2-Frame 3-Suction Nozzle 4-Vacuum Pump 5-Pushing Mechanism

The transfer mechanism, illustrated in Fig. 7.17, receives the product pushed by the pushing mechanism. It is driven and controlled by a servo motor (3) to convey the product along the conveyor track (2) to the next process.

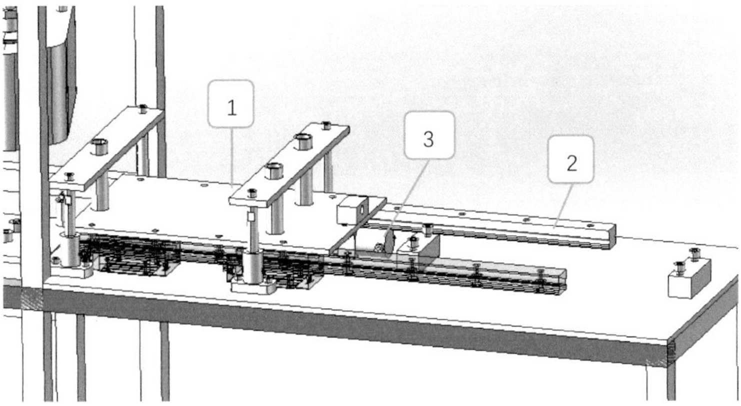

Fig. 7.17 Transfer Mechanism
1 - Transfer Mechanism 2 - Conveyor Track 3 - Servo Motor

IV. Specific Parameters (see Table 7.5)

Table 7.5 Automated Suction Transfer Device Parameters

Maximum Suction Weight:	100 kg
Maximum Suction Volume:	90 mm(H) × 165 mm(L) × 280 mm(W)
Working Efficiency:	400-600 pcs/h
Ambient Temperature:	0~55°C
Permissible Humidity:	10%~95%
Equipment Dimensions:	1460 mm(H) × 940 mm(L) × 340 mm(W)

V. Mechanical Design Highlights

The automated suction transfer device uses three suction nozzles positioned above and on the sides of the product, effectively maintaining stability during the transfer process. Moreover, compared to other mechanical handling methods, the vacuum nozzles can complete the entire conveying process without harming the product or raw materials.

Case 95: Electronic Equipment Handling Device

I. Case Description

The electronic equipment handling device is an integrated automation device that combines suction and movement. It is suitable for various sheet-shaped, spherical metal, or non-metal workpieces. The overall structure is shown in Fig. 7.18, comprising a frame (1), horizontal guide mechanism (2), conveyor mechanism (3), handling mechanism (4), and feeding mechanism (5). The electronic equipment handling device boasts high automation, high efficiency, and effective reduction of manual labor intensity.

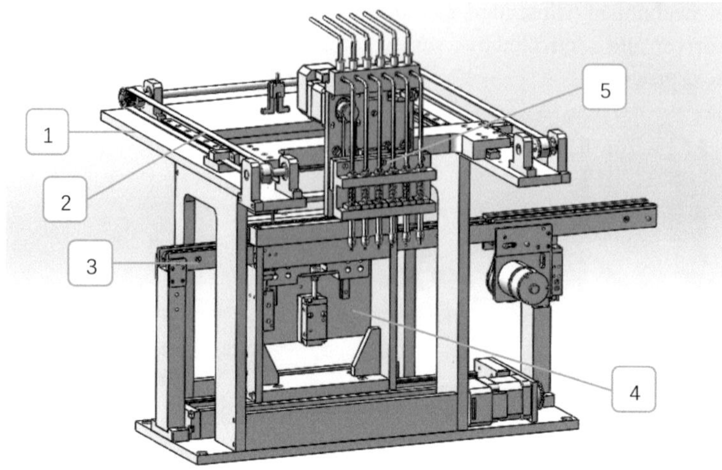

Fig. 7.18 Electronic Equipment Handling Device
1 - Frame 2 - Horizontal Guide Mechanism 3 - Conveyor Mechanism 4 - Handling Mechanism 5 - Feeding Mechanism

II. Working Principle

When the electronic equipment handling device is operational, the feeding mechanism suction picks the workpiece to be transported. After picking up the component, the feeding mechanism moves along the horizontal guide, transporting the electronic component to the handling mechanism. The vacuum generator on the feeding mechanism is closed, and the suction on the nozzle's adhesive end disappears, placing the electronic component on the handling mechanism's product frame. The handling mechanism is then moved to the next process by a synchronous belt, completing the automated handling of the electronic component.

III. Main Structure Introduction

The feeding mechanism, shown in Fig. 7.19, operates with a servo motor (2) driving the vacuum generator (1) through a pulley system (3). The nozzle (4) sucks the workpiece that needs to be transported.

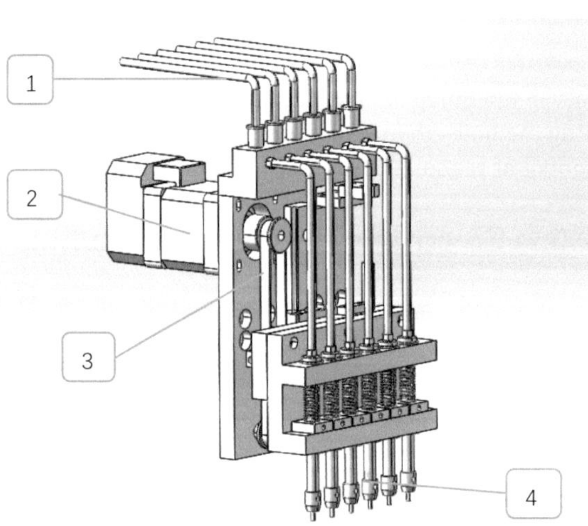

Fig. 7.19 Feeding Mechanism
1 - Vacuum Generator 2 - Servo Motor 3 - Pulley 4 - Nozzle

The guide mechanism, illustrated in Fig. 7.20, is driven by a servo motor (4), controlling the horizontal movement of the feeding mechanism (1) along the guide (2) using a pulley system (3).

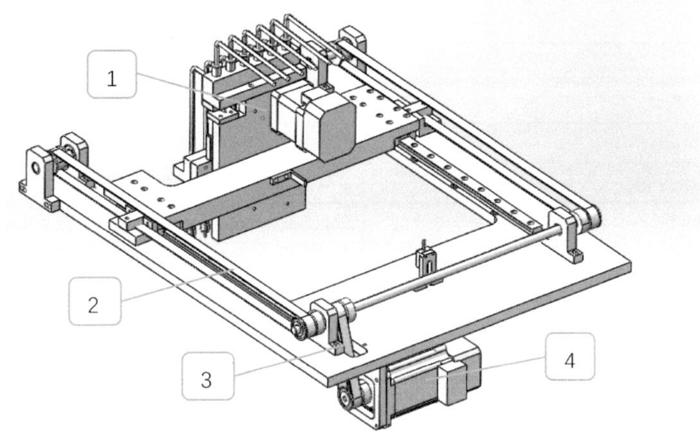

Fig. 7.20 Guide Mechanism
1 - Feeding Mechanism 2 - Guide 3 - Pulley 4 - Servo Motor

The handling mechanism, as depicted in Fig. 7.21, picks up the electronic component and places it on the product frame (1). Servo motors (4 and 3) are activated to control the lateral movement of the product frame along the synchronous belt (2).

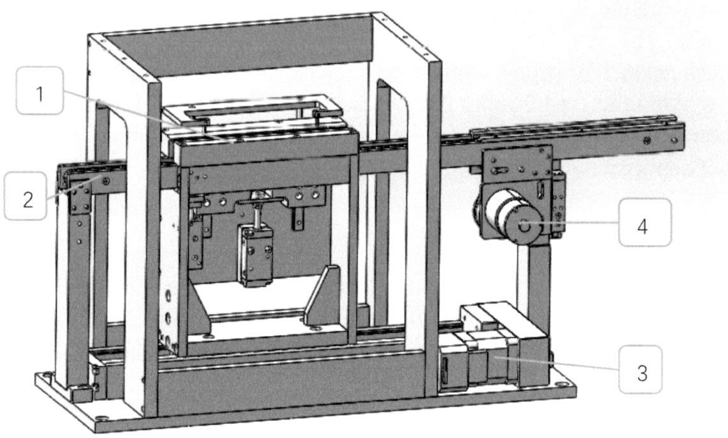

Fig. 7.21 Handling Mechanism
1 - Product Frame 2 - Synchronous Belt 3, 4 - Servo Motor

IV. Specific Parameters (see Table 7.6)

Table 7.6 Electronic Equipment Handling Device Parameters

Working Efficiency:	1500-2000pcs/h
Operating Rate:	≥93%
Yield Rate:	≥98%
Ambient Temperature:	0~55°C
Permissible Humidity:	30%~95%
Equipment Dimensions:	460 mm(H) × 575 mm(L) × 460 mm(W)
Applicable Products:	Sheet-shaped, spherical metal, or non-metal workpieces

V. Mechanical Design Highlights

Currently, in electronic and precision instrument industries, many small metal and non-metal workpieces need to be moved between processes. Due to the difficulty in manual handling, and to enhance production efficiency and reduce the workload on workers, automated production to replace manual operations is imperative. The electronic equipment handling device is an integrated automation device that combines suction and movement. It is suitable for various sheet-shaped, spherical metal, or non-metal workpieces. Through the coordination of the feeding mechanism and the driving mechanism, it achieves lateral movement after material suction, replacing manual material handling, especially for precision materials that are challenging for manual handling, significantly improving production efficiency.

Case 96: Punch Press Hydraulic Feeder Robot

I. Case Description

The punch press hydraulic feeder robot is primarily used for the feeding operation of punch press hydraulic machines. The overall structure is shown in Fig. 7.22, consisting of a rotary mechanism (1), rotating mechanism (2), lifting adjustment mechanism (3), frame (4), and feeder robot (5). The punch press hydraulic feeder robot can be used in conjunction with punch presses, hydraulic machines, and other equipment, offering advantages such as high production efficiency, low cost, and high product quality.

II. Working Principle

During the operation of the punch press hydraulic feeder robot, the lifting adjustment mechanism adjusts the feeder robot to the specified working height. The punch press hydraulic feeder robot is started, and it grabs the product, feeds it to the cutting die on the punch press, and after cutting, the feeder robot grabs the product again. Through the rotating mechanism, it is delivered to the unloading guide rail mechanism's unloading carrier, where manual stacking of materials is performed.

III. Main Structure Introduction

The feeder robot, as shown in Fig. 7.23, is controlled by cylinder (1) for vertical displacement and cylinder (2) for horizontal displacement. The rotary mechanism (3) is used to enhance the stability and precision of the feeder robot.

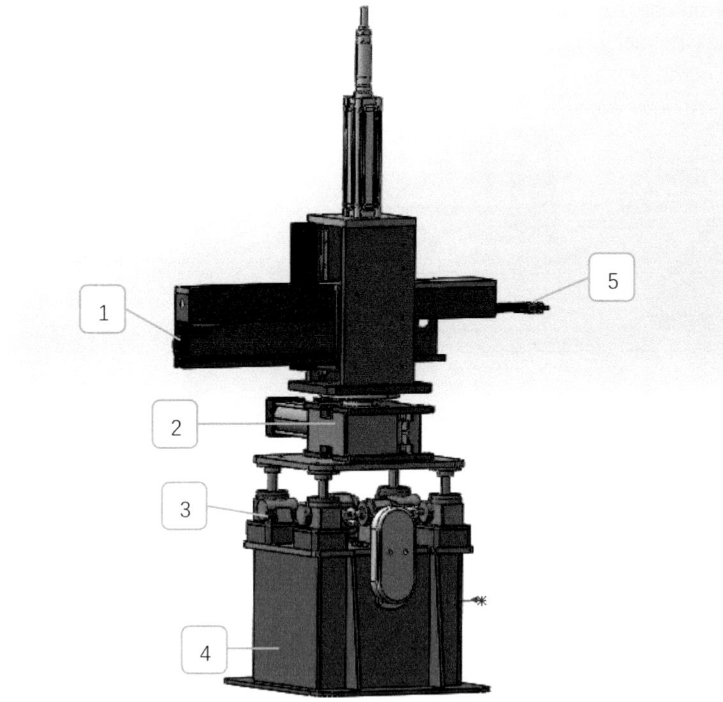

Fig. 7.22 Punch Press Hydraulic Feeder Robot
1 - Rotary Mechanism 2 - Rotating Mechanism 3 - Lifting Adjustment Mechanism 4 - Frame 5 - Feeder Robot

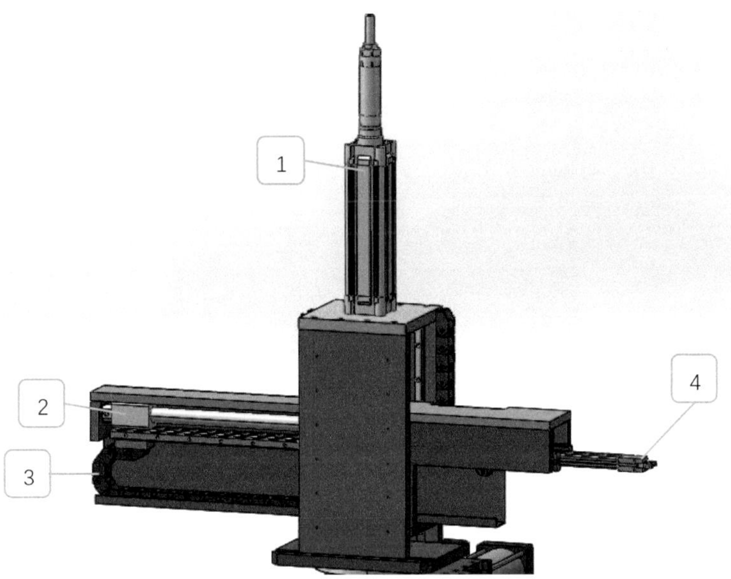

Fig. 7.23 Feeder Robot
1, 2 - Cylinder 3 - Rotary Mechanism 4 - Feeder Robot

The rotating mechanism, illustrated in Fig. 7.24, is driven by cylinder (1). During operation, cylinder (1) moves the rack (3) laterally, driving the gear (2) to rotate and, in turn, rotating the feeder robot around the central axis.

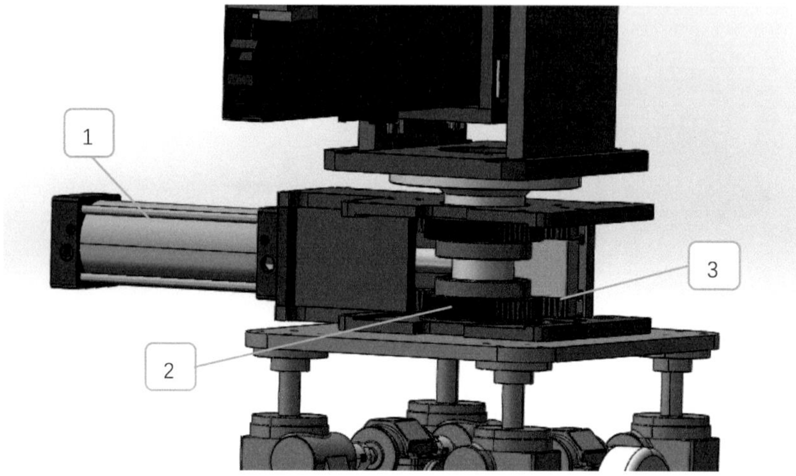

Fig. 7.24 Rotating Mechanism
1 - Cylinder 2 - Gear 3 - Rack

The lifting adjustment mechanism, as depicted in Fig. 7.25, is driven by cylinder (2), controlling the internal pressure to adjust the height of the working table (1).

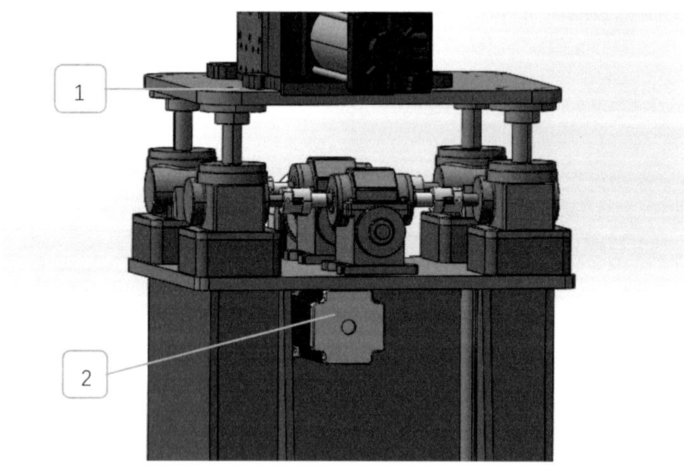

Fig. 7.25 Lifting Adjustment Mechanism
1 - Working Table 2 - Cylinder

IV. Specific Parameters (see Table 7.7)

Table 7.7 Punch Press Hydraulic Feeder Robot Parameters

Operating Pressure Range:	4-6 kg/cm^2
Positioning Accuracy:	±0.5 mm
Working Efficiency:	1500-2000 pcs/h

Contd.

Operating Rate:	≥93%
Yield Rate:	≥98%
Ambient Temperature:	0~55°C
Permissible Humidity:	20%~95%
Equipment Dimensions:	1680 mm(H) × 1045 mm(L) × 805 mm(W)

V. Mechanical Design Highlights

(1) The main drive is achieved by a high-performance servo motor.
(2) Quick positioning, lifting, and movement actions can be realized.
(3) Utilizes a specially designed lateral translation structure and auxiliary lifting structure.

VI. Precautions

(1) The machine speed should be moderate, preparing actions within the forming cycle.
(2) Be careful not to collide with the punch press robot when the upper and lower dies are engaged, as it may cause a short circuit or deformation of the machine.
(3) Regular maintenance and timely lubrication are required during machine operation.
(4) Avoid operating in dusty and oily environments.
(5) Do not operate in highly corrosive environments, as it may affect the service life of the punch press robot.

Case 97: Precision Vibratory Feeding Disk

I. Case Description

The precision vibratory feeding disk is designed for automatic feeding of precision components. The overall structure, as shown in Fig. 7.26, consists of transition feeding mechanism (1), vibratory feeding mechanism (2), vibratory base (3), linear feeding mechanism (4), and adjustment groove

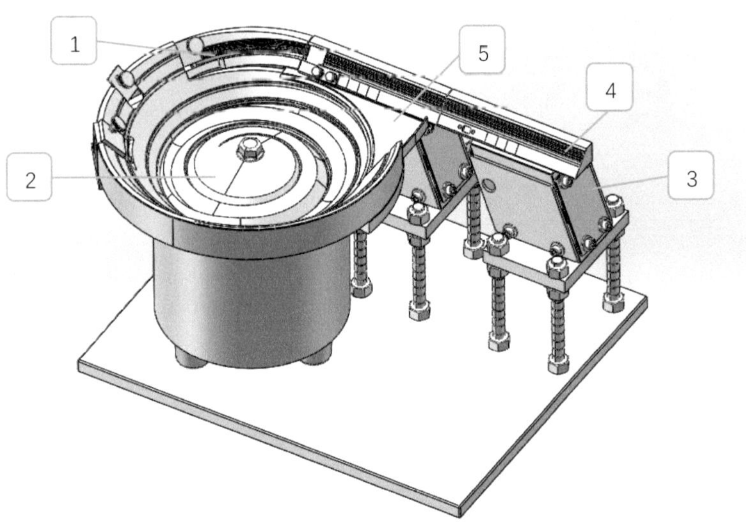

Fig. 7.26 Precision Vibratory Feeding Disk
1-Transition Feeding Mechanism 2-Vibratory Feeding Mechanism 3-Vibratory Base 4-Linear Feeding Mechanism 5-Adjustment Groove

(5). The outlet of the vibratory feeding mechanism is connected to the inlet of the transition feeding mechanism, and the outlet of the transition feeding mechanism is connected to the inlet of the linear feeding mechanism. The transition feeding mechanism (1) includes curved feeding track, straight feeding track, and block adjustment components. The precision vibratory feeding disk features high automation, stability, high efficiency, compact structure, and simple operation.

II. Working Principle

The precision vibratory feeding disk is an automatic feeding mechanism for precision components. During operation, the workpieces are placed in the vibratory feeding mechanism. The vibration of the vibratory feeding mechanism conveys the workpieces to the transition feeding mechanism and further guides them to the linear feeding mechanism.

III. Main Structure Introduction

The vibratory feeding mechanism, illustrated in Fig. 7.27, generates continuous vertical vibration through the servo motor inside the base (2), constantly feeding products into the feeding track (1).

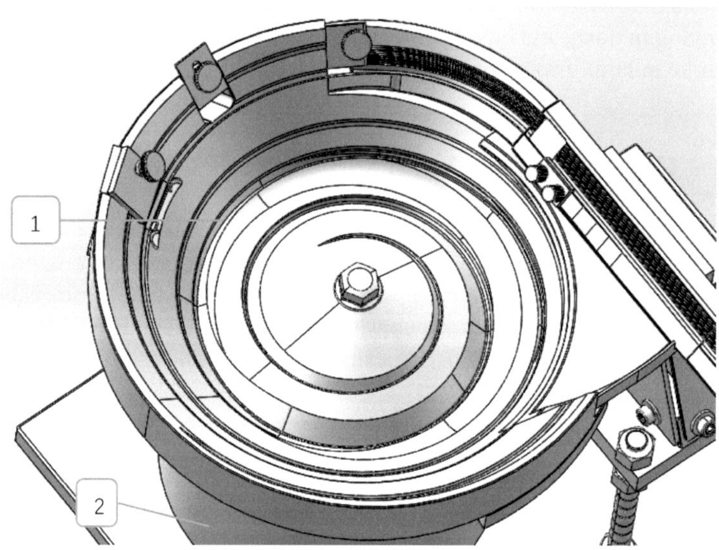

Fig. 7.27 Vibratory Feeding Mechanism
1 - Feeding Track 2 - Base

The transition feeding mechanism, shown in Fig. 7.28, includes curved feeding track (1), straight feeding track (2), and block adjustment components (3). The curved feeding track has a central bend groove line designed to adapt to the convex surface of the workpieces, allowing smooth operation. The block adjustment components (3) can adjust the arrangement of workpieces, reducing material jamming issues [127].

The vibratory base, as depicted in Fig. 7.29, drives the synchronized vibration of workpieces on the straight feeding track through vibration, smoothly conveying the workpieces to the linear feeding mechanism (1).

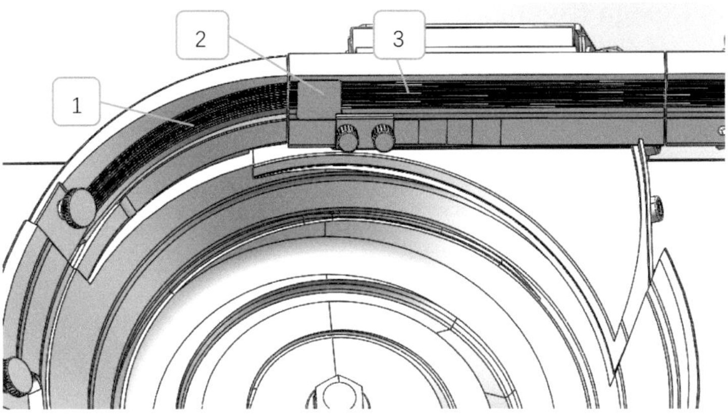

Fig. 7.28 Transition Feeding Mechanism
1 - Curved Feeding Track 2 - Straight Feeding Track 3 - Block Adjustment Components

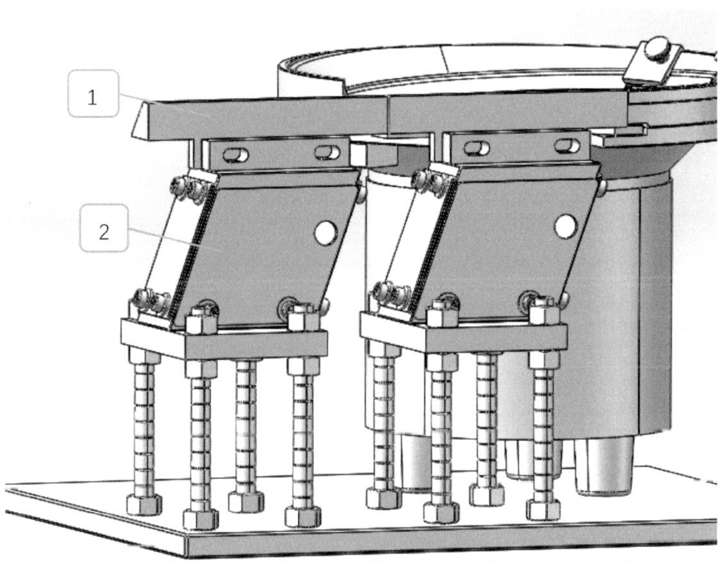

Fig. 7.29 Vibratory Base
1 - Linear Feeding Mechanism 2 - Vibratory Base

IV. Specific Parameters (see Table 7.8)

Table 7.8 Precision Vibratory Feeding Disk Parameters

Disk Diameter:	200 mm
Vibration Frequency:	50~140 Hz
Equipment Dimensions:	180 mm(H) × 260 mm(L) × 305 mm(W)
Lightweight and Small Parts Feeding:	Column Feeding
Parts Conveyance:	No Magnetic Influence During the Process

V. Mechanical Design Highlights

When workpieces pass through the curved feeding track, the block adjustment components can adjust the sorting positions of the workpieces, allowing them to move smoothly and orderly into the straight feeding track. The vibration of the vibratory base drives the synchronized vibration of workpieces on the straight feeding track, facilitating orderly and smooth conveyance to the linear feeding mechanism [128]. Through this process, the conveyance of precision components becomes more orderly and smooth. Additionally, the curved feeding track features a convex platform, preventing workpieces from sliding downward when they enter the curved feeding track.

Case 98: Screw Feeding Mechanism

I. Case Description

The screw feeding mechanism is designed for the feeding process of screws during production. The overall structure, as shown in Fig. 7.30, consists of the feeding area (1), guide plate (2), rotating mechanism (3), limit plate (4), and correction mechanism (5). The screw feeding mechanism features a simple structure, high working efficiency, and accurate positioning, significantly reducing manual labor costs.

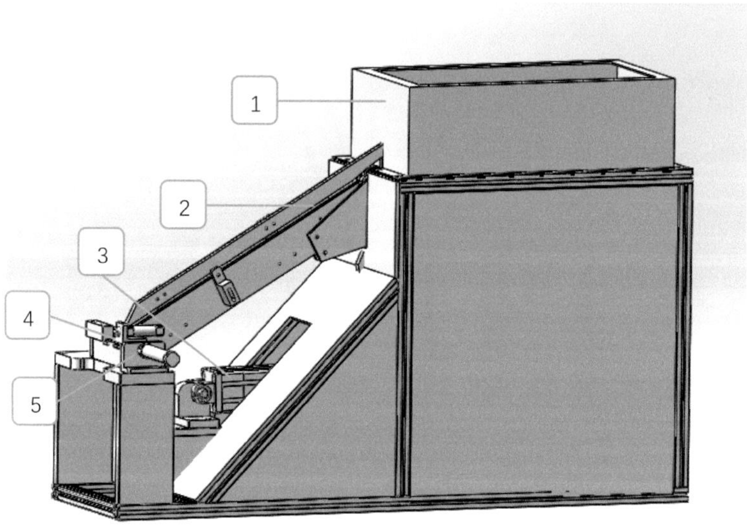

Fig. 7.30 Screw Feeding Mechanism
1 - Feeding Area 2 - Guide Plate 3 - Rotating Mechanism 4 - Limit Plate 5 - Correction Mechanism

II. Working Principle

During operation, screws are manually placed in the feeding area. Once confirmed in the feeding area, the piston rod of the cylinder in the rotating mechanism (3) is connected to a rotating wheel, driving the wheel to rotate. The rotating motion is used to intermittently pick up screws. The guide plate conveys the picked-up screws to the limit plate, and the correction mechanism adjusts the screw positions. After correction, the screws are transported to the next process [129].

III. Main Mechanism Introduction

The rotating mechanism, as shown in Fig. 7.31, consists of a cylinder (3), piston rod-connected rotating wheel (2), guide wheel (1), and a limiting groove (4). During operation, the cylinder (3) rotates, driving the rotating wheel (2) in reciprocating motion. This, in turn, moves the guide wheel (1) back and forth, continuously extracting screws from the feeding area and delivering them to the screw guide rail (5).

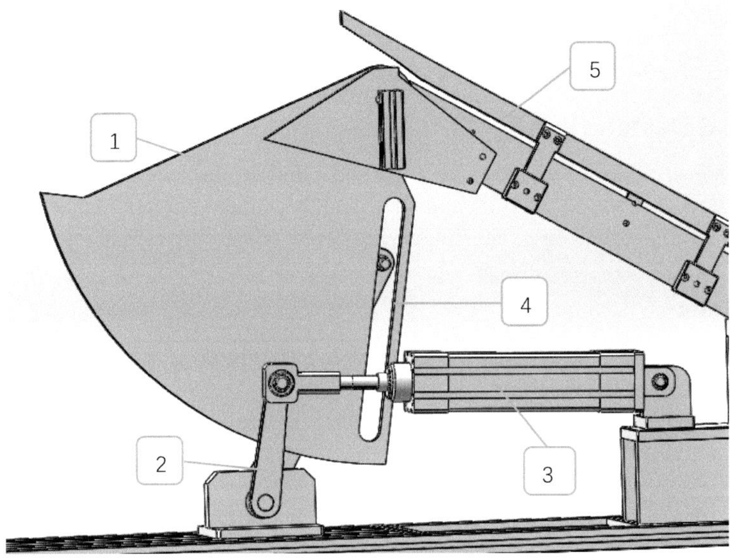

Fig. 7.31 Rotating Mechanism
1 - Guide Wheel 2 - Limiting Groove 3 - Screw Guide Rail 4 - Rotating Wheel 5 - Cylinder

The correction mechanism, illustrated in Fig. 7.32, is driven by a cylinder (3) to control the limit plate (2) to clamp the screw, preventing it from falling off.

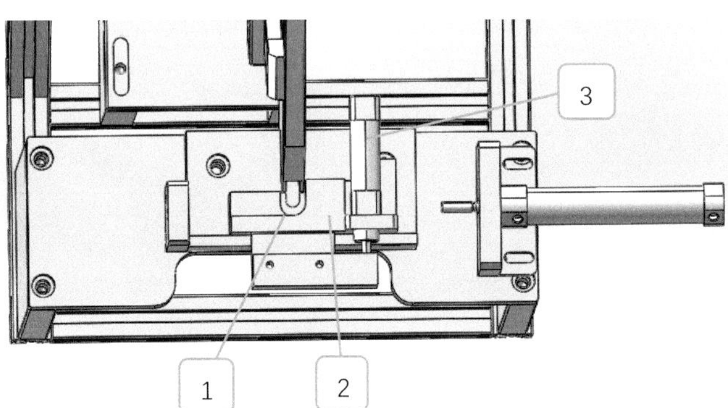

Fig. 7.32 Correction Mechanism
1 - Correction Groove 2 - Limit Plate 3 - Cylinder

IV. Specific Parameters (see Table 7.9)

Table 7.9 Screw Feeding Mechanism Parameters

Working Efficiency:	500-800 pcs/h
Operating Rate:	≥93%
Pressure Range:	4-6kg/cm^2
Operating Temperature:	0~55°C
Permissible Humidity:	10%~95%
Equipment Dimensions:	555 mm(H) × 820 mm(L) × 345 mm(W)

V. Mechanical Design Highlights

The screw feeding mechanism employs new materials and optimized structures to gradually pick up screws, ensuring accurate positioning and significantly reducing manual labor costs. This design also eliminates material jamming issues, improving screw conveying efficiency.

VI. Precautions

(1) Use screws with fewer defects such as foreign objects, mixtures, and defective heads to ensure smooth operation of the machine.
(2) Select commonly used screws with a large quantity and few varieties to improve the efficiency of the automatic screw machine.

Case 99: Aluminum Can Feeding Machine

I. Case Description

The aluminum can feeding machine is designed for the transportation of aluminum cans. The overall structure, as shown in Fig. 7.33, consists of a vibratory conveyor (1), bottle mouth correction mechanism (2), flat belt (3), and frame (4). The aluminum can feeding machine features simple operation and high working efficiency.

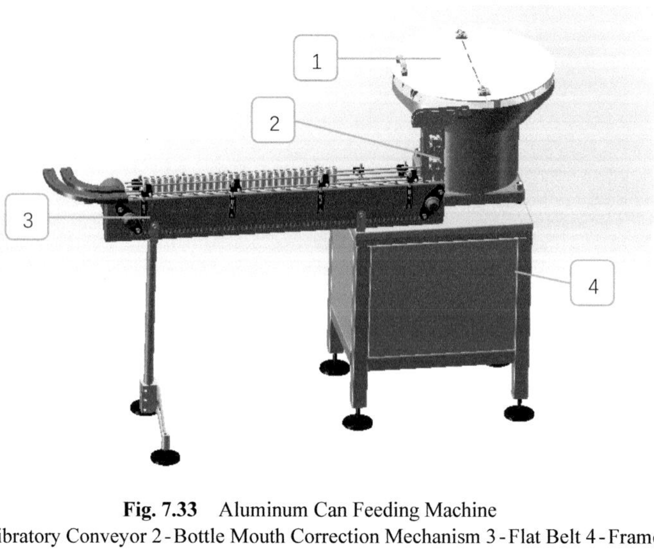

Fig. 7.33 Aluminum Can Feeding Machine
1 - Vibratory Conveyor 2 - Bottle Mouth Correction Mechanism 3 - Flat Belt 4 - Frame

II. Working Principle

During operation, aluminum cans to be fed are placed into the vibratory conveyor. The vibratory conveyor conveys the aluminum cans to the bottle mouth correction mechanism, which adjusts the cans to have their mouths facing upwards. The corrected aluminum cans are then neatly arranged and conveyed to the next process by the flat belt.

III. Main Structure Introduction

The bottle mouth correction mechanism, as shown in Fig. 7.34, uses vibratory conveyor (1) to convey aluminum cans to the correction mechanism (3). The correction mechanism (3) features a cam (2). If the mouth of the aluminum can faces the cam (2), the cam will extend into the mouth, causing the bottom of the can to fall onto the flat belt (4). If the bottom of the aluminum can faces the cam (2), the cam will directly about the bottom, causing it to fall onto the flat belt (4).

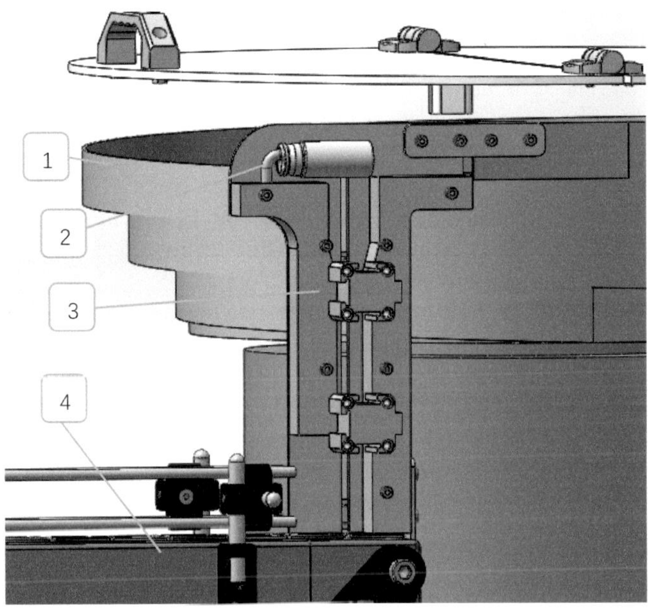

Fig. 7.34 Bottle Mouth Correction Mechanism
1 - Vibratory Conveyor 2 - Cam 3 - Bottle Mouth Correction Mechanism 4 - Flat Belt

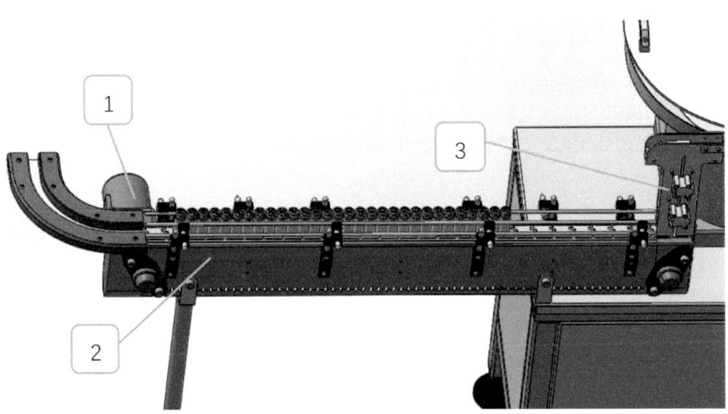

Fig. 7.35 Flat Belt
1 - Servo Motor 2 - Flat Belt 3 - Bottle Mouth Correction Mechanism

The flat belt, as shown in Fig. 7.35, is driven by a servo motor (1) to convey the neatly arranged aluminum cans to the next process.

IV. Specific Parameters (see Table 7.10)

Table 7.10 Aluminum Can Feeding Machine Parameters

Working Efficiency:	1800-2500 pcs/h
Operating Rate:	≥95%
Operating Temperature:	0~45°C
Permissible Humidity:	25%~75%
Equipment Dimensions:	1135 mm(H) × 1735 mm(L) × 770 mm(W)

V. Mechanical Design Highlights and Precautions

(1) The feeding of aluminum cans is directional, and it is necessary to control the cans to keep the mouths facing up. The aluminum can feeding machine achieves this by using the bottle mouth correction mechanism to ensure that the cans maintain an upward-facing orientation after being output from the vibratory conveyor.
(2) During the initial loading of the aluminum can feeding machine, the feed speed should be gradually increased to eventually reach the rated conveying capacity. The feeding should be uniform to prevent material accumulation and motor overload, which could damage the equipment.
(3) Regularly inspect all parts of the aluminum can feeding machine to check for loose fasteners. If any are found, they should be promptly secured to ensure the proper operation of the equipment.
(4) Before starting the machine each time, check the screws to see if they are loose. If any loosening is observed, correct it immediately.
(5) Do not remove the cover of the feeder while the equipment is running to prevent accidents.
(6) If any abnormal phenomena are noticed during the operation of the aluminum can feeding machine, they must be investigated and eliminated first. If no issues are found, the equipment can continue to operate.

Case 100: Chip Conveyor

I. Case Description

The chip conveyor is a mechanical device designed to crush solid raw materials to the required size and transport them. The overall structure, as shown in Fig. 7.36, consists of a flat belt (1), feed inlet (2), crushing mechanism (3), frame (4), and chip collection mechanism (5). The chip conveyor features good crushing efficiency and smooth transportation.

II. Working Principle

During operation, waste material is fed into the chip conveyor through the feed inlet. The crushing mechanism processes the waste material, reducing it into smaller particles. The crushed chips are then transported upward to the chip collection mechanism using the inclined flat belt. The flat belt has side baffles to prevent the chips from falling off.

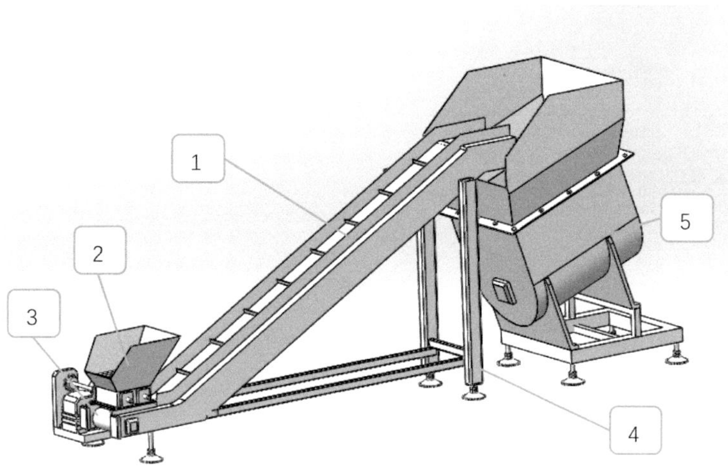

Fig. 7.36 Chip Conveyor
1 - Flat Belt 2 - Feed Inlet 3 - Crushing Mechanism 4 - Frame 5 - Chip Collection Mechanism

III. Main Structure Introduction

The crushing mechanism, as shown in Fig. 7.37, is powered by an electric motor (1). The rotation of two rollers (3) is controlled by a pulley (2), achieving the purpose of crushing.

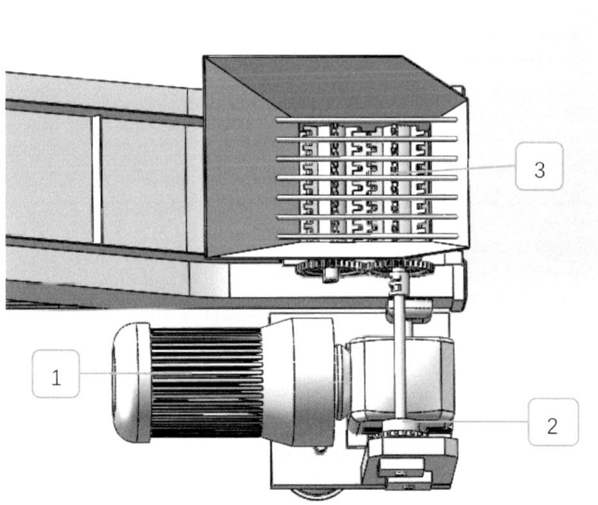

Fig. 7.37 Crushing Mechanism
1 - Electric Motor 2 - Pulley 3 - Rollers

The flat belt, as shown in Fig. 7.38, is synchronized by an electric motor (1). It features a scraper (2) on its surface and baffles (3) on both sides to ensure the smooth transportation of crushed chips to the chip collection mechanism.

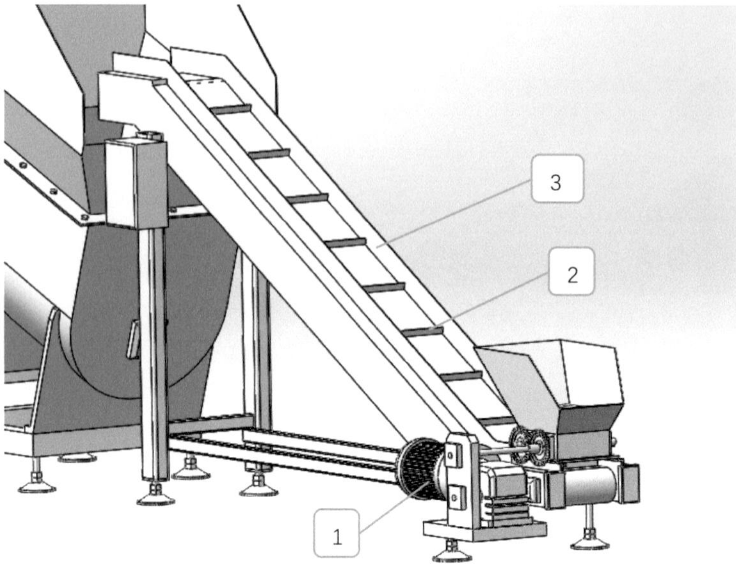

Fig. 7.38 Flat Belt
1 - Electric Motor 2 - Scraper 3 - Baffles

IV. Specific Parameters (see Table 7.11)

Table 7.11 Chip Conveyor Parameters

Scraper Chain Speed:	0.5 m/s
Scraper Chain Width:	30 mm
Production Capacity:	4 t/h
Electric Motor Power:	2.2 kW
Permissible Humidity:	10%~95%
Equipment Dimensions:	215 mm(H) × 500 mm(L) × 115 mm(W)

V. Mechanical Design Highlights

The chip conveyor not only crushes materials but also transports them. It provides effective crushing, and by adjusting the electric motor speed, the material flow can be controlled. The material is thoroughly crushed under the action of the rollers. The chip collection mechanism has a weight sensor at the bottom, allowing real-time monitoring of the material weight in the collection bucket. This helps in assessing whether the material flow is appropriate and facilitates material retrieval [130].

VI. Precautions

(1) The chip conveyor should be started without a load to reduce startup resistance. After running the conveyor without a load for 5 minutes, it can be loaded for a full operation.
(2) Full loading of the conveyor is generally not allowed. In special cases where full loading is necessary, clear the upper part of the flat belt of any material before starting [131].

(3) Avoid contact between the guide plate of the feed hopper and the surface of the flat belt to prevent wear on the flat belt surface.
(4) After stopping the feed, empty the waste on the carrying section before stopping the machine.
(5) When the conveyor equipment is placed outdoors, operators should ensure waterproofing, rain protection, and moisture prevention for the equipment's electric motor and distribution box. Alternatively, cover the machinery with a rainproof cover.

References

[1] Chen, X. 2020. Research on multi-mode calibration and rigid-flexible coupling error compensation method of industrial robot [D]. Jiangnan University.

[2] Jin, R. and Y. Geng. 2020. Cartesian trajectory planning for singularity avoidance in space robot dynamics [J]. Journal of Astronautics, 2020, 41(08): 989–999.

[3] Ni, R. 2020. Modeling and analysis of six-degree-of-freedom collaborative robots [D]. Qingdao University.

[4] Wang, J. 2020. Research on path planning and control methods for multi-joint manipulator [D]. Beijing University of Chemical Technology.

[5] Yang, M., R. Zhang, J. Zhang, H. Zhu, Y. Sun, Y. Chen, and D. Xu. 2020. Overview of SCARA four-axis robot control system [J]. Electric Drive, 2020, 50(01): 14–23.

[6] Xiao, S. 2019. Kinematics and dynamics research of six-degree-of-freedom welding robot [D]. Xinjiang University.

[7] Yuan, C., C. Yang, J. Wei, J. Xing, and G. Gong. 2021. Link parameters optimization of stacking robot [J]. Packaging Engineering, 2021, 42(13): 225–229.

[8] Gu, C., S. Liu, Y. Yao, and M. Zhao. 2021. Dynamics simulation of high-speed heavy-duty stacking robot [J]. Packaging Engineering, 2021, 42(07): 186–194.

[9] Huo, T., B. Zhang. 2021. Trajectory planning algorithm for stacking robot based on velocity optimization [J]. Mechanical Engineering and Technology, 2021, 50(02): 81–83.

[10] Liu, H. 2020. Design of industrial robot sorting system [J]. Forging Equipment and Manufacturing Technology, 2020, 55(06): 69–72.

[11] Xu, J., and Y. Wang. 2020. Finite element analysis and optimization research on joint structure of stacking robot [J]. Mechanical and Electrical Engineering, 2020, 37(09): 1089–1093.

[12] Chen, R., S. Dou, H. Wang, X. Zou, X. Zhan. 2020. Motion analysis and trajectory planning of stacking robot [J]. Automation Application, 2020(08): 77–79.

[13] Zhang, F. 2020. Cascade control simulation of parallel manipulator under the Internet of Things [J]. Computer Simulation, 2020, 37(11): 348–351+449.

[14] Wan, Z. and J. Zhang. 2020. Tracking control research of parallel hydraulic manipulator based on optimized pose control [J]. Machine Tool and Hydraulic, 2020, 48(21): 57–61+44.

[15] Cheng, R. and H. Lv. 2018. Design of special indexing and positioning fixture for hydraulic pump cylinder body [J]. Manufacturing Automation, 2018, 40(01): 108–109+115.

[16] Zhu, J., X. Yin, and Z. Li. 2015. Research on design of single-head rotary beveling machine [J]. Mechanical Engineering and Automation, 2015(02): 117–118+121.

[17] Cai, Q., W. Peng, C. Zhuo, and Q. Huang. 2019. Design of fully automatic batch cutting machine for triangular wood [J]. Forestry Machinery and Woodworking Equipment, 2019, 47(05): 35–39.

[18] Xiong, Z. 2019. Optimization design of steel pipe circular sawing machine fixture [J]. Journal of Baotou Vocational and Technical College, 2019, 20(04): 6–8.

[19] Lu, D. 2018. Development of gantry-type upright dual sawing stone cutting machine [D]. Shandong University.

[20] Wang, H. 2014. Design of control system for paper cutting machine based on PLC [D]. Dalian University of Technology.

[21] Ma, W., K. Wu, N. Yang, and J. Zhao. 2017. Application of servo electronic cam in paper cutting roll of paper cutting machine [J]. China Pulp and Paper, 2017, 36(02): 49–52.

[22] Wang, S. and P. Hao. 2013. Control strategy of paper cutting machine transmission system [J]. Paper and Paper Making, 2013, 32(06): 24–27.

[23] Jiang, Z. and Y. Yue. 2011. Research on the reflector and air cooling system of UV curing furnace under the Internet of Things [J]. Shanxi Electronics, 2011(08): 46–48.

[24] Ma, X., D. He, and W. Li. 2015. Research on intelligent control of UV curing machine based on Internet of Things [J]. Optics and Optoelectronics, 2015(07): 51–57.

[25] Chen, W. and J. Chen. 2018. Research on automatic control system of high-speed automatic UV curing machine [J]. Light Industry Machinery, 2018, 36(02): 9–12.
[26] Liu, X., S. Zhang, H. Wang, and H. Zhang. 2016. Design of automatic temperature control system for UV curing machine [J]. Industrial Control Computer, 2016(09): 57–58.
[27] Li, H., Q. Zhang, Y. Su, and Z. Xue. 2018. Research on intelligent temperature control system of UV curing machine based on Internet of Things [J]. Electronic Technology and Software Engineering, 2018(03): 98–99.
[28] Zhang, X., X. Wang, and X. Wang. 2019. Research on automatic control system of UV curing machine based on single-chip microcomputer [J]. Light Industry Machinery, 2019, 47(04): 23–27.
[29] Zhu, K. 2018. Research on the speed control system of high-speed UV curing machine [J]. Packaging Engineering, 2018, 39(04): 111–115.
[30] Wang, Z., and C. Wu. 2018. Design of intelligent control system for UV curing machine [J]. Modern Manufacturing Technology and Equipment, 2018, 47(05): 119–122.
[31] Liu, J., Y. Liang, W. Han, K. Sun, X. Zhou, and B. Wang. Research on the automatic control system of UV curing machine [J]. Packaging Engineering, 2018, 39(09): 137–142.
[32] Wang, H. 2018. Design of automatic control system for UV curing machine [J]. Journal of Guangxi Chemical Industry, 2018, 38(01): 53–55+65.
[33] Zhou, S., J. Liu, X. Wu, and X. Wang. 2017. Research on intelligent control system of UV curing machine based on Internet of Things [J]. China Mechanical Engineering, 2017, 28(24): 3282–3287.
[34] Qian, W., W. Liu, X. Liang, and S. Shi. 2017. Design of automatic temperature control system for UV curing machine based on single-chip microcomputer [J]. Packaging Engineering, 2017, 38(12): 54–58.
[35] Wang, Y., L. Zhang, and H. Liu. 2017. Design of intelligent control system for UV curing machine [J]. Packaging Engineering, 2017, 38(18): 82–86.
[36] Wei, M. and B. Wang. 2016. Design of automatic control system for UV curing machine based on PLC [J]. Light Industry Machinery, 2016, 44(02): 54–56.
[37] Huang, W., X. Liu, R. Zhang, C. Ma, and B. Yang. 2016. Design and implementation of automatic control system for UV curing machine [J]. Journal of Guangdong University of Technology, 2016, 33(04): 13-16+50.
[38] Li, X., and X. Wang. 2016. Design of automatic temperature control system for UV curing machine [J]. Packaging Engineering, 2016, 37(05): 80–84.
[39] Wang, X. 2015. Design and implementation of automatic control system for UV curing machine [D]. Xi'an University of Technology.
[40] Zhang, H., W. Zhang, C. Tian, K. Wang, F. Tian, Y. Zhang and L. Liu. 2015. Design of automatic control system for UV curing machine [J]. Mechanical and Electrical Information, 2015, 10(03): 59–60.
[41] Liu, X., W. Zhang, L. Zhang, X. Chen, M. Li, X. Bai, and J. Wang. 2015. Research on automatic temperature control system for UV curing machine [J]. Journal of Electronic Measurement and Instrumentation, 2015, 29(08): 1135–1140.
[42] Zhang, W., H. Jiang, X. Gao, X. Zhai, Z. Liu, Y. Zhang. 2015. Design of automatic control system for UV curing machine [J]. Mechanical and Electrical Information, 2015, 6(03): 59–60.
[43] Wang, C., and H. Chen. 2015. Research on automatic control system of UV curing machine [J]. Chemical Automation and Instrumentation, 2015, 42(03): 79–81.
[44] Li, Y., Y Wang, M. Gao and H. Liu. 2015. Design of intelligent control system for UV curing machine [J]. Light Industry Machinery, 2015, 43(05): 58–62.
[45] Wu, Y., F. Li, and Y. 2015. Research on automatic temperature control system of UV curing machine based on single-chip microcomputer [J]. Chemical Fertilizer Equipment, 2015(03): 63–66.
[46] Yu, Y., and H. Li H. 2015. Design of automatic temperature control system for UV curing machine based on PLC [J]. Chemical Fertilizer Equipment, 2015(02): 63–66.
[47] Wang, W., M. Zhang, and W. Zhang. 2015. Design of automatic temperature control system for UV curing machine based on PLC [J]. Chemical Fertilizer Equipment, 2015(01): 58–61.
[48] Liang, X., F. Wu, and Q. Tang. 2014. Design of automatic control system for UV curing machine [J]. Electronic Test, 2014(19): 70–71.
[49] Zeng, W., J. Zhang, and Zeng X. 2014. Design of automatic temperature control system for UV curing machine [J]. Electronic Measurement Technology, 2014, 37(12): 16–19+32.
[50] Zhang, Y., X. Gao, Y. Wang and Y. Wu. 2014. Research on automatic control system of UV curing machine [J]. Mechanical and Electrical Information, 2014, 5(02): 65–66.
[51] Li, Z., S. Fan, and Z. Fan. 2014. Design of automatic control system for UV curing machine based on PLC [J]. Packaging Engineering, 2014, 35(06): 115–119.
[52] Zhang, T., and R. Huang. 2014. Design of automatic control system for UV curing machine [J]. Modern Manufacturing Technology and Equipment, 2014, 43(05): 123–126.

[53] Zhang, T. and R. Huang. 2014. Design of automatic control system for UV curing machine [J]. Electronic Measurement Technology, 2014, 37(05): 22–25.
[54] Liu, L. 2019. Research and Development of Automated Production Line for Coaxial Assembly of Rotors for Multiple Models of Traction Machines [D]. Guangdong University of Technology, 2019.
[55] Yang, D. 2020. Research and Application of Intelligent Manufacturing Technology for Elevator Traction Machines[J]. China Elevator, 2020, 31(20): 17–19+26.
[56] Luo, X. 2019. Intelligent Manufacturing in Shanghai Mitsubishi Elevators—Design and Integration of Intelligent Assembly Line for Traction Machine[J]. China Elevator, 2019, 30(08): 37–39.
[57] Liu, Y., X. Wei, and L. Wei. 2012. Dynamic Simulation of the Installation Process of E-Type Snap Rings Based on ANSYS/LS-DYNA[J]. Mechanical and Electrical Engineering Technology, 2012, 41(09): 64–66+74.
[58] Yang, J. 2020. Analysis and Monitoring System for Automatic Locking of Screws [D]. Taiyuan University of Technology.
[59] Su, Q. 2020. Discussion on the Design of Screw Locking Automation Control System Based on PLC[J]. Computer Products and Circulation, 2020(03): 278.
[60] Sun, P. 2018. Development of Fully Automatic Screw Locking System Based on Machine Vision [D]. Soochow University.
[61] Gao, Y. 2012. Several Measures to Improve the Manufacturing Quality of AC Motor Stators[J]. Electric Motor Technology, 2012(06): 51–52+56.
[62] Wang, C., M. Cao, Z. Sun, Y. Hu, and J. Zhang. 2006. Research on Molds Based on Packaged Motor Stator Windings[J]. Coal Engineering, 2006(10): 81–82.
[63] Liu, W. 2018. Development of Visual System for Automatic Assembly Line of Brake Wheel in EMU[D]. Harbin Institute of Technology.
[64] Li, F. 2017. Research on Hybrid Assembly Process Design Technology for Solid-State Relays[D]. Harbin Institute of Technology.
[65] Chen, Q. 2019. Basic Principles, Current Status, and Development Trends of Automobile Shock Absorbers[J]. Metallurgical Management, 2019(17): 51–52.
[66] Shi, X., Y. Yao, L. Dong, W. Tian, and H. Wu. 2020. Design of Improved Dual-Station Screw Locking Robot[J]. Mechanical and Electrical Engineering, 2020, 37(02): 196–200.
[67] Shi, Z., and Y. Xiao. 2019. Analysis of the Composition of a Micro Automatic Screw Locking Workstation[J]. Modern Manufacturing Technology and Equipment, 2019(05): 131+133.
[68] Qin, X., Y. Huang, and B. Yang. 2018. Design of Desktop Fully Automatic Screw Locking Machine Control System Based on PLC[J]. Shandong Industrial Technology, 2018(07): 5–6+48.
[69] Chu, G., and M. Chang. 2017. Design of a New Type of Fully Automatic Screwdriver[J]. Equipment Manufacturing Technology, 2017(08): 259–260.
[70] Wang, R., W. Zhang, J. Xu, Y. Chen, and Z. Chen. 2021. Structural Analysis and Structural Parameter Optimization Design of Electrical Connector Contact Pieces[J]. Mechanical Manufacturing and Automation, 2021, 50(03): 136–139.
[71] Xun, Q., C. Yang, H. Wang, J. Yu, Z. Ye Z. 2009. Control System Design of Six-Station Combination Machine Tool Based on PLC[J]. Combination Machine Tool and Automation Processing Technology, 2009(07): 53–56.
[72] Nie, X., and J. Tan. 2018. Design of Six-Station Drilling and Tapping Machine[J]. Equipment Manufacturing Technology, 2018(11): 27–30.
[73] Wang, K., and H. Yin. 2020. Research and Application of Automatic Tapping Machine Control System[J]. Open Journal of Electric Power, 2020(03): 12–15+19.
[74] Zhou, P., F. Chen, and Q. Li. Design of an Automatic Inner Hole Grinding Machine for Round Pipes[J]. Manufacturing Technology and Machine Tool, 2017(06): 117–119.
[75] Wang, Z. 2017. Research on Control Technology of Grinding Robot System[D]. Anhui University of Engineering.
[76] Ge, X., M. Xu, J. Xie, Y. Wei, and B. Li. 2020. Design and Analysis of Inner Wall Grinding Machine for Long Cylindrical Pipes[J]. Science and Technology Wind, 2020(30): 169–170.
[77] Guo, M. 2018. Development of Three-Station Five-Axis Polishing Machine[D]. Jilin University.
[78] Yu, J. 2017. Key Technology Research on Adaptive Constant Pressure Grinding of Six-Axis Four-Station Polishing Machine[D]. Huazhong University of Science and Technology.
[79] Feng, Y. 2019. Development and Research of Automatic Dust Removal Equipment for Electronic Components[J]. Electronic World, 2019(16): 114–116.
[80] Shan, G., and P. Dong. 2019. Design and Research on Fully Automatic Cable Stripping Equipment Based on PLC Control[J]. Hydraulic and Pneumatic, 2019(08): 116–120.
[81] Sha, M, and J. Wang. 2011. Progressive Die Design and Application at the Terminal Connection Level[J]. Die & Mould Industry, 2011, 11(11): 25–27.

[82] Jiang, B. 2010. Research on Deviation and Control of Connector Terminal Parts in Precision Progressive Die Stamping[D]. Jiangsu University.
[83] Zhang, X. 2013. Research and Design of Special CNC Milling Machine[D]. Chongqing University.
[84] Zhang, J. 2016. Finite Element Analysis and Structural Optimization of CNC Milling Machine[D]. Changchun University of Technology.
[85] Ning, T. 2016. Design and Research of Fault Diagnosis System for CNC Milling Machine Servo System[D]. Changchun University of Technology.
[86] Zhao, Y. 2013. Development of Small High-Speed CNC Milling Machine[D]. South China University of Technology.
[87] Li, S. 2019. Design and Improvement of Fixtures for CNC Milling Machines[J]. Heilongjiang Science, 2019, 10(08): 84–85.
[88] Ren, C., X. Lou, C. Yang, Q. Miao, Y. Deng and T. Jiang. 2021. Research on PLC-based Infeed and Outfeed Table Control System for Small Diameter Wood Circular Sawing Machine[J]. Modern Electronics Technique, 2021, 44(01): 127–131.
[89] Zhang, Y. 2017. Research and Control System Design of 3D Printer Mechanical Structure Based on FDM Technology[D]. Changchun University of Technology.
[90] Zhang, Z. 2015. Design and Research of 3D Printer Based on FDM Technology[D]. Changchun University of Technology.
[91] Liu, Z. 2014. Design of Control System for Hydraulic Tube Expander[J]. Mechanical and Electrical Engineering Technology, 2014, 43(06): 93–95+143.
[92] Yan, H. 1999. Development of Fully Automatic Vertical Expanding Machine[J]. Machinery Design & Manufacture, 1999(05): 53–55.
[93] Xiong, C., Y. Zhang, X. Luo, J. Chen, and X. Tang. 2014. Determination of Localization for Welding Process of Domestic Material Wheel Rims[J]. Mechanical Engineering Materials, 2014, 38(12): 50–53+86.
[94] Wu, Z. 2015. Simulation and Die Optimization Design of Automotive Sealing Strip Extrusion Forming[D]. Guizhou University.
[95] Zou, N. 2013. Design of a New Type of Punching and Cutting Machine[J]. Mechanical Strength, 2013, 35(01): 116–118.
[96] Mo, J, M. Li, B. Zhou, Y. Wang, K. Zhao. 2017. Design and Implementation of Vision Detection System for Positioning and Bracket Bonding of Automotive Windshield[J]. Journal of Electronic Measurement and Instrumentation, 2017, 31(03). 343–352.
[97] Li, M. 2017. Research on Vision Positioning Gluing and Bracket Bonding Detection System for Automotive Windshields[D]. Hunan University.
[98] Zhang, S. and Y. Ji. 2012. Design of Processing Technology for Automotive Connecting Rods[J]. Hunan Agricultural Machinery, 2012, 39(07): 50–51.
[99] Feng, R. 2017. Research on a New Type of Engine Splitting Connecting Rod Boring and Milling CNC Machine Tool[J]. China High-Tech Enterprises, 2017(08): 21–22.
[100] Tian, S. 2013. Research on Aircraft Parts Cleaning Technology and Equipment Development[D]. Dalian University of Technology.
[101] Sītú, Y. and J. Yè. 2016. PLC based Control System for Small Parts Assembly Line Cleaning[J]. Automation Technology and Applications, 2016, 35(09): 62–65+74.
[102] Li, S.. 2006. Design of Pneumatic Disc Automatic Drilling Machine[J]. Machine Tool & Hydraulic, 2006(04): 146–148.
[103] Tang, L. 2018. Research and Implementation of USB3.0 Interface Testing Methods[D]. University of Electronic Science and Technology.
[104] An, Y. 2016. Design of a Pressure Vessel Airtightness Detection System Based on Machine Vision[D]. Zhejiang Sci-Tech University, 2016.
[105] Feng, Y. and J. Nie. 2019. Application of Airtightness Testing Technology in Production[J]. China Metrology, 2019(05): 108–110.
[106] Fan, L. 2014. Research and Design of Battery Performance Online Detection System[D]. Xi'an Technological University.
[107] Wang, J. and B. Yu. 2020. Brief Analysis on How to Improve the Detection Quality of Electronic Control Products[J]. Technology and Market, 2020, 27(08): 58–59.
[108] Wang, Z., L. Yu, and Z. Shi. 2019. Resistance Strain Measurement Technology[J]. Jiangsu Building Materials, 2019(03): 14–15.
[109] Li, P. 2019. Research on Improving Accuracy in Resistance Strain Measurement[J]. Metrology & Measurement Technology, 2019, 46(12): 62–64.

[110] Lan, X. 2013. Implementation of Automatic Soldering Machine Based on PLC Controller[J]. Manufacturing Automation, 2013, 35(08): 132–135.
[111] Xu, W. and Q. Ji. 2016. Design of Automatic Soldering Machine Control System[J]. Journal of Changshu Institute of Technology, 2016, 30(02): 56–59.
[112] Tang, B. and W. Liang. 2014. Design of an Automatic Soldering Machine[J]. Automation Application, 2014(07): 50–51+53.
[113] Xia, Y. 2016. Stability Online Evaluation of Mid-Frequency Resistance Spot Welding Process for Automotive Body[D]. Harbin Institute of Technology.
[114] Ning, Y. 2011. Hybrid Design and Planning of Car Floor Welding Line Based on Digital Factory[D]. Jilin University.
[115] Li, S. 2020. Standardized Design Discussion on Copper Terminal Structure Suitable for Ultrasonic Welding of Aluminum Conductors[J]. Automotive Components, 2020 (08): 15–21.
[116] Yang, W. 2015. Application Research of Vision Guidance in Robot Automatic Welding[D]. Northeast Electric Power University.
[117] Tian, Y. 2016. Key Technology Research on Robot Header Automatic Welding System[D]. Jiangnan University.
[118] Qiu, W. and R. Wang. 2019. Design and Application of Robot Automatic Welding Workstation[J]. Welding Technology, 2019, 48(03): 64–66.
[119] Yang, M. 2017. Design and Research of Automatic Feeding Device for Hot Melt Machine[J]. Southern Agriculture Machinery, 2017, 48(19): 36–37.
[120] Xu, C., T. Lv, G. Chen, and S. Yin. 2014. Dynamic Analysis of Toothed Rack Lifting Mechanism with Self-Raising Platform[J]. Journal of Mechanical Engineering, 2014, 50(19): 66–72.
[121] Shi, T. 2017. Stability Analysis and Structural Optimization of Scissors-type Lifting Mechanism[D]. Harbin University of Science and Technology.
[122] Wang, Z., X. Chen, X. Zhou, Z. Gao, G. Feng, and F. Du. 2014. Application of PLC in Palletizing Machine on Wine Filling Line[J]. Tianjin Agricultural Sciences, 2014, 20(03): 76–80.
[123] Hong, L., Y. Sun, and C. Cao. 2018. Mechanism Design and Motor Synchronization Control of Multi-Position Feeding Manipulator[J]. Forging Technology, 2018, 43(06): 89–93.
[124] Wu, F. 2017. Research on Coordination Control of Feeding Manipulator and Servo Press[J]. Hubei University of Technology.
[125] Ding, J. 2019. Design and Application of Continuous Stamping Automatic Feeding Manipulator[J]. Machine Tool & Hydraulic, 2019, 47(19): 100–103+124.
[126] Zhou, Y. 2016. Design and Simulation Analysis of Automatic Feeding Device for Stamping Production Line[D]. Chongqing University.
[127] Dai, S. 2015. Design and Implementation of Adaptive Piezoelectric Vibrating Disk Control System[D]. Hangzhou Dianzi University.
[128] Zhang, L. 2013. Structural Strength Analysis and Structural Parameter Optimization of Vibrating Disk[D]. Northeastern University.
[129] Lin, Z., Q. Wang, Z. Luo, B. Zhu, J. Ge, Z. Fu. 2021.Design and Research of Screw Automatic Feeding Device[J]. Mechanical Engineer, 2021(03): 104–107+110.
[130] Li, Z. 2010. Analysis and Research on Automation Logistics System in Heavy Machine Shop[D]. China Academy of Machinery Science & Technology.
[131] Jia, J. 2011. Research on Energy-Saving Optimization of Material Yard Stacking Conveyor Belt System[D]. Northeastern University.

About the Authors

Hao Zhang is a Chartered Engineer, Life Fellow of the Royal Society for the encouragement of Arts Manufactures and Commerce, Fellow of the Institution of Engineering and Technology, and an IPRA & Fellow Assessor & Registration Assessor of IET. He has numerous awards, such as the Outstanding Alumni of Jiangnan University & Tianjin University of Science and Technology, the Distinguished Alumni of Jiangsu Shazhou Middle School, the Distinguished Celebrity Medal of Zhangjiagang City, The IET Excellence and Innovation award, Geneva Invention Award, German iENA Gold Medal, First Prize of China Machinery Industry Science and Technology Awards, Silver Medal of Asia Exhibition of Innovations and Inventions, Science and Technology Achievement Award of China, Thailand Award for the Best International Invention & Innovation, Outstanding Author of China Power Press and China Machine Press, Outstanding Reviewer and Outstanding Paper of Emerald, the Best paper Award of IEEE and HCI. He is a highly cited researcher in intelligent manufacturing.

Adrian David Cheok AM is an internationally renowned academic, inventor, and engineer. He was appointed a Member of the Order of Australia by the Governor-General on behalf of Her Majesty Queen Elizabeth II for his outstanding contributions to international education and research. He is a Fellow of the Institution of Engineering and Technology, a Fellow of Engineers Australia, a Fellow of the Royal Society for the encouragement of Arts Manufactures and Commerce. Currently a Full Professor at Nanjing University of Information Science and Technology and Director of the Imagineering Institute. He is a Full Member of Sigma Xi. His innovative work in multisensory communication, mixed reality, and human-computer interaction has secured over $130 million in research funding and has been showcased globally, including at Ars Electronica and Wired NextFest. He is also the recipient of the World Economic Forum's Young Global Leader award and the Albert Nelson Marquis Lifetime Achievement Award.

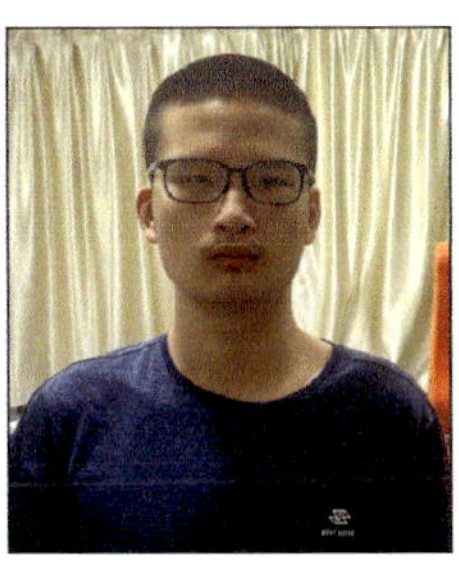

Yujun Zhang is studying at College of Artificial Intelligence and Automation, Hohai University. He won Second Prize in Jiangsu Provincial University Higher Mathematics Competition, First Prize in the Chinese Chemistry Olympiad Competition. He hosted and participated in the following key research and development projects: Double saw blade cutting machine, Automatic sugar beet harvesting machine, nonstandard taping machine, E-type circlip assembly machine, Automatic harpoon loading machine, Pipeline assembly machine, Relay automatic assembly machine, Multi-station automatic polishing machine, Special CNC milling machine, Parts cleaning machine, Pair-type feeding wire welding machine.

Tianyong Zhang is Assistant Editorial Manager of Journal of Intelligent Manufacturing and Special Equipment; Chief Administrative Officer of Hangzhou Huajian Intelligent Technology Research Institute. He has hosted and participated in the following key research and development projects: Exquisite hot melt machine, Auto parts packaging machine, Electronic components automatic feeding machine, Design and improvement of chip conveyor, Single-power stripping and twisting line equipment, Buckle terminal sorting machine, Switch terminal bending machine, Double-turret non-standard assembly machine, Capacitor foot cutting and bending molding machine, 4-head automatic liquid filling machine, Jelly food filling machine, Highly automated filling production line and other equipment, Participate in editing "Design of Common Intelligent Elevation Integrated Drive Controller". He received the International Young Scientist Award. He formulated two national standards.

Index